Wonder & Science

Imagining Worlds in Early Modern Europe

Mary Baine Campbell

CORNELL UNIVERSITY PRESS   *Ithaca and London*

Copyright © 1999 by Cornell University

All rights reserved. Except for brief quotations in a review, this book, or parts thereof, must not be reproduced in any form without permission in writing from the publisher. For information, address Cornell University Press, Sage House, 512 East State Street, Ithaca, New York 14850.

First published 1999 by Cornell University Press
First printing Cornell Paperbacks, 2004

Printed in the United States of America

Library of Congress Cataloging-in-Publication Data

Campbell, Mary B., 1954–
    Wonder and science : imagining worlds in early modern Europe / Mary Baine Campbell.
      p.   cm.
    Includes bibliographical references and index.
    ISBN 0-8014-3648-6 (cloth : alk. paper)
    ISBN 0-8014-8918-0 (pbk. : alk. paper)
    1. Europe—Intellectual life—16th century. 2. Europe—Intellectual life—17th century. 3. Europe—Intellectual life—18th century. 4. Imaginary places—Early works to 1800—History and criticism. 5. Cosmography—Early works to 1800—History and criticism. 6. Ethnology—Early works to 1800—History and criticism. 7. Philosophy and science—Europe—History. 8. Wonder (Philosophy)—History. I. Title.
CB203.C36   1999
001.1′094′0903—dc21   99-29103

Cornell University Press strives to use environmentally responsible suppliers and materials to the fullest extent possible in the publishing of its books. Such materials include vegetable-based, low-VOC inks and acid-free papers that are recycled, totally chlorine-free, or partly composed of nonwood fibers. For further information, visit our website at www.cornellpress.cornell.edu.

Cloth printing   10 9 8 7 6 5 4 3 2 1
Paperback printing   10 9 8 7 6 5 4 3 2 1

For Jason S. McLachlan

BLAKE: Pray, Mr. Taylor, did you ever find yourself, as it were, standing close beside the vast and luminous orb of the moon?

TAYLOR: Not that I remember, Mr. Blake; did you ever?

BLAKE: Yes, frequently; and I have felt an almost indescribable desire to throw myself into it headlong.

—William Blake, *Blake Records* (*Supplement*)

CONTENTS

		List of Illustrations	ix
		Acknowledgments	xi
	I	Introduction	1
		Wonder, and Science	2
		Worldmaking	9
		A Map	13
PART I		IMAGINATION AND DISCIPLINE	23
	II	Travel Writing and Ethnographic Pleasure: André Thevet and *America, Part I*	25
		Pleasure	26
		The Isles of André Thevet	30
		Anecdote	44
		Thomas Hariot, John White, and Theodor de Bry: The *Briefe report* and *America, Part I*	51
	III	The Nature of Things and the Vexations of Art	69
		Francis Bacon and the *Novum Organum*: Repression	71
		Catalogues and Tables: Iteration	78
		Sir Thomas Browne: "Ironicall Mistakes"	85
		Knowledge Is Power	96
		Robert Plot: Cybernetics	101
PART II:		ALTERNATIVE WORLDS	111
	IV	On the Infinite Universe and the Innumerable Worlds	113
		Sublimity: Bruno's *De l'infinito universo et mondi*	116
		The Empirical: Galileo and the "Enchanted Glass"	123
		Lunar Astronomy	133
		Fiction: The Minuet of Fontenelle	143

V	A World in the Moon: Celestial Fictions of Francis Godwin and Cyrano de Bergerac	151
	The Man in the Moon: Imaginary Gardens, Real Toads	155
	Cyrano de Bergerac's Other World — Mimesis and Alterity	171
VI	Outside In: Hooke, Cavendish, and the Invisible Worlds	181
	Truth Even unto Its Innermost Parts: *Micrographia* and the Details of the Corporeal Eye	183
	Margaret Cavendish's *Blazing-World*	202
	Other World	202
	Inner World	206
	What the Microscope Missed	213
PART III	THE ARTS OF ANTHROPOLOGY	221
VII	Anthropometamorphosis: Manners, Customs, Fashions, and Monsters	225
	Anthropometamorphosis	233
	Categories and Misfits, or, What Is That Naked Monster Doing in a Book about Clothes?	250
VIII	"My Travels to the other World": Aphra Behn and Surinam	257
	Fashion Plates in Surinam	260
	Exposed, Excised, Experienced	272
	Fiction, Science, and Agency	278
IX	*E Pluribus Unum*: Lafitau's *Moeurs des sauvages amériquains* and Enlightenment Ethnology	285
	On the Brink of Anthropology	286
	Father Lafitau and the *Moeurs des sauvages amériquaines*	289
	"La Religion influoit en tout"	293
	Ventriloquism and Possession: Empty Bodies, Moving Souls	295
	Torture	299
	Psalmanazar, Formosa, and the Other Self	310
	Coda: The Wild Child	319
	Works Cited	325
	Index	353

ILLUSTRATIONS

1	Marguerite de Roberval, castaway, from André Thevet's *Cosmographie universelle*	22
2	Pietro Coppo's woodblock map of the New World	42
3	Adam and Eve, from Theodor de Bry's frontispiece to the final section of his *America, Part I*	54
4	"The Pyne fruite," watercolor by John White	56
5	"Scorpions," watercolor by John White	56
6	"Theire sitting at meate," watercolor by John White	58
7	Theodor de Bry's engraving of John White's "Theire sitting at meate," from *America, Part I*	58
8	Algonquin woman and child, with European doll, from Theodor de Bry's *America, Part I*	59
9	Pict carrying European head, from Theodor de Bry's *America, Part I*	64
10	Initial "P" in the first edition of Francis Bacon's *Novum Organum*	86
11	Waterworks of the "artificial island" at Enston, from Robert Plot's *Natural History of Oxfordshire*	107
12	Grotto at Enston, from Robert Plot's *Natural History of Oxfordshire*	108
13	The Copernican universe, from Leonard Digges's *Prognostication everlasting*	110
14	Domingo Gonsales flies to the Moon, from Francis Godwin's *Man in the Moon*	157
15	Airborne by means of evaporating dew, from Cyrano de Bergerac's *L'Autre monde*	157
16	Point of needle, full stop, and razor's edge, from Robert Hooke's *Micrographia*	193
17	Eye of a gray drone-fly, from Robert Hooke's *Micrographia*	197
18	Detail from drone-fly's eye, from Robert Hooke's *Micrographia*	199
19	A patch of the Moon and "a few stars," from Robert Hooke's *Micrographia*	200
20	Frontispiece to John Bulwer's *Anthropometamorphosis*	220
21	Matron, virgo, femina, from Jean-Jacques Boissard's *Habitus variarum orbis gentium*	229
22	Cross-cultural cosmetics, from John Bulwer's *Anthropometamorphosis*	237

23	Cyclops, from the *Recueil de la diversité des habits*	252
24	Woman from Brabant in mourning, from the *Recueil de la diversité des habits*	255
25	"America," woodcut by Jan van der Straet in *Nova reperta*	263
26	American fashions, from Joseph Lafitau's *Moeurs des sauvages amériquains*	284
27	"Tortures," from Joseph Lafitau's *Moeurs des sauvages amériquains*	307
28	Formosan fashions, from George Psalmanazar's *Historical and Geographical Description of Formosa*	311
29	The Formosan alphabet, from George Psalmanazar's *Description of Formosa*	315
30	Psalmanazar's manuscript copy of the Lord's Prayer in Formosan	316
31	Manuscript copy of the alphabet of "Javasu," probably by Caraboo	324

ACKNOWLEDGMENTS

This book has been gathering since I first designed an interdisciplinary course in some of its materials at Columbia University in 1986, where I was a Mellon Fellow in the Society of Fellows in the Humanities. Necessarily after so much time there are many institutions and persons I want to thank, and there will not be room here to express all the gratitude that has been gathering along with the book. So my first acknowledgment is of all the friends, colleagues and strangers who provided me with wonderful facts, amazing lies, names and definitions, and moments of understanding or challenge. Even more important to me than the collegiality of a world of wonder-mongers has been the inspiration of seven scholars and thinkers, all of whom have also supported me in large or small pragmatic ways but for whose example and capaciousness I am most centrally thankful: Lorraine Daston, Daria Donnelly, Allen Grossman, Anne Janowitz, Katharine Park, Eve Kosofsky Sedgwick, and David Wilson.

Institutional support has been plentiful, and I feel my luckiness in this time of dwindling support for intellectual work. I thank the Mellon Foundation for having funded two fellowships, at the conception and conclusion of the project, and the research centers at which I held them: Columbia University's Society of Fellows and the National Humanities Center. Along the way I have also been abetted by fellowships from the American Council of Learned Societies, Brandeis University (the Perlmutter Research Fellowship), and the NEH. I have depended on a number of libraries: Harvard's Widener, Houghton, Cabot, and Tozzer libraries (with extra gratitude to the Houghton, especially to Tom Ford); Columbia's Butler Library; the John Carter Brown Library at Brown University; the British Library; Lambeth Palace Library; the Goldfarb Library at Brandeis; Boston University's Mugar Library and Science and Engineering Library; the library staff at the National Humanities Center (especially Eliza Robertson) and the various libraries on which that center draws, particularly in my case the University of North Carolina Library at Chapel Hill (whose North Carolina Collection provided the photographs of John White's drawings, though permission to reprint them comes from elsewhere), and Duke University's Perkins Library and Rare Book Library.

The most important institution of all has of course been my own university, which even in these hard times has not lost its commitment to scholarship and has supported, in addition to the Perlmutter Fellowship and two Mazer Research Grants, considerable leave time for me. Particular thanks are due to Dean of Arts and Sciences Robin Feuer Miller, to the inimitable Arlyne Weisman, and to Susan Staves, the eighth wonder of the world and frequently chair of my department. Past and present Brandeis colleagues besides those mentioned elsewhere who have made differences include Lennard Davis, Steven J. Harris, Erica Harth, Tom King, Kathleen Perry Long, Paul Morrison, Sylvan S. Schweber, and Gary Taylor. Past and present graduate students of particularly provocative presence who have studied these and related materials with me and/or provided research assistance are Michael Booth, Elizabeth Hale, Tonya Krouse, Mary Leader, Brandie Siegfried, Judith Tabron, Peter Wogan, and Laura Yim. And I must also thank the example and instruction of four people who were my teachers or mentors at Boston University in matters that sparked my thoughts on the wonders and the sciences herein: Laurence A. Breiner, Michael McKeon (whose course in the "Origins of the English Novel" left a crucial mark), and the late Celia Millward, sorely missed. Robert Levine patiently nurtured my Latin for many years after I left graduate school.

Chemist and dancer François Amar, Lorraine Daston, and Peter Hulme invited me to participate in particularly stimulating seminars on such matters as "Wonder, Horror and Curiosity," "Writing Travels," "Constructing an Object for Science," and on astronomy in an NEH curriculum-designing project for teaching sciences and arts together. Other impresarios organized and galvanized conferences, seminars and special sessions which have provided me with an intellectual community: thanks to Christopher Baswell, Glenn Burger, Eugene Goodheart, Roland Greene, John Headley, Glenn Hooper, Jhana Howells, Claire Jowett, James Kloppenberg, Jodi Mikalachki, Katharine Park, James Romm, Katherine Rowe, Janet Whatley, and Tim Youngs for their intellectual energy and for their hospitality.

Anyone writing between and across disciplines depends on the kindness of strangers, as well as on friends, family, pets, and stimulants. I have been especially fortunate in my two residential fellowships to meet people with generous hearts as well as interesting minds, and fortunate too in my friends. Kathleen Biddick, Laurence A. Breiner, Mary V. Dearborn, John Headley, Anne Janowitz, Marina Leslie, Katharine Park, Marie Plasse (who taught me the corporeal), Nicole Rafter, and above all Helaine Razovsky have read parts of the manuscript at various stages; at the National Humanities Center, Laura Engelstein, Karen Hanson, Judy Klein, Lydia Liu, Michèle Longino (especially), George Saliba, and Richard Trexler all read and commented on sections relevant to their own work or expertise, and I got help in matters philosophical and French respectively from Marilyn Frye and Claude Reichler. Karen Carroll's copy editing rivaled Kent Mulliken's photography. P. de Rivière gave endlessly of his time. From the gestational days at the Society of Fellows I thank Richard Andrews, Roger Blumberg, Janet Johnson, Dilwyn Knox, Loretta Nassar, Mark Rollins, Nancy Stepan, and Gauri Viswanathan.

Parts of Chapters 2 and 7 have appeared as articles in *Studies in Travel Writing* (first issue, 1997), *Monster Theory* (ed. Jeffrey Jerome Cohen [© The Regents of the University of Minnesota, 1996]), and *The Work of Dissimilitude* (eds. Robert A. White and David Allen). Permission to reproduce them, revised, is gratefully acknowledged to *Studies in Travel Writing*, the University of Minnesota Press, and the University of Delaware Press. A few paragraphs from Chapter 5 appeared in an article in *Literature and History* (6[2] [Autumn 1997]), borrowed here by permission of the Manchester University Press. Permission to quote from Marianne Moore's "Poetry" was granted by Marianne Craig Moore, Literary Executor of the Estate of Marianne Moore. All rights reserved. It was reprinted with the permission of Simon and Schuster from *The Collected Poems of Marianne Moore* (© 1935 by Marianne Moore; copyright renewed © 1963 by Marianne Moore and T. S. Eliot) and, in Britain, Faber and Faber. Louise Glück's "Vita Nova" is from *Vita Nova* by Louise Glück (© 1999 by Louise Glück), reprinted by permission of the author and of The Ecco Press. Permission to quote from a letter in *No One May Ever Have the Same Knowledge Again*, (ed. Sarah Simons), was granted by the Museum of Jurassic Technology. Figures nos. 10–24, 26–29, and 31 are reproduced by permission of the Houghton Library, Harvard University; nos. 1, 3, and 7–9, by permission of the John Carter Brown Library at Brown University. Figures nos. 4–6, John White's watercolors (copyright the Trustees of the British Museum, British Museum Press) are reproduced by permission of the British Museum Press. Figure no. 2 appears by permission of the British Library, no. 25 by permission of the Smithsonian Institution Libraries, Smithsonian Institution (© Smithsonian Institution), and no. 30 by permission of the Archbishop of Canterbury and the Trustees of Lambeth Palace Library. Where no source is cited, translations are my own.

At Cornell University Press my editor, Bernie Kendler, has been peerless in wit, brevity, and good advice about writing; it was a pleasure, too, to work with Candace Akins. The Press's reader, the historian of science (and of jokes) Paula Findlen, was generous with information and materials, and thrillingly simpatico. My father, Kenneth L. Campbell, offered every kind of support, from editing to my new xerox machine. I have been especially grateful for the patience and encouragement of my friends who are ill. The greatest aid of all came from Jason McLachlan, wonderful scientist, who has been in on this adventure from the start and read the whole manuscript judiciously and often. This book is for him, its Muse.

Wonder and Science

I • INTRODUCTION

The second of July, we found shoal water, where we smelt so sweet, and so strong a smell, as if we had been in the midst of some delicate garden by which we were assured that the land could not be far distant: and keeping good watch, and bearing but slack sail, the fourth of the same month we arrived upon the coast.

—Philip Amadas and Arthur Barlowe, in Richard Hakluyt's
Principal Voyages . . . of the English Nation (1589)

THIS IS HOW I learned the adventure of the New World at Hudson Elementary School: as a moment of wonder, compounded of dream, surprise, delight, the trope of paradise realized on earth. It has been presented that way by many writers. Some of them were early explorers, others recent critics of the rhetorical contraption in the making among such reports and "discourses" as that of Philip Amadas and Arthur Barlowe. On an ocean voyage from New York to Southampton as a child, in 1963, I sniffed eagerly for the coast of England when it came in sight but had to conclude that only America could produce that flowery odor. On the way, I had dropped a tightly sealed bottle overboard; it contained a note about an island I'd discovered in the middle of the ocean, signed "Christopher Columbus." My instruction had been clear: the distance between these shores marked the space of wonder, fantasy, invented worlds.

It has been a long and weary awakening to some of the New World's other spaces—immemorial homeland of many now extinguished or deracinated nations, for example, or new home to a nation based not only on freedom but on slavery, home to an imperial upstart made dangerous to smaller and weaker nations by the myth of its own paradisal innocence and welcome.

But it is never possible to waken all the way, nor is it wholly desirable. The longing for another world seems to have been a *real* pressure on the construction of the Edenic narrative of America. Though "Eden" came to obscure terrible exploitations of people, nations, land, and resources, and came to it soon, that does not give us leave, as curious historians of culture, to dismiss the element of true desire in the false consciousness of colonial empire. The pressure of that same desire produced many kinds of other worlds, less usable, less phenomenal, and although they are all stained with the original sin of the conquests, they represent efforts of the imagination to see and yearn past the bounds of the known and approved, past the spiritual oppression of the "dark existing ground."[1]

Wonder and Science joins that effort. Without closing its ears to the din of the real, the book wants to render an account of wishes, pleasures, excitements, sublimities, and, above all, possibilities. Although one can see, in the stream of worldmaking texts to be encountered here, a narrative that terminates in Goya's dark "Dreams of Reason," the moments the texts encapsulate and the possibilities of vision they suggest are rich with ambivalence and undecidability. These texts make up, in part, a history of imaginative literature, a history of science, a history of their mutually determining emergence, a history of cognitive transformation and the means it expresses and is expressed by. The entanglements of these texts with one another and with the history of early colonial empires make up a fabric knotty with significance for all these histories, and for the characters of the forms and genres on which they are woven. Their future remains unwritten.

WONDER, AND SCIENCE

> To recognize in paranoia a distinctively rigid relation to temporality, at once anticipatory and retroactive, averse above all to surprise, is also to glimpse the lineaments of other possibilities.... To read from a reparative position is to surrender the knowing, anxious, paranoid determination that no horror, however apparently unthinkable, shall ever come to the reader *as new*.... Because there can be terrible surprises, however, there can also be good ones.... Because [the reparative reader] has room to realize that the future may be different from the present, it is also possible for her to entertain such profoundly relieving, ethically crucial possibilities as that the past, in turn, could have happened differently from the way it actually did.
>
> —Eve Kosofsky Sedgwick,
> "Paranoid Reading and Reparative Reading" (1997)

This book was conceived at the height of "paranoid" achievement in the fields of literary and, as it had begun to be called, cultural studies.[2] As a literary historian myself, of a European corpus (travel writing) intimately complicit in the inception and expansion of imperial colonialisms, I was moved by the subtlety and fervor of the new critiques, particularly critiques of de-

---

1. From Louise Glück, "Vita Nova": "By the tables, patches of new grass, the pale green / pieced into the dark existing ground" (*Vita Nova*, 2).
2. For the rest of the text from which my epigraph is drawn, as well as my energy in this Introduction, see Eve Sedgwick's exhilarating introduction to *Novel Gazing*.

velopments in the historical periods with which *Wonder and Science* is concerned.[3] As a poet, however, I felt not only moved but implicated; I wanted to defend the value of the "cognitive emotion" associated at least since Aristotle with the writing and reading of poetry, and associated increasingly with the manipulation of the colonized, the selling of the colony to backers back home, the exoticizing of whatever could be (or seem to be) subdued.[4] One cannot defend such a sentimentalized and caricatured concept as "wonder" against persons invested in despising or fearing its power (which is real — consider the origins of fireworks, and the message there). But one may certainly argue for the value of a pleasurable emotion, or relation to knowing, that requires the suspension of mastery, certainty, knowingness itself. And no one can deny the actual functionality of such a relation to knowing, at least no one who has ever observed a seminar or a lab in motion. The animus that relegates wonder and other cognitive pleasures to the trivial or sentimental sidelines of life's serious business is a historical product of the transformations under observation in this book. Other transformations are under way now.

I will consider wonder, then, not as the sentimentally ennobling sensation of worldly conquistadors, stopped in their bloody tracks for a moment by the surplus of novelty or its temporary absence of visible use value. In this book, wonder might first be seen as a register opposed to that of "paranoid reading" — one which embraces surprise, enjoys the excess and alteration which generate it, is constitutively open to the rewriting of the past as well as the future, the making of new worlds. In short, wonder will be taken on its own terms, not as or not only as a rhetorical masking or a deflection of "reality."

"Wonder" is a form of perception the necessary conditions of which have undergone radical changes between the highly rationalist but fact-poor Middle Ages and the end of the current millennium: evaluation of it has a history, recently rehearsed by Katharine Park and Lorraine Daston in their *Wonders and the Order of Nature* and Caroline Walker Bynum in her 1996 address to the American Historical Society; earlier, by the poet and critic J. V. Cunningham in *Woe or Wonder*, a study of Shakespearian dramaturgy. All of these works address wonder, explicitly or not, as a historical phenomenon

---

3. I am thinking, in particular, of Edward Said's *Orientalism*, Stephen Greenblatt's *Renaissance Self-Fashioning* (chap. 4, "To Fashion a Gentleman"), Bernadette Bucher's *Icon and Conquest* (*La sauvage aux seins pendants*), Peter Hulme's *Colonial Encounters*, Patricia Parker's *Literary Fat Ladies*, Mary Louise Pratt's *Imperial Eyes*, Jonathan Goldberg's *Sodometries*, Michel de Certeau's "Ethnography, or The Speech of the Other."

4. See especially Stephen Greenblatt, *Marvelous Possessions*, and Louis Montrose, "The Work of Gender in the Discourse of Discovery."

differently valenced and valued (and experienced) in different times and places. Its status in the last stages of modern Western culture in its dominant form is ambivalent—it is eschewed but also craved, wildly popular and markedly absent as a value, in discourses of power such as those of business, government, the sciences, and "rigorous" scholarship.

The period under examination here extends from the middle of the sixteenth century (by which time the novelty of the New World was starting to provoke innovations in philosophical and narrative genres) to the early eighteenth. It was a period of intense intellectual, technological, religious, and economic transformation in western Europe—not to mention the even more drastic changes it brought to the Americas and Africa.[5] Transformations so multiple and simultaneous can interfere drastically with the sense of mastery that knowledge confers; wonder may be rife without being necessarily pleasurable. Quotations for this period from the Oxford English Dictionary's article on "wonder" tend to emphasize its associations with stasis and incomplete understanding. From Thomas Hoby's translation of Castiglione's *Courtier*, for example, we get this: "then he turning about, and beholding him . . . with a wonder, stayed a while withouten any word." A seventeenth-century citation speaks of Galen "hushed into a wonder by certain anatomical observations." Speechlessness and a kind of paralysis are major symptoms of the state of wonder, as brevity and isolation often are, or were, of its rhetorical presentation (especially in the Middle Ages, when systematized knowledge—of most things besides theology—was less common). But this was an age of discovery, invention, venture capital, conquest; the active, not the contemplative virtues were in the ascendance. Increasingly, wonder was suspect, inconvenient, needed to be put in its (eventually aesthetic) place.[6]

From a fully modern point of view, a datum stripped of the ramifications of system, whether theological or scientific, is enjoyed in the same way that a fetish, a mantra, an icon are enjoyed. It arrests the gaze, the intellect, the emotions, because (consciously at least) it leads nowhere, reminds us of

---

5. I have not chosen to start this account with discussion of Columbus, Vespucci, and Peter Martyr, although their texts would have made another sensible place to begin, because I and so many others have already attended to that first wave of wondering reaction and conventionalized dissemination; for work in English in the last ten years, see my own *Witness and the Other World*, Greenblatt's *Marvelous Possessions*, Valerie Flint's *Imaginative Landscape of Christopher Columbus*, and Anthony Grafton's *New World, Ancient Texts*. And we are not alone.

6. For a stimulating analysis of wonder as an aesthetic phenomenon central to mathematics and optics as well as to painting and architecture, see Philip Fisher, *Wonder, the Rainbow, and the Aesthetics of Rare Experiences*.

nothing. It has no use value. As a result, wonder is a form of perception now mostly associated with innocence: with children, the uneducated (that is, the poor), women, lunatics, and non-Western cultures. And of course artists. It is often, in our time and in the "developed" countries, a form of perception artificially contrived by the political obstruction of access for many people to the systematic explanations we call the sciences, or else by an overt evasion or rejection of those explanations.[7] It is a form far less well studied than its elite relative, the sublime, for reasons that have something to do with the social status of those persons we conceive as most susceptible to it. And if literary critics and theorists have slighted the phenomenon on the bases of class, race, gender, and age, historians of science until Daston and Park have ignored it as well, in connection with its function as epistemological drag. The relation of wonder to knowledge is crucial but largely oppositional — "broken knowledge," Francis Bacon calls it in *The Advancement of Learning*. No one has yet written a book called *The Advancement of Wonder*.

The "science" of my title refers to the epistemological innovation that eventually constituted "wonder" as a drag. It was a long struggle (not over yet), and what the seventeenth and eighteenth centuries called natural philosophy brought as many marvels into the light as did the explorers and poets —

---

7. For one of many contemporary works that could seriously complicate that statement, see the collection of letters written to observers at the Mount Wilson Observatory between 1915 and 1935, *No One May Ever Have the Same Knowledge Again* (ed. Sarah Simons). The "artificiality" of modern conditions of wonder is certainly problematized in a letter like the following:

> "I once was taking a minister of the gospel to his appointment to preach . . . and as I spake to him of the wonderfull creation of God how God had Created millions of other suns and worlds, says he what do you mean by saying other suns and worlds, I answered by saying the other suns and systems of worlds, for God would no doubt not make suns to shine without each sun had a system of worlds such a sun would be to no purpose it would not be of use to anyone not even its Creator one had Just as well to say I am going to build a big costly mansion but I am not going to let anyone live in it . . . , so no doubt from a reasoneble standpoint, each sun has its system of worlds and the worlds of each system are for the purposes of habitation for Gods inteligent creatures to lie on and be hapy. . . . then he sais to me is there an other world and where in the Bible do you find where it speaks of other worlds: well says I to him the Bible says in many places that God made the worlds and Jesus said himself that he mad the worlds St. John 1–3 all things were made by him and many other bible texts prove that God has made the worlds by millions and yet the minister did not believe that there was any other world only this one on which we live and it is as flat as a pancake round as a pancake but not round as a ball and that the sun did actualy go around this pancake every 24 hours (17–18). (Some of [my friends] think that [the stars] are only little sulpher balls up in the sky . . . but if I could only have a scope to convince them of theyr ignorance I should be Glad [19].)

often in highly figural or sensational styles. Nonetheless, the tension was significant, and generative.[8] Even Bernard de Fontenelle (lifetime secretary of the Académie des Sciences), in his exquisite and fictional 1686 *Conversations on the Plurality of Worlds*, pooh-poohs the preference for ignorance as mystery: "Most cherish a false notion of mystery wrapped in obscurity," he tells his female interlocutor, "the Marquise," in the first of their moonlit conversations. "They only admire Nature because they believe she's a kind of magic, and the minute they begin to understand her they lose all respect for her" (12). This sounds a little like Carl Sagan, whose corny sublime proselytized for science to the same kind of audience Fontenelle here scolds. Fontenelle was, like Sagan, a scientific popularizer, and the genre as we know it is one of those provoked to birth in the long period under discussion here. The medieval equivalent to popularizing (not a very close equivalent, as the populace wasn't literate) tends to further isolate and fragment the data it purveys, in the interest of exciting the reader, while the ostensible impulse of early modern popularizing is to rationalize strangeness and correct error. John Wilkins's "Epistle to the Reader" in his 1638 *Discourse Concerning a New World and Another Planet* advises us "to come unto [the book] with an equal mind, not swayed by prejudice, but indifferently resolved to assent unto that which upon deliberation shall seem most probable unto thy reason" (A3r). Such prefatory appeals to "indifference" are normal by Wilkins's time, if often disingenuous. They may remind us of the conventional claims to historicity with which the fictions of the time, especially early novels, begin: we see in both conventions a form of repression that hints at an identity between the experiences of fiction and "wonder," despite the fundamental opposition between the imaginary and what is offered as, however marvelous, the factual. But fact and fiction are in fact etymological brethren, both children of *facere*, to make or fabricate. "Les faits sont faits," says Bachelard.

Wonder and Science concentrates on the mutually exclusionary process of development, in the brethren discourses of natural philosophy (which includes the cosmography and anthropology crucial to any history of "worlds") and of fiction, of Truth seen as constitutively distinct from Beauty. Both narratives are subsumed in the larger narratives of "the Age of Discovery," the age, epistemologically speaking, of the plethora. More poured in than had been dreamed of in anyone's philosophy, and wonder poured out

---

8. See John Bender, "Enlightenment Fiction and the Scientific Hypothesis," which defines the generativity as, by 1750, a dialectic, in which "the guarantee of factuality in science increasingly required the presence of its opposite, a manifest yet verisimilar fictionality in the novel" (6).

to meet it. There were deluges, plenum upon plenum. By the time of the compromised Enlightenment/siècle des Lumières/Aufklärung/Iluminismo/Ilustración there was order again, or system, institutionalized science and a recognizable new prose genre that offered for sale to small but growing literate publics the "true histories" of imaginary people and places, finally distinguishable from the false histories, travels, "relations" of real ones.[9] Wonder was still an available experience, as in some sense it always is (see Charles Darwin), but its sources were more likely to be sequestered at public demonstrations of electricity, or behind the proscenium arch at the Comédie Française.[10] It was no longer a tidal wave; it no longer pressed upon people to devise systems of catchment and diversion. It no longer threatened.

The story encompassed by the chapters of this book, then, is one that ends in the world of C. P. Snow's "Two Cultures" (more recently and more analytically summed up by Bruno Latour as "the modern Constitution").[11] It is also a story that points to the infiltration of the two "cultures" by the ghosts and powers of cultures destroyed or "reduced" to Christianity and colonial dependence along the way. These ghosts are at last being called up, at the end of the millennium or the end of a modernity constituted by divisions such as Snow's.[12] Coptic roots for much of Greece's sacrosanct literary language have been suggested, the Islamic sources unearthed of some of Copernicus's

---

9. See especially William Nelson, *Fact or Fiction*; Percy G. Adams, *Travel Literature and the Evolution of the Novel*; Lennard Davis, *Factual Fictions*, and Michael McKeon, chap. 1 ("The Destabilization of Generic Categories") in *Origins of the English Novel*. J. Paul Hunter's *Before Novels* is less crucially focused on that matter but has much to say about it with regard to the English novel.

10. In *The Expression of the Emotions in Man and Animals* (1872) Darwin provided a gestural definition of wonder (raised eyebrows, open mouth, etc.) and a function for it in situations of danger—the gestures make for better vision and more oxygen intake—that would bring it ironically into the vigilant and defensive mode of "paranoia" as described by Sedgwick.

11. "These separations [of humans and nonhumans . . . and between what happens 'above' and what happens 'below'] could be compared to the division that distinguishes the judiciary from the executive branch of the government. This division is powerless to account for the multiple links, the intersecting influences, the continual negotiations between judges and politicians. Yet it would be a mistake to deny the effectiveness of the separation. The modern divide between the natural world and the social world has the same constitutional character, with one difference: up to now, no one has taken on the task of studying scientists and politicians in tandem, since no central vantage point has seemed to exist" (Latour, *We Have Never Been Modern*, 13).

12. For a stunning example of this invocation on a popular scale, see Brackette Williams's "Dutchman Ghosts and the History Mystery," a study of the spirits of long-dead (and historically actual) Dutch colonists who lost to the English in the struggle for control of Guyana. These spirits possess many current Guyanese descendants of the African slaves whose 1763 Berbice rebellion laid the ground for the intra-European struggle over a country that belonged to none of the actors. The Dutchman ghosts speak with English accents.

key mathematical formulations, the West African cosmological symbolism hidden in the familiar motifs of nineteenth-century American quilts, the twelfth-century philosophical-allegorical romance of Grenadian Muslim Ibn Tufayl behind Defoe's construction of *Robinson Crusoe*.[13] The *invisibilia* and occult properties of the Neoplatonists are resurfacing in Boyle's air pump, Newton's laws, and the haunted courtrooms of colonial Massachusetts, the fine detail of pornography in the microscopical anatomies of plants (or vice versa).[14]

Most of us are old enough to feel something akin to wonder at the crossing of boundaries erected so long ago in the process of institutionalizing sciences, academic fields, nations. It is a different sort from the raw, even frightening *admiratio* of sixteenth- and seventeenth-century people of enough leisure and education to concern themselves with New World societies, heliocentrism, plural worlds, or fiction. Then, wonder had a valence with horror and terror. Our epistemologically compartmentalized world of the intelligible, with its departments and dissociations, was the cure for that unmanageable excess. It was also a way of denying some of the depths, guilty or not, of representation. The breakup of the old colonial system in the second half of this century has turned out to be as well the breaking up of the fields, the walls between the compartments, indeed the whole *Titanic* with its equally regulated compartments of class, its designers who could not let themselves believe the power of a mute, unarmed, and merely "natural" iceberg.

It is interesting that the postcolonial is also the postdisciplinary, in an unmoored muddle of "literary kinds" whose ontological distinctions are melting. What held all that together? Why does it all go down at once? What had Jane Austen or Balzac to do with West Indian sugar plantations, the Moon, the microscope, the *Journal des Sçavans*? This book does not provide an answer, although it assumes a model in which the actual (if very much pre- and postfabricated) encounters of European explorers with "new" genealogies of humans in the New World provoke, infiltrate, inhabit other forms of discovery, reportage, storytelling, worldmaking. Such explorations and encounters

---

13. See Martin Bernal, *Black Athena*; George Saliba, *A History of Arabic Astronomy*, 26–31 and 245–305; Gladys-Marie Fry, *Stitched from the Soul: Slave Quilting in the Ante-Bellum South*; Len Evan Goodman's introduction to his translation of Ibn Tufayl's *Hai ibn Yaqzan*.

14. Latour on Hobbes's horror at Boyle's masterpiece, the air pump: "Worse still, this new coterie [experimental philosophers] chooses to concentrate its work on an air pump that once again produces immaterial bodies, the vacuum—as if Hobbes had not had enough trouble getting rid of phantoms and spirits!" (20). For micrography and pornography see Chapter 6.

quickly become long-lived metaphors, clichés for use in other prospecting ventures and texts, or sites of parody and fictional imitation. We might best see the picture of European imaginative thinking between the times of Columbus and Lafitau or Linnaeus in musical terms, as a many-voiced fugue whose first motif appears when Christopher Columbus writes his remarkable "Letter" to Queen Isabella's Keeper of the Purse, instantiating a new earth, another paradise, even (as he would come prophetically to claim) "a new heaven."[15]

But that would be to leave out the narrative truth: this fugue does not end where it began, though it ends, in the early eighteenth century, in a newly stabilized worldview shared by most literate Europeans despite sharp differences of class, gender, and belief. And its telos, from my point of view, is neither its modern end, nor (as nostalgic ideal) its medieval beginning, but its chaotic middle, where inventiveness and possibility, greed and hope propounded many new worlds beyond the one in which, always already, a nightmare of slavery began to forge its chains of association.

WORLDMAKING

> If I ask about the world, you can offer to tell me how it is under one or another frame of reference; but if I insist that you tell me how it is apart from all frames, what can you say? We are confined to ways of describing what is described. Our universe, so to speak, consists of these ways rather than of a world or worlds.
>
> . . . The many stuffs—matter, energy, waves, phenomena—that worlds are made of are made along with the worlds. But made from what? Not from nothing, after all, but *from other worlds*.
>
> —Nelson Goodman, *Ways of Worldmaking* (1978)

Many, many worlds were made and unmade between André Thevet's voyage to Brazil and Joseph Lafitau's to Canada, between the late masterpieces of Neoplatonism and the first of a theorized anthropology. I will not pay much attention to the construction of new religions, new nations, new polities, new colonies, or new empires, except as those condition or enable the new sciences, particularly cosmography, planetary astronomy, and anthropology, and the new genres of representation, particularly ethnography and the novel, though also the printed scientific report (an avatar of "writing to the minute" and an important developmental arena for the verisimilitude of de-

---

15. "Of the new heaven and of the new earth, which Our Lord made, . . . He made me the messenger and He showed me where to go" (Jane, *Select Documents* 2:48).

tail). For someone trained in literary studies these genres and disciplines are more accessible and easily analyzed than the more deeply mathematized physics or optics of the period; more importantly, they are ways of making *other* worlds, which are what especially interests me, and should interest anyone who cares about the powers of the imagination and the consequences of its productions.[16]

In my use of the word "world" I will not go so far as Nelson Goodman, who seems to imply that for every proposition or axiom or even semantic pattern there is an implied world for which it is true. For my category of "worlds" I would like to retain as an attribute the social concept of the habitable or inhabited. This category would include what anthropologists call (or used to call) a culture, as well as what a literary critic might call the "world" of a particular novel (which requires far more than one axiom to found—we judge the success of realistic fiction in part on how many interlocking axioms or laws it manifests coherently). Perhaps more strangely, or less "thickly," it would also include the unspecified habitability of the "innumerable worlds" of Giordano Bruno's controversial speculations, and the extension and non-human sentience with which the microscopic "world" is represented in the first decades of its accessibility. It would be a mistake to require that a "world" be inhabited by human persons, at a time of such ontological insecurity about that category, and such jealous hoarding of its privileges.[17]

Modern cultural anthropology, or ethnology as I will usually be calling it, is the major arena of *inhabited-world*making (especially *other-world*making), at least in terms of its explicitness of focus and of its historical consequentiality. From within the borders of the culture of science it articulates entire and distinct webs of possibility for human relations, actions, imagination, meanings.

---

16. It might be objected that botany, too, especially as seen through the lens of Mary Louise Pratt's treatment of it in *Imperial Eyes*, is a kind of "worldmaking," and that is true. But it is one that, like Enlightenment anthropology (see Chapter 9), participates in the great universalist program of the imperial nations that eventually begins to regulate the imaginative welter this book attends to. The Enlightenment is the end of my story, the closure. On the Renaissance roots of the universalist vision, see John Headley, "The Sixteenth-Century Venetian Celebration of the Earth's Total Habitability," and "Geography and Empire in the Late Renaissance."

17. The New World, for instance, was only debatably inhabited by humans, and the matter of privileges there, one of life or death. See Lewis Hanke or Anthony Pagden on the sixteenth-century debate at Valladolid, between Sepúlveda and Las Casas, over the humanity of the native Americans. Sepúlveda argued for the Aristotelian category of natural slaves, which would in practice have ensured that the Spanish would not even have to go through the motions of "negotiating" the seizure of American property and persons. On a world inhabited by bees, see Fontenelle, *Entrétiens* (from the 1740 edition, quoted in the Notes to Hargreaves's translation, 78–80).

Anthropology in its larger sense considers these cultural webs in pursuit of a more general and unified description of the human, per se. The ethnographies underpinning anthropological knowledge of cultures are subject to the limitations of human vision, especially vision of novelty, and human language (inevitably culture-bound as even the technical lexicons of the sciences are). The magnetism of the ethnographer's own cultural assumptions curves her descriptions of other cultures into globes that tend to function as versions—better, worse, or merely wondrous in their difference—of the home globe.[18] This process is even easier to watch before the development of masking vocabularies, when descriptions are made in "ordinary language" and without allegiance to an institutionalized megainvestigation. The earliest European ethnological societies all date from the first half of the nineteenth century, but one can see a century earlier, in the Jesuit missionary Lafitau's comparative ethnology, as well as in lesser works of the early Enlightenment comparing ancient and exotic religions, the staking out of the territory for investigation, the "invention of culture."[19]

Another modern genre, the fictional novel (whose very name recalls origins in the management of novelty and its associated wonder), bears a close resemblance to the structural features of ethnography and shares some liter-

---

18. The literature on the problems associated with these inevitabilities has grown to huge proportions in recent years. Critiques from the fields of literary studies and history include Edward Said's groundbreaking polemic, *Orientalism* (1979), James Clifford and George E. Marcus's collection, *Writing Culture* (1986), George Stocking's collection, *Observers Observed* (1983), and his *Victorian Anthropology* (1987). Influential examinations of ethnography from within are Talal Asad's collection, *Anthropology and the Colonial Encounter* (1973), Johannes Fabian's searing *Of Time and the Other* (1984), the special issue of *Ethnohistory* on travel writing (1986), that of *Cultural Anthropology* (introduced by Arjun Appadurai) entitled "Place and Voice in Anthropological Theory" (1988), Clifford Geertz's *Works and Lives* (1988), and Richard G. Fox's forward-looking collection, *Recapturing Anthropology* (1991). Many important thinkers unnamed here can be found in the collections listed above, but even then we are talking about the tip of the iceberg—and the iceberg of authors writing in English only, and mainly about the problematics of ethnography and the "field" encounters in which it is grounded.

19. For an almost ahistorical theoretical account of this relativisitic concept ("cultures"), generated in the same intellectual moment as Goodman's *Ways of Worldmaking*, see the second (1981) edition of Wagner's *Invention of Culture*. Wagner contrasts the "objective" invention, based on "observing and learning" (4) with the "uncontrolled fantasies . . . of Herodotus, or the travelers' tales of the Middle Ages," which "we can scarcely speak of [as] a proper relating of cultures" (3). The difference is indeed a difference of kind, in at least some respects; the concept of individual "cultures" was alien to medieval European thinking (though obviously medieval travelers noticed religious and behavioral differences of "rule" and "custom" between Christian, Muslim, Jewish, and Byzantine peoples). But when the Flemish monk William of Rubruck, at the end of the thirteenth century, likens the Mongol territory he is entering to "another world," we are not as far apart as Wagner claims.

ary genealogy with the enterprise we now call ethnographic. Exotic travel writing had equally intimate relations with the future science of anthropology and the art of the novel, both of which phenomena represent encounters with alien or exotic people or envision them collectively as cultures, and both of which seemed to emerge in recognizable forms at about the same time. The turbulent and transitional "early modern" period in European culture saw in fact any number of attempts to understand, trade places with, spy on, or see "inside" beings definitively abnormal: Tupinamba Indians, moon-people, flies, monsters, itinerant rogues, and, above all, women. The period saw, one might claim, the *creation* of an "inside," characteristically located in those others crowding the world "outside" the sensorium, or nation, or species of the writer, and generating for its literary environment the novel. *All* the genres occupied in such exoticist probings were sensational in one way or another, as long as the turbulence lasted, and even the great natural historian Buffon was a rhetorician and a sentimentalist. But the systematizing effects of the life sciences in the Enlightenment seem finally to have formalized the exclusion of sensibility from elite scientific practice and the exclusion of information-value from mainstream fictional narrative — exclusions that until recently functioned as constitutive for the enterprises of the so-called "Two Cultures."[20]

The interesting textual worlds examined in the following chapters are rarely novels — even Cyrano's *Voyage to the Moon*, Margaret Cavendish's *Blazing-World*, and Aphra Behn's *Oroonoko* have been denied that formal affiliation in some histories.[21] These disparate texts (cosmography, colonial report, empirical and theoretical works of natural philosophy and natural history, dialogues, fantastic voyages, utopias, exotic fictions, confessions) are interesting in many ways. One of these ways is in their preinstitutional uses of

---

20. Important recent works of literary history that take into account the epistemological relations of early modern fiction to other kinds of representation include Nelson, Adams, Davis, McKeon, and Hunter (see n. 8), and Ilse Vickers, *Defoe and the New Sciences*; for comparative and French studies, see Timothy Reiss, *The Discourse of Modernism* and Natalie Zemon Davis, *Fiction in the Archives*. Although not directly concerned with the history of early novelistic fiction per se, Karl Guthke's *The Last Frontier* cannot be left off this list. Denise Albanese's examination of the connected discourses of science and colonialism, in *New Science, New World*, involves much discussion of poetic texts. Claire Jowett's *Real and Imaginary Worlds: English Politics, Utopianism, and Travel Literature* will be a valuable addition.

21. They are simply absent from the influential histories of F. R. Leavis, Ian Watt, and Arnold Kettle; Hunter, in the ambiguously titled *Before Novels*, includes Behn, but not Cavendish or Cyrano de Bergerac (whose *Autre Monde* is defined by the literary historian Erica Harth as a work of "scientific popularization" [*Cyrano de Bergerac*, 4]).

now-familiar novelistic conceptual structures and language habits: the first-person protagonist-narrator; the value of individual experience; minutiae of description; the rhetoric of facticity; the reportability of bodily sensation and the reader's discovery of the sensational as pleasure; "verisimilitude" ("Truth's greatest enemy," said Richard Flecknoe in 1656 [*Relation*, 29–30]); the importance of identification in the reading experience; the "suspension of disbelief"; the exoticization of distance and the ratio of distance to the importance of belief; the salience of individual persons (characters) unaffiliated genealogically with protagonist, author, or implied reader; the fascination with internal states unmarked or disguised by describable exteriors of costume, carriage, or behavior (and the faith in their knowability); the conceivability of alternative cultures and environments (the assumption of what we might call the "versionicity" of the universe). All this and much more is present, in bits and pieces, among the texts and kinds of text to which this book solicits attention. These features in their eventual transfer to the zone of make-believe and aesthetic irreality seem to me likely to come trailing clouds of history, from the long period of heightened novelty and wonder which is their home, or homes. And these are the materials of worldmaking as we know it now.

A MAP

> Modernizing progress is thinkable only on condition that all the elements that are contemporary according to the calendar belong to the same time. For this to be the case, these elements have to form a complete and recognizable cohort. Then, and only then, time forms a continuous and progressive flow. . . . This beautiful order is disturbed once the quasi-objects are seen as mixing up different periods, ontologies or genres. Then a historical period will give an impression of a great hotchpotch. Instead of a fine and laminary flow, we will most often get a turbulent flow of whirlpools and rapids. Time becomes reversible instead of irreversible.
>
> —Bruno Latour, *We Have Never Been Modern* (1993)

For the sociologist of science Bruno Latour, the past itself (as conventionally/narratively understood—as "over") is yet another invention of the imaginary "modern" dispensation ("as if there were a past!" [*We Have Never Been Modern*, 72]). After a brief discussion of how the "irreversible arrow—progress or decadence" (73) has been laboriously constructed, starting in the seventeenth century, into a "continuous and progressive flow" in our historical imaginations, Latour introduces his "quasi-objects," the innumerable "hybrids" of nature and culture which modernism, as he sees it, constitutively denies in its drive towards the "purification" and separation of kinds and pursuits and modes of consciousness: "The beautiful order is disturbed once the quasi-objects are seen as mixing up different periods, ontologies or genres."[22] Latour is writing in defense and celebration of the new discipline of science studies, but the present work of literary history and criticism attends to just this "hotchpotch," with its capacities for "reprise, repetition or revisiting" (74). The hotchpotch came to my attention as a set of textual hybrids, which gradually revealed mixings, as well, of ontologies and "periods," at least as I had been taught them. The value of their study is essentially an imaginative one, a liberation from rule-bound generic confines without a loss of context or density of resonance, and a reunion of disparate enterprises under the rubric of making new or other worlds—a job of the mind as necessary as clearing the air of the one(s) we live in.

Two features of exclusion and inclusiveness require some comment before I go on to describe the book's arrangement of its materials. Enthralled by something closer to Foucault's "heterotopia," I have paid almost no attention to the concept of "utopia," though more than one work analyzed or at least discussed here is commonly classed in the genre that goes by that name.[23] This may seem perverse after so much reference to the liberatory powers of alternative textual worlds, but the pedagogical moralism associated with that category renders it alien as a literary idea to the main focus of this book: utopia is not really a version of "other world." Nowhere else can we see so

---

22. One of Latour's clearest examples of quasi-objects (other than Boyle's air pump as described by Steven Shapin and Simon Schaffer in *Leviathan and the Air Pump*) is: "a slag heap in northern France, a symbol of the exploitation of the workers, that has just been classified as an ecological preserve because of the rare flora it has been fostering" (2). The term seems usually to refer to intermediaries and systems that belong to both the human/social and natural spheres (artificially divided by "modern" thought). An even clearer example might be Christopher Wren's "robot scientist," described in Chapter 3.

23. In Foucault's beautiful and important discussion of what he calls *heterotopias*, "something like counter-sites" ("Of Other Spaces," 24), he offers a distinction I can adapt to my purposes: though he is only interested in territorially or spatially "real" places in constructing his term ("places that do exist and that are formed in the very founding of society," 24—cemeteries, for instance), he contrasts it to the more mimetic *utopia* ("sites that have a general relation of direct or inverted analogy with the real space of Society," 24). "Places of this kind are outside of all places, even though it may be possible to indicate their location in reality. Because these places are absolutely different from all the sites that they reflect and speak about, I shall call them, by way of contrast to utopias, heterotopias" (24). Like Foucault, I'd put "certain colonies" in this category (or his subcategory, "the heterotopia not of illusion, but of compensation . . . I wonder if certain colonies have not functioned somewhat in this manner" [27]); unlike him, I would also add certain genres and even the pre-institutional singular work of imagination—Kepler's *Somnium*, Gertrude Stein's *Tender Buttons*.

plainly what Nelson Goodman means when he speaks of worlds being "made . . . *from other worlds*" (*Ways of Worldmaking*, 6). Normative examples of utopia exert a pressure on readers to alter their own worlds in its direction — one drawn, like satire's, from the central conscious values of its author's social time and place. Its potential or intended otherness, difference, alternativity, novelty, and capacity for invoking wonder are seriously compromised by its ethical density and conventionality.[24] (For a similar reason, among others, I have chosen not to write a chapter on *Gulliver's Travels*, despite its having so obviously been made from the very "*other worlds*" presented in the following chapters.)

The borderless inclusivity of my choice of texts as far as nations of origin go might also need explaining. Although I *focus* on only two societies, English and French, I discuss Italian, German, and (more briefly) Spanish works as well, and include both Latin and vernacular works in my hotchpotch. This internationalism responds to the fact that, although the centuries from which my texts were drawn were important times of national consolidation for all but Italy, the book culture within which the management and eventual compartmentalization of wonder was perhaps most energetically taking place was one of polyglots and, as time went on, of translation. Personal libraries in the seventeenth century tended to be divided in auction catalogues by language (starting with Latin works and often ending with untitled and linguistically unspecified pamphlets, jest books, and "curious" or "philosophical" literature — that is, pornography); clearly early readers were not as picky as current academic departments about the language in which their reading was done.[25] As we move toward the eighteenth century, catalogues tend to consist more and more of books in the vernacular language of their owners,

---

24. For a stimulating, nuanced, and more generous view of the genre in the English sixteenth and seventeenth centuries, see Marina Leslie, *Renaissance Utopias*.

25. There were other kinds of border crossing as well; *Wonder and Science* treats some of the many books that were printed in countries where printers' and compositors' vernaculars were not those of the author (de Bry's *America, Part I*; Bruno's *De l'infinito*), not to mention books first published in languages not the authors' own (e.g., Psalmanazar's *Description of Formosa*) — or in such imaginary countries as Utopia (Godwin's *Nuncius Inanimatus*). I have looked at dozens of auction catalogues in the British Library from the sale of seventeenth-century private libraries (including some French libraries); there are also some in print (e.g., John Dee's, Walter Ralegh's, and Richard Burton's, all cited elsewhere) and a number of studies assessing circulation and readership through other means (printers' and booksellers' inventories, stationers' registers, citations, and so on). The literature amounts to a field and cannot even be representatively cited, but I should note my indebtedness to the work of Roger Chartier, Miriam Usher Chrissman, Robert Darnton, Sears Jaynes, and Margaret Spufford, as well as that of Dory Black (in her thesis, "Working Women's Writing in Early Modern England").

but translations are also increasingly common, and faithful, so the mixing of national intellectual cultures remains high—higher than it is at present (in America anyhow). Although the events of the Reformation and Counter Reformation certainly split Europe ideologically, these splits were intranational as much as international; and the long dominance of Latin as the language of philosophy and knowledge, as well as of serious poetry, had maintained a European culture of letters amidst the vernacular diversity, despite the hardship of travel before the centralization of nation-states began to provide connected infrastructures. Of equal importance for my relative indifference to national boundaries is the fact that the seagoing nations of Europe were almost uniform in their possessive and exploitative response to the geographical expansions of the known world; they all wanted and almost all created colonies, satellites—moons of their own.

Wonder and Science is divided into three chronologically overlapping parts. The two long chapters of Part 1 ("Imagination and Discipline") set the stage by charting the process of shifting frameworks for representation and comprehension in the late sixteenth and earlier seventeenth centuries, especially as this relates to culture and cultural groups. The first of these chapters examines both the increasingly readerly and phantasmatic cosmography and the increasingly scientized ethnography of sixteenth-century France and England. The works of "Royal Cosmographer" André Thevet mark a successful intersection of both old-fashioned wondermongering and new-fangled fictional rhetoric, particularly in his accounts of such self-enclosed "worlds" as islands represent and his manufacture of an experiencing narrator. His English near-contemporary, the mathematician and colonial investor Thomas Hariot, lent a report on Virginia, marked by now-familiar features of scientized ethnographic representation, to the first publishing *coup* of Theodor de Bry and his sons, who published the popular illustrated series *Les grands voyages* for over forty years (1590–1634).[26] The second chapter is concerned with the attempted (and fitful) exclusion of "wonder" and the subjective from the codification of learning, and discusses such wonder-ridden and stylistically beautiful manifestoes and polemics of the seventeenth century as Francis Bacon's *Novum organum* and Thomas Browne's

---

26. The final volumes were printed by their successor, Matthias Merian (the Elder), father of the botanist, engraver, and proto-ecologist Maria Sybilla Merian, about whose adventures in Europe and particularly in Surinam Natalie Zemon Davis has recently written in *Women on the Margins*.

Pseudodoxia epidemica, as well as an exemplum of the new orientation in Robert Plot's microcosmography of Oxfordshire.

Part 2 ("Alternative Worlds") addresses the wonderful, the sensational, and the sublime in several seventeenth-century works, many of them overtly fictional, that set out to represent new and/or imaginary worlds, inevitably habitable and usually inhabited: lunar fictions and Margaret Cavendish's feminist quasi-utopia, *The Blazing-World*, but also Galileo's *Sidereus Nuncius*, Fontenelle's *Entrétiens sur le pluralité des mondes*, and Robert Hooke's *Micrographia*, among others. Although pursuit of the "real character" and other universal (denotative) language schemes was alive throughout the century, philosophers and astronomers were often exploiting the powers of excitation as much as were writers of fictions (Steven Shapin's term for their efforts is "virtual witnessing"), and these powers were largely released through language. Henry Stubbe called the new philosophers "novellists."[27] One can offer only so many people a direct look through one's telescope, and indeed many refused Galileo's offer or saw nothing when they looked. These powers are now the special province of a "dissociated" literary institution, but they gained much in the way of technical development from the linguistic and communicative pressures borne by early modern natural philosophy. Stepping back for a very long view of the European world of "letters," one might almost see the imaginative prose of the modern West as the persistent repetition and recall of those moments when "a new planet" swam into our ken. Keats thought so, at any rate.

The three chapters of Part 3 ("The Arts of Anthropology") concentrate on one sixteenth- and several diverse seventeenth- and early eighteenth-century avatars of anthropology as both a discourse and a quickly changing set of genres. They take up the relations of ethnographic fashion plates and antifashion polemic to both anthropology and novelistic narrative. From this grounding in attention to the body and its culturally significant adornment, the discussion moves on to the shared and opposed aims and techniques of ethnography and realistic fiction, via a reading of Aphra Behn's fashion-conscious and body-conscious novel, *Oroonoko*. The final chapter concerns the same generic tension, but in the form of a dialectical relation between two texts, Lafitau's monumental opus of comparative ethnology, *Moeurs des*

27. On Stubbe, see McKeon, 71. Much of the sensational and wondrous power of Hooke's *Micrographia* comes from the illustrations, but the illustrations neither stand alone nor excite alone; one has to be made aware of the scale of the depicted objects before they are strictly speaking "wonderful." Drawings of the moon such as those with which Galileo illustrated the *Sidereus Nuncius* are not even abstractly interesting without the text that identifies them.

sauvages amériquains, and "George Psalmanazar's" false but initially credited ethnographic memoir (read and absorbed by Swift), *The Geographical and Historical Description of Formosa*.

The early texts on fashions (including body fashions) convey a sense of other cultures (including European ones) as aggregates of performance, and in fact, early ballet often consisted of performance of these fashions and their associated props, both by "exotic" people themselves and by Europeans dressed in their borrowed finery or imitations of it.[28] Behn's *Oroonoko*, once charged with plagiarizing a contemporary travel account of Surinam, and referred to sometimes as a "voyage," is a perfect instance of the hybrid—as ethnographic as it is novelistic, and thereby offering us surprising insights into the germination of both enterprises. Lafitau's *Moeurs* not only records voluminous ethnographic detail from his six years of fieldwork with the Iroquois of southern Quebec, but compares most of its data on the Iroquois with what he can gather from other reports on American peoples and with the ancient civilizations of Eurasia, forming from these materials a body of ethnological theory about the universal religious instinct in human societies (as well as describing systematically the kinship system of the Iroquois). Paired with Psalmanazar's risky ethnographic hoax of a few years earlier, Lafitau's work can show us the continuing but now unacknowledged interpenetration of "objective" and "subjective," epistemological and poetic texts, as well as illustrating the universalizing tendency of the emerging science of anthropology. Psalmanazar, the wonderful "Formosan," especially in his life as temporary exotic toast of London town, literalizes to the point of criminality the notion of culture as performance, while Lafitau merely puts it to use in the process of acclimating himself to the social world in which and about which his studies were carried out.

The destination of *Wonder and Science* as originally conceived had been a treatment of *Gulliver's Travels*, that *summa* of discourses, but Psalmanazar's wonderful lie brings more clearly into focus the aggression and desperation (and underground success) of the outnumbered but undefeated forces of

---

28. As Behn reminds us in the text of *Oroonoko*, the feather skirt and headdress she brought home from Surinam were used in the opening performances of Dryden and Howard's play, *The Indian Queen* (later—1695—Purcell's opera). See chapter 8. On exotic dress and the performance of foreign cultures in ballet and opera, see my notes in Chapter 7. (See also Michèle Longino, "Staging of Exoticism," for a critical study of French classical drama and its major "Oriental" characters; Julia Douthwaite, *Exotic Women*, on the roles of non-European women in late seventeenth- and eighteenth-century French fiction; Tom King, *Queer Articulations*, on theatrical and transnational queer performance in the Restoration.)

"the thing which is not." *Wonder and Science* concludes with a coda looking ahead from lonesome Robinson Crusoe on his island to two resistant and anomalous aliens of the turn of the nineteenth century, the English working girl who called herself "Princess Caraboo" and Victor, the presumably French (?!) "Wild Child" of Aveyron, who was constructed and represented to the world first by Jean Marc Gaspard Itard, a physician who worked for the Imperial Institute for the Deaf and Dumb in Paris, and later by the French filmmaker François Truffaut. The interest for me in these two aliens lies partly in their disturbance of the ethnological and anthropological graphs laid down by Enlightenment science and colonial capitalism over the elusive diversity of human behaviors and intentions. Where are these creatures "from"? Not from the imaginary "Javasu," exactly, but is Devonshire a better answer for the Princess Caraboo? And in what sense is the mute and feral Wild Boy French? Or human?

As tiny black holes in the map of cultural space they stand not only for the incomplete assimilability of the human world into the world of the systematically "known"—for the human person herself as a "wonder"—but also for the poet and poetry as it came to be understood under the Romantic dispensation; "Caraboo" and "Victor" are inventors and refusers of language, an envelope wholly inadequate for them as it stands. Of course they are not texts (though texts have been made of them), but their salience is a salience in the history of representing other worlds. "Caraboo" created herself from such representations, and "Victor" evaded them altogether. And both of them could say, revising Milton's Satan: "I myself am a world."

Timothy Reiss's Foucauldian classic, *The Discourse of Modernism* (1982), treats many of the same texts as *Wonder and Science* does in light of the rise to dominance of a new "discourse class" (or episteme) that he calls "analytico-referential." One underlying hope of my account is that it can portray a persistence—or at least a latency—of ways of knowing less firmly exclusionary and dichotomized than this dominant modern discourse class. A feminist literary historian cannot be satisfied with a concentration on the "dominant" mode because it is constructed, in part, to exclude or objectify women and their intellectual fellowship. This particular framework of understanding was constructed to objectify and maintain as imaginary *many* forms of being occupied by the disenfranchised—women, monsters, colonial subjects, even the living "things" now understood as the proper objects of the human and life sciences. In trying to describe the expansion and fragmentation of representational prose in early modern Europe one must take into account the involvement of the disenfranchised, the obscure, and the imaginary, in whose ambivalent hands the features of a dominant discourse are usually undermined. Some of these imaginary beings were *agents* in the production of the separated and mutually exclusive domains. So my sense of the hegemonic is weaker than Reiss's, as is my desire to find a systematic coherence in the events and objects gathered here.

Bruno Latour's invigorating recent polemic on the intellectual developments of this period, already cited, downplays the hegemony and monotony of the *mentalité* Reiss explicates (as well as its "revolutionary" severance from an imaginary past). I don't agree, though, with Latour's solution to the improbable separation of spheres assumed by "modern" thought, that anthropology should not only "come home from the tropics" (Latour, 100) but expand its scope to become our way of knowing everything (though the prospect of a science that considered all cultures evenly, even its own, has its allure).[29] According to Freud, "A room illuminated to its furthest corners would be uninhabitable." *Wonder and Science* agrees with him. And it is the troublesome heart of anthropology to be about faraway places. The discipline has functions other than those of serving an imperialist state or putting together a carefully organized portrait of "the human." Or rather, it is a genre as well as a discipline and purveys an Imaginary as well as a rationality. Anthropology brought "home" has provided those of us in the unexamined groups with many fascinating new objects of attention and assuaged the guilt of those who feel understandably awkward about watching and explaining dominated cultures for a dominating one. But it continues to estrange and exoticize what it looks at; as the global village loses (for some of us) the dimension of space through the omnipresence of mass media, the internet, and air travel, it is less imperative that anthropological objects be located physically outside the home territory of the anthropologist (an increasingly difficult phenomenon to define, anyway). That does not mean that anthropology is home from the tropics, any more than are writers of science fiction who locate their plots in New York City.

The ethical problem of the poet (which category includes the novelist) is related to the closeness, still functionally resonant, between poetry and an-

---

29. In his introduction Latour praises anthropology for its ability to hybridize the object of its descriptions: "Once she has been sent to the field, even the most rationalist ethnographer is perfectly capable of bringing together in a single monograph the myths, ethnosciences, genealogies, political forms, techniques, religions, epics and rites of the people she is studying. . . . In works produced by anthropologists abroad you will not find a single trait that is not simultaneously real, social and narrated" (7)

thropology and between the travel relation and the expansion of European states. *Wonder and Science* means to find the entangled textual fibers of these increasingly differentiated modes of looking or understanding, as manifested in early modern texts of ambiguous provenance or genre, in part to restore a dimension of historicity to the particular language habits of emerging modern genres, in part to discover functions besides colonialist complicity for poetic estrangement and the cognitive emotion, wonder, which is its Final Cause.

Fig. 1. Marguerite de Roberval, niece of the French explorer Roberval, castaway on a multiply imaginary island. From André Thevet's *Cosmographie universelle* (Paris, 1575). By permission of The John Carter Brown Library at Brown University.

Part I
IMAGINATION AND DISCIPLINE

THE CHAPTERS IN PART 1 illustrate the war of impulses carried on between the times of the French traveler and cosmographer André Thevet (author of a major popular work on the New World, the *Singularitez de la France Antarctique* [1557]), and the methodical account by Robert Plot, the Ashmolean Museum's first curator and Oxford's first professor of chemistry, of a more manageable territory in his *Natural History of Oxfordshire* (1677). From "Singularities" to "Natural History," the century saw an effortful transformation of epistemological methods and tastes throughout western Europe.

Thevet's large, exuberant, increasingly imaginary representations of worlds apart, especially the Americas, meet a kind of comeuppance in the chaste and technically accurate account of the mathematician and colonial investor Thomas Hariot; Hariot's ethnographically tilted *Briefe and true report of the New-found land of Virginia* (1588) still commands respect from ethnohistorians, while Thevet gets called a liar. In Thevet, on the one hand, we can see early stirrings of the all-consuming first-person narrator (already glimpsed in the narratives of Columbus and Cabeza de Vaca) that will come home to roost in the bourgeois novel. In Hariot, on the other hand, we have an instance of the more easily credited and utilitarian "author-evacuated" delivery of an ethnographically present-tense local culture. Hariot's approach will blossom eventually as a fieldwork-based ethnology, fully visible in Father Joseph Lafitau's 1724 masterpiece of comparative ethnology, the *Moeurs des sauvages amériquains*. What is jettisoned from Thevet's approach to other worlds will gradually be converted to the uses of writers who look for imaginative opportunity *as opposed to* the management of knowledge (consider the equally well-traveled Lemuel Gulliver).

The divisions of labor and conflicts of desire represented in this dynamic belong to the larger story once anachronistically conceived as the "Scientific Revolution." With conscious clarity and institutionalizing fervor, natural philosophers (there were no "scientists" yet to revolve), such as Francis Bacon, René Descartes, and the smaller fry who catalogued and categorized, took it upon themselves to bring rigorous method to the information explosion of their times. Chapter 3 examines some of the manifestoes and exemplars of the new mode, with its impulse of expurgation as well as of expansion. Bacon comes first, as the most influential of the expurgators, as well as one of the most resonant and beautiful writers among them; he knew well what powerful pleasures he was rejecting in his program of philosophical renewal and purification. The chapter ends with Plot's dutiful attempt to put his idol Bacon's prescriptions to work, on a local culture less challenging than Hariot's Algonquin "Virginia."

By the time of Plot's *Natural History*, Descartes and Bacon have come and gone, Gresham College has been founded — as well as the Académie des Sciences in Paris, the Royal Society in London, the Accademia del Cimento in Florence, and several others. Public museums exist, scientific journals are published in vernacular languages, herb-women and out-of-work soldiers act as research assistants to the scientific virtuosi (see Keith Thomas, *Man and the Natural World*, 73), and "arts" such as chemistry and geography are taught in the universities (see Phyllis Allen). Religious authority has been questioned at gunpoint by the wars that wracked Europe; in England a divinely anointed king has been beheaded and replaced by a commoner. The political and especially the commercial structures of most western European nations have adapted to the expansion of colonialist enterprise, the traffic in slaves has more than tripled since Thevet's time, and most of a transformation in intellectual culture has taken place.

My purpose in Part 1 is to characterize that transformation by examining some crucial world-ordering texts, not to hypothesize its causes, which are the material of social and economic history. It is a transformation that brings the literary culture of western Europe much closer to the individuated, divided relation it now has with the "disciplines" of the sciences. But along the way of change are visible many a strange hybrid and wonder-ridden testimony.

II • TRAVEL WRITING AND ETHNOGRAPHIC PLEASURE
André Thevet and *America, Part I*

> For things whereof the perfect knowledge is taken away from us by antiquity, must be described in history, as geographers in their maps describe those countries whereof as yet there is made no true discovery; that is, either by leaving some part blank, or by inserting the land of pigmies. . . . To which purpose I remember a pretty jest of Don Pedro de Sarmiento, a worthy Spanish gentleman, who had been employed by his king in planting a colony upon the straits of Magellan: for when I asked him, being then my prisoner, some questions about an island in those straits, which methought might have done either benefit or displeasure to his enterprise, he told me merrily that it was to be called the Painter's Wife's Island; saying, that whilst the fellow drew that map, his wife sitting by desired him to put in one country for her, that she, in imagination, might have an island of her own.
>
> —Walter Ralegh, *History of the World* (1614)

WRITERS ADORE A VACUUM. During the long period covered by this book "science" was mostly still being written by writers, only some of them writing *as* scientists (a word not invented till the 1840s); this is true especially for the kind of material later to be codified as ethnographic, under the aegis of the science of anthropology. As any number of recent works have demonstrated, the "New World" in the sixteenth century provided a writerly occasion second to none, unless perhaps to that of the "new heaven" emerging simultaneously from a more sedentary exploration. Ethnohistorians now often look to sixteenth-century texts for information in their attempts to form an accurate picture of the American histories and social structures fatally interrupted during what they call the "contact period," but these texts were produced in answer to a complex of desires and needs less regulated and less utilitarian than is the average modern ethnography. Indeed, some of the desires spoken to by books such as André Thevet's or Theodor de Bry's *America* are much more like those of the Painter's Wife than those of Ralegh; I will focus here on the part such desires had to play in the forging of a textual New World for European readers.[1]

In this chapter I look at the relations of ethnology and fictional narrative in some books written before the successful establishment of French and English settler colonies in the New World—before, that is, the geographer's map became too crowded with verifiable place names to bear the insertion of "the land of the pigmies." I look mainly at the major works of the traveler André Thevet, a Dominican friar and the Royal Cosmographer of France for most of the second half of the sixteenth century, and more briefly at *America, Part I* (1590), the famous collaboration of the English scientist and colonialist Thomas Hariot with the painter John White and the Dutch Protestant engraver and publisher Theodor de Bry.

PLEASURE

Sixteenth-century travel accounts such as these bear the mark of "discoveries" not only personal and scientific but fully, globally, political. They constitute at once the origins of the modern science of ethnology and the textual justification of several governments' policies of colonial appropriation. That the developments of colonial empire and prose fiction were also chronologically parallel is in itself evidence of nothing (since the emergence of the printing press and commercial publishing were simultaneous with the marked increase in quantity and sophistication of prose fiction), but I think that in fact there are significant connections to be made. The salient factor common to these and other *sequelae* of national consolidation and expansion during the early modern period might be formulated most generally as pleasure.

But "pleasure" is an unwieldy category—as is for that matter "travel writing," which for our purposes will have to include any ethnographic writing (no matter how fraudulent) grounded in personal experience. We will be interested in several specific forms of delight, wonder, frisson, satisfaction, existing as objects of demand in a suddenly expanded marketplace and also as aesthetic effects reliably, characteristically, available in a number of representational genres. Ex cathedra, Michel de Certeau pronounces, "What travel literature really fabricates is the primitive as a body of pleasure" ("Ethno-Graphy," 226); Marianna Torgovnik's recent meditations on Mali-

---

1. In *The Creature in the Map* (32–40), Charles Nicholl puts the passage from Ralegh (*Works* 4:683–85) into the context of his search for information on the whereabouts of El Dorado (and the pragmatic rhetoric of hiding it).

nowski's field diaries (*Gone Primitive*) substantiate this claim for modern ethnography. I have been thinking about the pleasure not only of the producer but of the reader or consumer, and in relation to texts from the predawn of ethnography. The pleasure of looking, whether at subjugated naked women on a tropical isle or at elegant engravings in an expensive illustrated book, was, in the sixteenth century, a pleasure on the march. To what pain did it correspond? What profit did it serve?

Most of the pleasures Freud discusses in *Civilization and Its Discontents* function anaesthetically, as forms of escape from the suffering inflicted by "all the regulations of the universe" (76); the last form he brings up before psychosis and religion is the creation and enjoyment of "art," "making oneself independent of the external world, by looking for happiness in the inner things of the mind." "As a goal of life," he says of the enjoyment of beauty (whose sources are scientific and "natural" as well as "artistic"), "this aesthetic attitude . . . offers little protection against the threat of suffering, but it can compensate for a great deal" (82). He complains a few sentences farther on of the mysteriousness of beauty's function: "Beauty has no obvious use; nor is there any clear cultural necessity for it. Yet civilization could not do without it" (82). Like Darwin, Freud gives up on the question of beauty's function, which I cannot answer either, posed at such a high level of generality. But if we draw it closer to the grimy context of imperialist expansion under capitalism, it is rather less of a conundrum. Beauty is, among other essences, the attribute that desire projects on what it is pleased to consume, and the production of desire and consumption are prerequisites to the growth of markets.

The most obvious case among sixteenth-century literatures of mixing business with pleasure — colonial business, erotic pleasure — is that of Hariot's inventory of Virginian "commodities" (including Algonquin people and their customs), illustrated by White (later Governor of the Roanoke colony), and engraved and published by de Bry as the first volume of *America*, a series of proto-ethnographic illustrated travel accounts that began the series of series, the *Grands voyages* (1590–1634). Before the inauguration of de Bry's *Great Voyages*, the most important sixteenth-century illustrated texts containing ethnographic data and composed by European observers are Fernández de Oviedo's *Historia general de las Indias* (Seville, 1535; composed in Mexico), Hans von Staden's account of his captivity among the "cannibals" of Brazil (*Warhafftige Historia und beschriebung einer landtschafft der Wilden* . . . [Marpurg, 1557]), Thevet's *Singularitez de la France antarctique* (Paris, 1557) and *Cosmographie universelle* (Paris, 1575), Jean de Léry's account (written partly in angry response to Thevet's works) of his experience in Brazil, the *Histoire d'un voyage fait en la terre du Brézil* (La Rochelle, 1578), and Girolamo Benzoni's *Historia del mondo nuevo* (Venice, 1565).[2] These commodities offered for sale all over Europe and regularly translated are important, not only to reenvisioning the early history of ethnography, but to understanding the history of reading for pleasure—a history which must include novellas, prose romances, voyages (real and imaginary), cosmographies and natural histories alongside the more commonly studied history, poetry, and philosophy favored by humanists.[3] (The gentleman-scholar John Morris's library, to take a random instance, included individual voyage accounts by Léry, Marc Lescarbot, Samuel de Champlain, and Thomas Gage, as well as Strabo's *Geographia* and Ramusio's *Viaggi*, the rough Italian equivalent of Richard Hakluyt's *Voyages*.)[4]

Both Thevet's and de Bry's publications exploit the potential of "manners and customs" for providing (to writer and artist as well as reader) erotic and narcissistic pleasure. But Hariot's *written* text, in its antinarrative, taxonomic, and utilitarian rhetoric, points the way toward a scientific discourse in which authority will ultimately derive from experience verifiable because *repeatable* (the historical dimension of human life and lives will have to be left out).

---

2. For an annotated bibliography of early illustrations on maps and broadsides as well as in printed books, see William C. Sturtevant, "First Visual Images of America."

3. Ethnography, as we more soberly understand the practice, did not begin with Thevet and Hariot or the other books listed here. Its most recognizable forerunners were two Franciscans, Andrés de Olmos and Bernardino de Sahagún, who spent years in Mexico interviewing native informants in Nahautl for the purposes of compiling what amounted to conversion handbooks. The Codex Tudela (presumably de Olmos's work) and especially the twelve books of Sahagún's "General History of the Things of New Spain" (see Sahagún, *Florentine Codex*) are manuscript sources of great importance (and danger) in the modern scientific effort to piece together a description of Aztec culture and social structure. Sahagún was interested in most of the same phenomena that interests modern professional ethnologists—kinship relations, religious beliefs, rituals and ceremonies, medicine, proverbs and folkways—but his point of view was pragmatically evangelical. He collected data from his informants so he and his fellow missionaries could better discern heresy and backsliding among their Mexican flocks. The works, which remained in manuscript and fell into disuse and disrepute after a change of imperial policy in New Spain, were beautifully illustrated, mostly by Mexican artists, even as most of the verbal text was dictated by Mexican informants. Because of their lack of circulation, they are more important now for ethnohistory than for the history *of* ethnology or of ethnography. On Sahagún in the context of genre see Walter D. Mignolo, *Darker Side of the Renaissance*, chap. 4.

4. For the complete catalogue, see T. A. Birrell, *Library of John Morris*. The catalogue of Secretary of State Thomas Smith included Sebastian Münster's *Cosmographia* and Richard Eden's translation of Peter Martyr's *Decades of the New World*; the charming Captain Cox of Coventry owned, at his death, thirty-three romances, four plays, seven ballads, and three madrigals. See Sears R. Jayne, *Library Catalogues of the English Renaissance*.

Though he bases his authority on witnessing, Thevet's experiences are likely to be repeated only in the imaginations of his readers (and are more often than not themselves repeated from the texts of other voyagers and cosmographers). The "knowledge" expressed in his works tends to be rather a representation or imitation of knowledge, as the novel imitates "true history." For Thevet, native Americans are frissons, *topoi*, characters—a medium for the textual rhythms of his dynamic, associative consciousness—and the function of his books is to articulate settings for the imaginary adventures of well-to-do and leisured readers.

In discussing Thevet I will be looking most closely at the kinds of private satisfaction provided by the cosmographies of the unscrupulous climber, "un homme nouveau," as Frank Lestringant labels him in his definitive study (*André Thevet* [1991]). But the notoriety and vendibility of Thevet's work suggest these pleasures added up to a mass phenomenon, with implications for the social as well as the strictly literary history of reading. To quote Freud one last time on the subject of (an)aesthetic pleasure:

> Another procedure [of securing happiness] . . . regards reality as the sole enemy and as the source of all suffering. . . . The hermit turns his back on the world and will have no truck with it. But . . . one can try to recreate the world, try to build up in its stead another world in which its most unbearable features are eliminated and replaced by others that are in conformity with one's own wishes. But whoever, in desperate defiance, sets out upon this path to happiness . . . becomes a madman, who for the most part finds no one to help him in carrying through his delusion. . . . A special importance attaches to the case in which this attempt to procure a certainty of happiness and a protection against suffering through a delusional remolding of reality is made by a considerable number of people in common. (*Civilization and Its Discontents*, 81)

Freud, of course, is talking about religion. But I am talking about the cosmographer whom Lestringant credits with "l'invention du Nouveau Monde" (104) and his many readers. Was the reality of Europe in the late Renaissance in fact (and not only for Americans and Africans) a "source of all suffering," "an enemy" from which a reading and writing population constructed, "by a delusional remolding," a new heaven and a new earth in the Southern and Western hemispheres? If so, this aesthetic function of cosmography (which produces, not a mere "body of pleasure," but a *world* of pleasure) could be seen (hyperbolically, perhaps) as one aspect of a comprehensive phenomenon: the Age of Discovery, composed as the psychotic transformation of the Age of Religious Wars. The rest of this chapter will examine the technical features of this transformation and their valences with the emerging institutions of ethnography and fiction.

THE ISLES OF ANDRÉ THEVET

The dissemination of first-hand information from the voyages involved from the start the work of professional writers and encyclopedists who collected and often recast in their own words the journals, letters, and oral reports of such early explorers as Columbus, Vespucci, Magellan, Pigafetta, Verrazzano, Cartier, and so on. European readers, scientists, investors, and would-be adventurers got their vicarious experience of the world outside Europe as often as not from the big books of Ramusio, Martyr, Benzoni, Eden, Thevet, Hakluyt, de Bry; the atlases of Mercator, Ortelius, Blaeu, von Linschoten; and the natural histories of Gesner, Topsell, Belon, Aldrovandi, Acosta, and Fernández de Oviedo. The stuffing of this paragraph with lists is apropos; one of the major effects (and pleasures) of the anthologies and collections was that of abundance and endless choice — precisely what the actual experience of distant exploration did not offer, with its "many sorrows, . . . labor, hunger, heat, sickness, & perill" ("Epistle dedicatorie" in Ralegh, *Empyre of Guiana*). From before the beginning, ethnography was to participate, at least rhetorically, in that embarrassment of riches which expansionist Europe never perhaps found embarrassing enough, offering to reader-buyers an apparently unlimited panorama of exotic morphology — both physical and cultural — that seemed to exist for their wandering eyes alone.[5]

No one wrote bigger books than André Thevet (c. 1516–1594), the Royal Cosmographer to the last four Valois kings (a largely honorary post) and the author of the popular and much-translated *Singularitez de la France antarctique* (1557). Thevet's several illustrated works of cosmography reveal a connection between wealth and natural history, manifesting the acquisitive and accumulative impulse that would flower as well in trade as in science, through a vulgar possessiveness that serves to clarify the better-bred gigantism of his fellow text-collectors (whom he often attacked and from whom he just as often stole). Although he himself had traveled as far as America,[6] most of his

---

5. In a discussion of English-Indian trade, James Axtell describes what to the native Americans was the "Englishmen's implausible preference for the greasy beaver robes they had worn for a year or more in their smoky lodges. Far from the hatters' workshops and high fashions of London, they could not appreciate that prolonged wear removed the long guard hairs from the downy, barbed underfur used in felting. Only the English could savor the delicious irony of how 'foule hands (in smoakie houses) [had] the first handling of those Furres which are after worn upon the hands of Queenes and heads of Princes'" ("English Colonial Impact on Indian Culture," 253; interior quotation from Roger Williams, *Key into the Language of America*).

6. Falling ill in Brazil after about ten weeks, he was forced to cut his visit short and probably never set foot in North America, though he "came very close to Canada" on his way home (*Singularitez*, 148r [*Thevet's North America*, 132]). I will be citing from the French first editions (or, in

American material is cribbed — from such sources, oral as well as printed and manuscript, as the pilot Alfonse de Saintonge, the explorer Jacques Cartier, the cosmographer François de Belleforest (his one-time collaborator and later his most hated rival), René Laudonnière (whose account later became part 2 of de Bry's *America*), Giovanni Ramusio, Francisco López de Gómara, the *Codex Mendoza*, Andrés de Olmos, John Cabot, and others collected by some of the historians named above.[7] Like the charming medieval plagiarist, Sir John Mandeville, Thevet wrote himself into the records of other men's experience; unlike Mandeville, he did so in a print culture where plagiarism and literary property had come to mean something (as his own accusations against Belleforest, Hakluyt, and even Rabelais [!] attest).[8]

A particularly stark instance of Thevet's writing himself in occurs in his discussion, in the late and still largely unpublished "Grand insulaire," of the geographical location of "France Royal" in New France: "the sun has a meridian [there] as high as the meridian of Angoulême the place of my birth, and has its noon when the sun is in the south-southwest.... And when in Angoulême [it] is noon, at France Roy and on our isle of Orleans it is only nine-thirty A.M." (406v). Roger Schlesinger's note informs us that "this passage is practically word for word from Alfonse [de Saintonge], but the lat-

---

the case of the *Singularitez*, the second [Paris, 1558], of which a modern facsimile exists [Baudry, 1981]). The only contemporary English translation of any of Thevet's work, a selection of chapters on North America (Schlesinger and Stabler, *Thevet's North America*) includes foliation from these editions throughout. This book also includes the only published text of the Canadian portions of a manuscript work, the 1586 "Grand insulaire et pilotage d'André Thevet."

7. See "Sources" in Schlesinger's introduction to *André Thevet's North America*.

8. In fact, it hadn't quite come to that yet, at least in English where "plagiarism" had not yet become part of the language (though *plagiaire* does appear in sixteenth-century French). None of the related entries in the *OED* is dated before the turn of the seventeenth century; though Thomas Browne claimed (in 1646) that "Plagiarie had not its nativitie with printing, but began in times when thefts were difficult," the word in our modern sense clearly belongs to the history of printing. Thevet published his works "avec le privilège du Roy," in the days before copyright — such privileges (limited term monopoly rights over the reprinting of a work) were afforded to about 5 percent of works printed in France in the first quarter of the sixteenth century. See Armstrong, *Before Copyright*. Thevet became in fact quite litigious as an author, in a literalization of the fictional phenomena of possession this chapter focuses on. The degree to which writers in the period of transition could potentially own knowledge, or sources, despite the absence of the copyright, was in some ways greater than it is today. Thevet, who for a while owned the sole manuscript copy of the *Codex Mendoza*, invites the reader "famished" for a look at this gorgeous account of Aztec history and customs, put together in Tenochtitlán for a Spanish viceroy, "to come see me and I will show him something that will be able to satisfy him" ("Grand insulaire" 181v). Thevet got the book from some French privateers who intercepted the Spanish ship bringing it back to the Emperor Charles V. Later Hakluyt bought it from Thevet, and Samuel Purchas published it in *Purchas His Pilgrimes* (1625). Like many other treasures displaced or precipitated by the voyages it ended up in the Bodleian Library.

ter's comparison of times is between 'Canada' (the town or province) and La Rochelle.... Thevet... substitutes the name of his birthplace, Angoulême, for La Rochelle and adds the Isle of Orleans for good measure" (Schlesinger and Stabler, 118, n. 146). In this rather poignant case of writerly narcissism, Thevet, while making no claims to authoritative eye-witness experience of New France (they are made elsewhere), orients America by means of his own obscure and irrelevant origin.[9] Not only does Angoulême here play a role once delegated to Jerusalem (the properly "Oriental" orientation point of medieval cosmography), but the notion of distance itself is personalized, individualized, in an extraordinary specification of the mental habit we now call Eurocentrism.

This is Thevet's perpetual habit, at all textual levels, and it offers a nice allegory of the probable or at least possible satisfactions of cosmographic reading in his avid but still relatively untraveled century. The distance that is so inherently pleasurable to a reader of voyages is the distance from oneself, with all that implies of gaps one might (or might not) close, differences one might absorb, spaces one might annex. When Robert Frost speaks scoffingly of Pascal's fear of the vast spaces of his newly enlarged cosmos and boasts of having, "so much nearer home," his "own desert places," he speaks perhaps from the perspective of a culture for which that absorption and expansion have already been internalized and to which those vast spaces—American as well as astronomical—are quite literally "home."[10]

Rather than offer, then, as Samuel Purchas would a century later, "a World of Authors" (*Purchas His Pilgrimes*, vol. 1, bk. 1, p. 1) Thevet offers us himself as a container of the world; this self *universelle* is not divided into authors but into *Singularitez*, and *plusiers isles*, to borrow from his titles.[11] The chapters of his *Singularitez* are verbal equivalents to the medallions that would increasingly come to surround the title on the title pages of books like his, or the borders of engraved world maps and maps of continents. The plea-

9. The organizing principle of Lestringant's 1991 biographical study is grounded in the view that "to the very end his career carries the trace of his initial obscurity ... without title or genealogy, unequipped with privilege or family history, Thevet was led to cosmography ... [,] which apprehends instantaneously ... the vast spaces of a reinvented world" (11).

10. See "Desert Places." Lestringant (1991) speaks of Thevet's ability "a tirer le meilleur parti d'un capital symbolique" (35), aligning his compositional practice with the practices of the venture capitalists and colonialists with whom he shared his New World. "To accumulate is to conquer the world of social wealth," says Marx (*Capital*, chap. 24.3); Thevet, like Bourdieu, knows this principle in terms of intellectual prestige.

11. *Cosmographie universelle* (1575); "Description de plusiers isles par M. André Thevet" (1588).

sure in plurality and multiplicity is not to be denied (and has a glorious future ahead of it in both the science and the fictions of seventeenth-century astronomy). It is to be, however, encapsulated—so many islands, so many singularities—so many jewels, *objets d'art*, exhibits, tableaux in the cabinet of the book. The mutual isolation and display of the many wonders and data collected in Thevet's books participate in the apparently compulsive aestheticization of Renaissance geography, ethnography, and natural history—at levels deeper, as I have suggested, than that of book production or private collection.[12]

The divisions of books such as Thevet's into isles, wonder-nuggets—in short, chapters—is not new in relation to exotica. The *Odyssey* produces an island for every marvel, just as Rabelais does; Marco Polo's chapters surround and contain individual cities or provinces (sometimes called isles) of Asia; the twentieth-century aesthetes Italo Calvino and Jorge Luis Borges have continued the practice in *Invisible Cities* and *Atlas*.[13] The salient factor in Thevet's case is that his books operate in another register than that of romance; they have, self-consciously, a quite pragmatic immediate context available (if not necessarily to him) in the dissemination of usable information to the potential exploiters and conquerors of the "isles" and marvels his books display. In fact, Thevet is in some ways most directly aiming at that audience (even when copying his data from works already in print)—he refers to "pilots and sailors" among his readers as often as not when he is articulating his motives, "so that [they] might have occasion to be grateful to me and avoid the perils, which threaten them from all sides" (1586, 157^v). The status of coherent and practical information is (already) higher than that of marvels, however elegantly conceived. The reader who needs longitudinal direction is more important to Thevet than the idle aesthete or dilettante who merely enjoys distance. But Thevet's ideal reader is only another case of the absorption of contemporary practical dramas of navigational science and "adventure capitalism" into the sphere of the aesthetic commodity.

Indeed, Thevet and Cartier and Rabelais are entangled together in a strange knot of mutual dependencies and epistemological dissonances. Both

---

12. On Renaissance collecting and museums, see Paula Findlen, "Jokes of Nature" and *Possessing Nature: Museums, Collecting and Scientific Culture in Early Modern Italy*; Lorraine Daston and Katharine Park have written on wonder cabinets and wonder generally in medieval and early modern natural history (*Wonders and the Orders of Nature: 1150–1750*). Stephen Greenblatt's *Marvelous Possessions* discusses some ramifications in empire-building of the rhetoric of the marvelous in voyage accounts.

13. On imperium/emporium in Calvino and Polo, see Laurence Breiner, "Italic Calvino."

Rabelais and Thevet, according to some historians, seem to have paid long visits to Cartier's home in Saint Malo in the process of researching their quite different books. Claims have been made that not only did Cartier give Rabelais the nautical vocabulary for Book 4 of his satires, but Rabelais was the real author of the sailor's otherwise inexplicably polished and even literary *Voyages*.[14] Meanwhile, it is often clear that Thevet has plagiarized Cartier's (Rabelais's?) *Voyages*, and in both the *Cosmographie* and again more elaborately in the "Grand insulaire" he lambastes Rabelais (in the form of "these Panurgical garblers" ["Panurgiques grableurs," 147ᵛ]) for stealing from him—for having "profited" from and "taken advantage of" [147ʳ] his account of the variously labeled "Isle of Demons" or "Isle of Roberval."[15] Stabler's examination of the evidence in *Études Rabelaisiennes* suggests rather that Thevet took advantage of Rabelais in this matter, or that both depended on a third source, the *De orbis terra concordia* (1544) of Thevet's traveling companion in the Levant, Guillaume Postel. What is most interesting from our point of view is simply the fact of the confusion. Oversensitive and hypocritical as Thevet indubitably was, we should still note that it was rhetorically possible for him to take (or pretend) offense at a writer working so different a field as was (to modern eyes) the author of *Pantagruel*. The confusion suggests an essential fictionality in the European imagination of extra-European actuality—which we will see elaborated later in other representational genres concerned with distances of scale and time as well as of undifferentiated geographical space.

The feature in question in this literary tug-of-war is Rabelais's "paroles gelées" (easily enough derived, in fact, from Postel, or the fourth-century

---

14. See Arthur P. Stabler, "Rabelais, Thevet, L'Ile des Démons," and Marius Barbeau, "How the Huron-Wyandot Language was Saved." The name listed parenthetically after the first entry for Cartier in John Alden's bibliography *European Americana* (vol. 1) is "Jehan Poullet." According to Lestringant (*André Thevet*, 72–83) the "conference of St. Malo" hypothesized by D. B. Quinn is almost certainly a "fiction," but even Lestringant hears resonances among the careers and writings of the putative participants.

15. "De tel recit, que i'ay autrefois publié, il y a certains grabeleurs ignorans, acazanés a leurs maisons, conteurs d'histoires tragicques, quj ont fait leur profit, se lavans leurs gorges des phantastiques apparitions qu'ilz disent estre par moy supposeés au dedans de ceste Isle. D'autres, mettans toutes pieces en besoigne, ont bien enflé la matiere, tellement se sont avantages qu'il y en a aucuns deux [d'eux], quj, pesle-meslant les tintouins d'Islande [Iceland] avec les tracas de ceste Isle desolée ont distillé dans l'alembic de leur cervelle mal disposée certaines voix gelées dont ilz font si tresgrand cas, qu'à lire les bayes, qu'ilz ont avancé, diroit on, veu la dexterité d'esprit, dont on tient qu'ilz estoient habillés, qu'ilz ont eux mesmes vire-volté parmy l'air, pour digirer ces voix saugrenées, dont ilz presument repaistre les oreilles du Lecteur" ("Grand insulaire" 147ʳ).

Neoplatonist Iamblichus). These frozen words were understood by the fabulously autocentric Thevet as a distortion bred from his own account of the Canadian "Isle of Demons." Thevet identifies the latter as the island where the Sieur de Roberval (commissioned by Francis I to found the first colony in New France) marooned his niece Marguerite in punishment for her shipboard affair with a member of the "good company of gentlemen and artisans of all kinds and several women" who accompanied him in this attempted settlement (Thevet, *La Cosmographie*, 1019ʳ). The "Isle of Demons" gets its name, according to Thevet's account in book 23, from "the great illusions and phantoms which are seen there" (1018ᵛ), for which reason it is, though "the largest and most beautiful" island off the coast of Newfoundland, uninhabited. "Pilots and mariners . . . [who] passed by this coast, when they were plagued by a big storm, . . . heard in the air, as if on the crow's nest or masts of their vessels, these human voices making a great noise, without their being able to discern intelligible words, only such a murmur as you hear on market-day in the middle of a public market" (1018ᵛ). Thevet says he interviewed Marguerite on her return to France and asked her about these voices. She said she had been troubled by them only at the beginning of her twenty-seven-month ordeal, after the deaths of her lover, her servant, and her child left her even more utterly alone (1019ʳ).[16]

In Rabelais (*Quart livre*, chap. 55), the voices turned up "as we were banqueting, far out at sea, feasting and speechifying and telling nice little stories." As in Thevet there is some question whether the voices are real or the protagonist's auditory hallucination, but at Pantagruel's repeated affirmations "we decided that either we could hear them too, or else there was a ringing in our ears. Indeed, the more keenly we listened, the more clearly we made out voices, till in the end we could hear whole words." Pantagruel offers an erudite occult account, but the favored explanation is found farther along the spectrum of materiality that Rabelais has already set up in the narrative: the captain tells them that they are at the edge of "the frozen sea," where last year "the frightful noises of battle [between the Arimaspians and the Cloud-riders] became frozen on the air. Just now . . . these noises are melting, and so you can hear them" (chap. 56).

---

16. Of Marguerite's voices, Thevet says, in the "Grand insulaire": "It is not improbable that if she had had considerable company the demons would rather have selected some retreat other than this ill-fortuned isle. And to tell the truth, solitude greatly reinforced the effect of these apparitions, as truly there is nothing in the world more inimical to the assurance which one could wish to maintain than to be alone in a place, deserted and abandoned by everybody in the world," etc. (146ʳ). See also note 22 of this chapter.

What a wonderful realization, *avant la lettre*, of the scientific ideal articulated by Thomas Sprat a century later in his *History of the Royal Society*: "so many *things*, almost in an equal number of *words*" (2.20.112–13)! In any case, whether or not Rabelais's episode is a parody of Thevet's marvelous Isle of Demons, it offers modern readers a sly commentary on Thevet's "demons." Unlike Rabelais, Thevet fails to identify *his* demons as native people engaged with one another—which means failing to admit their absence from his scene.

Instead of the material/linguistic traces of vividly absent natives provided by the satirist Rabelais, Thevet the royal cosmographer offers the quintessential colonialist adventure, a gloomy precursor of both *The Tempest* and *Robinson Crusoe*, the fantasy of European society reproducing itself in viral miniature among castaways on a "desert" island. It is a tale worth telling and full of historical instruction for us now. What does it mean, for instance, that the first French colonizer of the modern world felt compelled to *maroon* his own niece for having formed an erotic attachment with a member of his company? What significance might we find in the *order* of the deaths of the castaways—first the "poor gentleman," Marguerite's lover (who had voluntarily joined them and died "of sadness and disappointment") (1019^v), then Marguerite's servant, an old woman who had been the lovers' go-between, and finally the infant conceived and born on the island? What lessons in the close call of Marguerite's escape? (The Breton fishermen who found her took her smoke signal at first for one of the "illusions of demons which deceive passers by" and almost left her there [1020^r].) What of the strength, competence, and resourcefulness of poor Marguerite, who later assured Thevet that "on one day . . . she killed three bears, of which one was white as an egg" (1019^v, see figure 1) and admitted that (like Elizabeth Bishop's Crusoe) "she wished she was still there"? The positioning of the waywardly erotic European woman in relation to the European "gentleman of good birth," the old Norman servant-woman, the bastard infant conceived on the island (none of whom survive), and the indigenous demons (with which supernatural phenomena she is at first identified but then dramatically differentiated) would seem to locate her in the closest possible relation to Nature. At the same time she is all that remains of Culture (gun-bearing and signal-making) in this early instance of the desert-island trope.

Naturally the tale was destined for use. Already in his first telling of it, Thevet is grumbling at those "hare-brained wags [who], after hearing me tell this story . . . have added to it follies and lies which they have inserted into their fables and tragical stories, plagiarized from here and there" ("desrobées deça & dela," 1020ʳ).[17] Its usefulness suggests that despite the real curiosity Europeans of many sorts felt about the native Americans and their "manners and customs," the potential charm of an *uninhabited world* was at least as great.[18]

"Filling up the blanks," as Ralegh puts it, would be a major European pastime, both aesthetic and epistemological, for a long time to come. In the colonial/geographical context, this sometimes involves *making* a blank first— like Ralegh's painter, Thevet (who may have illustrated the *Cosmographie* himself, and beautifully—see figure 1) makes up an island of his own—two, in fact—to fit blank spaces near the mainlands of both South and North America, calling both the "Isle of Thevet."[19] This is another version of "writing himself in," of course—and points to one of the primary psychic uses of "islands" and "new" lands, a use that has complicated earnest attempts to "know" their geography and ethnography from that day to this. Real or unreal, an island is a good place (a *topos*) for imagining the expansion or unfettered dominion of the self. But more significantly as a representative Renaissance "origin" for the as yet unarticulated discipline of ethnology, the twice-nonexistent "Isle of Thevet" (never seen on any other map) illustrates the tendency to erase (which sometimes meant exterminate, though it often simply meant ignore) what was there and fill the "blank" with what one wants or already knows—to go to the end of the earth in search of "news" or "new lands" and find, as Thomas Ellis did, a European "box of nailes" (Hakluyt, *Divers Voyages*, 630–35), or an isle named "Thevet" or—not a demonic magician after all, no Sycorax or Caliban, but Marguerite de la Rocque de Roberval.

The castaway narrative then is the *anti*-ethnographic genre of voyage literature. Its emphasis on isolation and the "deserted" setting bespeaks an uneasy (un)consciousness that European settlement and cultivation of the "new

---

17. By which "wags" Stabler thinks Thevet means Rabelais and Thevet's hated rival Belleforest, though the story had also appeared in Marguerite of Navarre's collection of tales (the *Haeptameron* [1558]). Marguerite of Navarre was, like Thevet, from Angoulême.

18. It also suggests what later, overt fictions will confirm, that a European or "white" woman can function as a kind of native *manqué*, fully exoticized by her distance from home, her anomalous presence in the wilderness, and perhaps above all her acquisition of weaponry. Some memorable seventeenth-century fictions will pursue this frisson, close inverted relative of the similarly exoticizing narrative of the servant who enters her master's erotic and social world, and related as well to the captivity narratives of so many female American settlers in the seventeenth and eighteenth centuries.

19. See folios 260ᵛ and 143ᵛ of the "Grand insulaire," reproduced in Schlesinger and Stabler, *Thevet's North America* from the autograph manuscript in the Bibliothèque Nationale.

lands" is inappropriate where the land is inhabited, especially where the inhabitants could be perceived (as nomadic hunters like the Iroquois generally were not) as having a culture.[20] The castaway narrative is also the perfect plot-type for justifying the fundamental pleasure of inventing the island *mise-en-scène*, and will take off as a fictional device almost immediately (though it was not brand new even in Thevet's time).[21] Inevitably, as it seems to us now, it will culminate in the agonizingly revealing fantasy of *Robinson Crusoe*, the bourgeois capitalist who builds a colonial-mercantile fortune from the almost absolute zero of the solitary castaway. Generally read as a fiction inspired by the case of Alexander Selkirk, marooned for four years on an island off the coast of Chile, *Robinson Crusoe* is also worth considering as a transformation of Marguerite de Roberval's story. The voluntary action of her lover in joining her is described with a resonance especially suggestive in this connection: "The gentleman, seeing this cruelty and fearing lest they do the same to him in some other isle, was so beside himself that forgetting the peril of death into which he was hurling himself and fearful tales he had been told about this land, took his arquebus and clothing, with a firesteel and a few other commodities, some measures of biscuit, cider, linen, iron tools, and a few other things necessary for their use, and precipitated himself onto the isle to accompany his mistress" (1019r).

It is wonderful to imagine the clatter with which the gentleman must have landed on his desolate new home. Thevet clearly enjoys what Defoe will later elaborate to its most satisfying proportions, the list of "commodities" and civilized necessities with which his character proposes to maintain himself in the blank space of the desert isle. There is a long lapse of cultural and economic history between the Royal French cosmographer of the sixteenth century and the English literary hack of the eighteenth, but the retentive fantasy of stockpiling necessities in the face of exile and wilderness is only more pow-

---

20. European belief in the Wild Man constituted an ability to imagine people literally *without* culture, even people in groups. This belief dwindled among the educated by the early nineteenth century to a fascination (still scientifically respectable) with isolated cases of feral children like the famous "Wild Boy of Aveyron." We will return to the phenomenon of the feral child at the end of this book; with particular reference to *groups* of people encountered by the French in the New World, see Olive Dickason, *The Myth of the Savage*. For the legal consequences—which include the establishment of international law—see Dickason and L. C. Green, *The Law of Nations and the New World*, and James Muldoon, *The Americas in the Spanish World Order*.

21. Not only is it familiar in similar form to the *Odyssey*, but it had received an elaborate treatment by a twelfth-century Arab, Ibn Tufayl, whose work would be translated into Latin and then English in the seventeenth century, achieving considerable notice in the widening readership of a not yet plentiful prose fiction. See Len Evan Goodman's translation of ibn Tufayl's *Hayy ibn Yaqzan*.

erful by Defoe's time (and reaching perhaps a climactic literalization in the fanatical right-wing "survivalism" of our own).[22]

The story of Marguerite seems to provoke a kind of demonic confusion of "voices" in Thevet as well—some of which he identifies as external (the "Grand insulaire's" "Panurgical garblers" and "quintessentialists" [147ᵛ], the *Cosmographie*'s "hare-brained wags" [1020ʳ]), but some of which are in one or another sense his own, or at least internal to his texts. In his first account, the island (not yet Marguerite's) is called "The Isle of Devils" and no mention is made of Roberval's niece (*Singularitez*, 161ʳ). In his second account, an Isle of Demons is identified and the story of Roberval and his niece attached to it—with one of his many attacks on other writers for having provided the same narrative or data (*Cosmographie*, bk. 23, ch. 6). By the time of the "Grand insulaire," the island has fragmented into the Isle of Roberval, the Isle of Demons, and the Isle of the Damoiselle. Thevet identifies the Isle of Roberval as also called "the Isle of Demons because the Demons raise a terrible uproar there" (145ᵛ), but then in another chapter, headed "The Isle of the Demons," he expostulates angrily that "it is impossible . . . to take away the stupid fancy from men . . . that in this isle . . . there are demons and devilry . . . as I told you about another isle which I named Roberval rather close to this one" (153ʳ). He locates it at the "elevation . . . where the sea is frozen" (153ʳ; cf. Rabelais, *Quart livre*, chap. 56). The Isle of the Damoiselle is "so called because it was on it that was exiled the niece of the captain Roberval" (404ᵛ). However, it turns out that this latter isle is one of *two* on which Roberval's niece was exiled—the other (also the "other" Isle of Demons?) is "about seventeen leagues distant from this one as I have rather fully described in my book of the Singularities." But the *Singularitez* is the book (the only book) where *no* mention is made of Roberval, his niece, or even (except in the allusion of a name) the cacophonous demons.

The question whether a voice or its speech is internal or external, his own

---

22. What doesn't last is the heroic presence of the woman who actually did survive the experience. Thevet's Marguerite is a female hero who gradually loses companions, Defoe's Crusoe a male who gradually attracts them. Marguerite's "demons" are represented as the voices of her own emotional distress, the content of her threatened subjectivity in a situation of enforced isolation. Crusoe's demons turn out to be other people, frequent visitors to the island on which he had laboriously constructed his artificial isolation for twenty-five years. Marguerite does have at least one direct female descendant, in the heroine of Margaret Cavendish's utopian *Blazing-World* (1666), in which the castaway's greed is not, like Crusoe's, material, but epistemological, and who is left on an ice floe without a single penknife or biscuit. What she acquires is almost a surfeit of natural history (as well as occult access to knowledge of certain private lives). See Chapter 6.

or another's, seems to be as pressing to Thevet as to his castaway — in both of them we can perhaps observe the disorientation and reformulation of subjectivity in this time of unprecedented, and unequal, encounter — the "contact period." In a multiplication of entities analogous to the case of split personalities, Thevet has differentiated as putatively separate islands (personalities, entities) different and apparently contradictory kinds or *sources* of knowledge about a single island. In order to maintain the illusion of a single voice (his own) amid the welter of his sources, Thevet has had to multiply the entities of which the voice speaks — we count at least three islands. (As in a David Hockney collage, more meets the eye than is there.)[23]

The words "island" and "isle" and their identical French cognates were used more loosely in the Renaissance (and before) than they are now, with meanings ranging from "city block" (a meaning of Latin *insula*) to the Biblical "lands across the sea." What the usages have in common is the theme of detachment, isolation, insularity. "No man is an Iland," said John Donne in his *Devotions* (no. 17, 1624); "each man is an iland, a little World," said "L. S.," in *Nature's dowrie: or, the people's native liberty asserted* (1652).[24] At any rate, self and island were associated with each other, as we see literally in Thevet's "Isle[s] of Thevet," the name of which "came to pass in that I was the first who set foot on it" ("Grand insulaire" 145r). Like Columbus on *his* "undiscovered," seemingly uninhabited isles, Thevet thinks his isle is Paradise:

> The inhabitants and governors of it were only birds of diverse plumages and sizes in great numbers, but also beautiful fruit trees of several kinds and colors. And when we thought to go into the isle, because of its thick woods, I perceived some hills and on these I discovered that there were some grape leaves. As for the trees I never saw any like them, and you would have judged this place to be a second earthly paradise. (*Cosmographie*, 1015r)[25]

---

23. For instance, the "Isles of the Devils" in his earliest work the *Singularitez*, are at latitude 56 degrees (161r–v). They become a singular isle in the *Cosmographie*, where Thevet adds, following Cartier (from Ramusio's *Viaggi*) and Roberval, the story of Marguerite. In the "Grand insulaire" the isle moves to 58 degrees and changes its name to Roberval; the Isle of the Damoiselle splits off as Thevet follows the *Routier* of the Portuguese pilot Jean Alfonse de Saintonge, who in turn is responsible for the doubling of the Isle of the Damoiselles, "so called because it was on it that was exiled the niece of Captain Roberval. Afterwards she was transported with her lover to another isle" ("Grand insulaire" 404).

24. See article on "island" in the *OED*.

25. Cf. Columbus's "Letter to Sanchez" (1493) for a similar fantasia: "This island and all the others are very fertile to a limitless degree, and this island is extremely so. . . . Its lands are high, and there are in it very many sierras and very lofty mountains, beyond comparison with the island of Teneriffe. All are most beautiful, of a thousand shapes, and all are accessible and filled

The cosmographer's topos shrinks easily to the dimensions of yet another allegorical self-dramatization. But it is important to the structure of Thevet's cosmographies, as well as to the nature of his culture's categories of knowledge and interrogation, that the unit of the island is so permeable to fantasies of abandonment and detached self-sufficiency, private dominion, and utopian invention—and so easy, in fact, to invent.

Thevet's personal fascination with the island-unit may be illuminated for us by his claim that he knew twenty-eight languages (a language is an island, too), and that his only noncosmographical work was a collection of selves, as it were: *Les vrais pourtraicts et vies des hommes illustres grecz, latins, et payens* . . . (1584). He was a collector and accumulator, a seeker of fantastic alternatives to his own limited self and life—a bourgeois reader of novels *avant la lettre*, making do with genres of fact. But Thevet was not simply idiosyncratic in his transformation of the "New Found World" (as his 1568 English translator had it) into bite-sized tableaux—clearly not, as he was such a plagiarist and interviewer of other voyagers, and such a popular and widely translated author himself.[26] The island, however defined, is an obvious kind of package—a projection onto geographical space of the aesthetic desire for titling and framing. As an island is (etymologically and usually) something surrounded by water, a literary island—a geographic or ethnographic tableau—is a text implicitly surrounded by the (increasingly vast) formlessness of mere space, mere motion, the invisible machinery of one's own or an alien culture producing interaction by the yard.[27] In the case of continents like the Americas, especially North America, so little mapped as yet, the generation of closed

---

with trees of a thousand kinds and tall, and they seem to touch the sky. And I am told that they never lose their foliage, as I can understand, for I saw them as green and as lovely as they are in Spain in May, and some of them were flowering, some bearing fruit, and some in another stage, according to their nature. And the nightingale was singing and other birds of a thousand kinds in the month of November there were I went" (Jane, *Select Documents* 1:5–6).

26. The *Singularitez* came out seven times between 1557 and 1584; it had three French editions, three Italian, and one English edition. In the early seventeenth century, parts of both the text and the illustrations reappeared in works by such notable authors as Edward Topsell, Ambroise Paré, and Gaspard Bachot. (This is especially impressive in that many of these editions were large quarto volumes and many, though not all the same ones, were illustrated.) The gigantic *Cosmographie* came out only once, unsurprisingly—it was a double-volume folio, replete with maps, illustrations, and portraits and numbering 1,025 folio leaves in addition to masses of front matter! Two-thirds as long and just as big, *Les vrais pourtraicts* was republished three times in the seventeenth century, twice in English.

27. The phrase is adapted from an unpublished poem by Laurence Breiner: "Vivaldi by the yard, by the acre, by the ton /—that's love, true love."

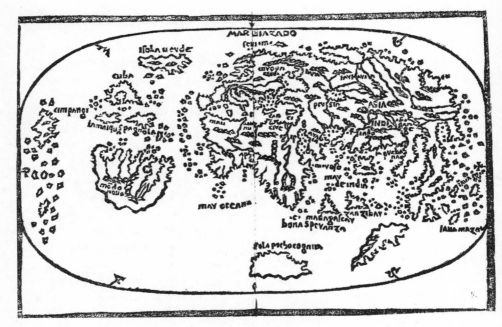

Fig. 2. Pietro Coppo's woodblock map, New World at left (Venice, 1528). (British Library G7292, Coppo Portolano, 1528, MAP.) By permission of the British Library.

and definite (and little) forms would be particularly desirable.[28] (See figure 2, Pietro Coppo's woodblock "portolan world chart" [Venice, 1528].)

The issue of origins, so problematic for "l'homme nouveau," is a problem often invoked or dramatized in desert-island texts. The fantasy may wish the island or its inhabitant(s) to be *sui originis*, but that would be an untenable contradiction in narrative terms or, as Thevet points out, biological ones. In the chapter on the Isle of Demons in the *Grand insulaire*, where Thevet accuses everyone but himself of fancying that "in this isle there are demons and devilry," the absence of demonic *invisibilia* is filled in, almost as if by a conversion formula, with flesh-and-blood visiting natives who "go often to our isle of demons . . . [and] never return to their huts without bringing back a great number of wild animals" (153ᵛ). That demonic absence also seems to

---

28. On this and other issues related to the pragmatic European colonialist interest in islands, see Richard Grove, *Green Imperialism*, chap. 2, esp. 24–32. "The tropical islands which were often the first landfalls and navigational points of reference, and which later became the first colonies, encapsulated an alternative kind of world as well as offering economic opportunities. An insular environment was one which was knowable in terms of the ease with which the island could be mentally circumscribed" (32).

leave the narrator room for a very cogent biogeographical question, elaborated at some length:

> I should like to consult some excellent personage who would explain to me the pros and cons of how it is possible that in this place and on a thousand other isles, which as I said never were inhabited any more than this one, you see teeming such a diversity of beasts and birds, and in what way they were procreated and engendered.... Now sentient beings must necessarily come from this commingling of semen, and for this reason I am in doubt (speaking according to natural reason) how these beasts were engendered in this isle.... Just as man can only be engendered from the effusion of seed and nourishment which he takes from his mother, likewise animals cannot be procreated from the mere humor of putrid earth, this being prevented by the lack of the radical warmth which through the blood causes this nourishment and life." (153ᵛ)

Although Thevet has gone scientific on us here, it is still possible to pick up, in his tentative responses to his own question, eerie echoes of the earlier narrative/fantastic functions of Marguerite de Roberval and the well-equipped lover who mysteriously "precipitated himself" onto the island. "One must not be surprised if there is an abundance of beasts there since these savages, in order to support and maintain themselves and to wear their skins after eating their flesh, load up their small boats with them." (Perhaps he assumes here that the natives raise food animals on the island, but the phrasing makes for a typical Thevetian logical reversal: the island is full of beasts because nearby people like to hunt them and carry them off; wealth or plenty is caused rather than simply matched by desire.) "As for the birds, since their bodies are more tenuous and participate of the nature of the air, even if they may not have been there since the beginning of the said isle, still they may have come and still be *coming from elsewhere*" (153ᵛ–154ʳ, emphasis mine).

We have then in this piece of speculative island biogeography, as in the tale of Marguerite and her lover, a provision of the island with supplies from the continent, mysterious landings "from elsewhere," and thus the necessary "matrix of [the] mother" (153ᵛ) — fertile among the oddly imported "beasts" as well as in the case of Marguerite. The issue is the same both for readers of exotic narrative and for those "intelligent people" to whose more philosophical "contentement" Thevet's "little digression" is dedicated (154): how is life (especially human life) originated there? This was of course the titillating and enormously disorienting question about the Americas in general — how could the implications of an inhabited world separated by oceans from the *orbis terrarum* be squared with the account of human (or even animal) origins in Genesis? In combination with other factors of a more immediately pragmatic nature, the question of origins would help to generate the science of anthropology (as well as that of geology) and begin, according to the great historian of science Paolo Rossi, the long, slow disengagement of scientific thought from theological analysis and explanation of nature.[29]

It is also a version of the perhaps more immediately titillating and disorienting question of one's own personal and biological origin, which sheds a strange little shaft of light on Thevet's citing, as support for his claim to having given his name to the Isle of Thevet by being the first to set foot on it, his own previous account of the event in the *Cosmographie universelle* ("Grand insulaire," 145ʳ)! For those who can muster narrative evidence, autochthony resolves anxiety at many levels, as Freud and Lévi-Strauss have shown us. In another account of the Isle of Roberval/Demons, Thevet defends the presence of demons by reminding his readers of the autochthonous creatures *par excellence*, the gnomes found by "all those who work with metals . . . in the recesses of mountains and mines." References to this belief can also be found, for example, in Sebastian Münster's *Cosmographia universalis* (1544, ccclxxix–ccclxxxi) and Georg Agricola's *De re metallica* (1556). It was good *scientia*, at the time; it was perhaps also a comforting stopgap as an image of how at least some sentient life is generated spontaneously, in the entrails of Mother Earth. A personal fantasy of spontaneous origin may reverberate in one's culture generally as a disinclination to fraternity (at least fraternity outside the territory of home); certainly it is congenial to an island-based conception of cultures or "nations." One of the major tensions in the early literature of anthropology is that between the religious and philosophical urge to enlarge the scope of the human (and ultimately Christian) family and the atavistic desire for self-sufficiency—genealogical as well as economic and political.

ANECDOTE

This chapter has been focused on the narcissism of Thevet's writing, structured to satisfy not only the writer's greed for singularity and authority but

---

29. See Rossi, *Dark Abyss of Time*, chaps. 5 and 8. Here is Linnaeus, trying to solve the problem of dispersal geologically and square his analysis with Genesis, 150 years later: "If we therefore enquire into the original appearance of the earth, we shall find reason to conclude, that instead of the present wide-extended regions, one small island only was in the beginning raised above the surface of the waters. If we trace back the multiplication of all plants and animals . . . we must stop at one original pair of each species. There must therefore have been in this island a kind of living museum" (*Museum S.R.M. Adolphi Friderici Regis* [1754], quoted in Janet Browne, *Secular Ark*, 18).

also that of his acquisitive reader. I have been looking at the books' mirror-functions, in which they blank out the vast tangle of American alterity, filling the emptied space with tableaux of the European self—receiving gifts, discovering abundances, forging dominions. But Thevet did include ethnographic material on the various Americans, South and North and Central as well; his corpus has a valence with anthropology as well as with bourgeois fiction. It is an extraordinarily contradictory accumulation of facts, names, anecdotes, and generalizations, for a long time believed to be, as Henry Harrisse put it, "a tissue of lies" and ignorant borrowings from the voyage literature of more experienced (at least in America) and sober travelers (Harrisse, *Découverte et évolution*, 156). Thevet is being rehabilitated in light of more recent historians' belief that much of his primary material is taken directly from oral sources and is not found elsewhere. The most interesting features of the texts from our rather more formal standpoint are their frequently anecdotal quality and the persistent emphasis (as everywhere) on analogous European origins—cultural aetiology seen, as it will be for centuries, as an invariable process of *maturation*—thus the eventual term "primitive":

> "So these Canadians fight with arrow-shots, round clubs, square clubs, lances, and wooden pikes.... So did the ancients like the savages: they fought with fist blows, kicks, biting, grabbed each other by the hair and other such actions. Later they used stones to fight, as you can read in the Holy Bible.... The Thebans and Lacedemonians revenged themselves on their enemies with blows of crowbars and huge wooden clubs. (*Singularitez*, 157 r–v)

Even closer to home,

> "Some have written that Hercules of Libya, coming to France, found the people living almost in the manner of the savages both of the East Indies and America [Brazil], completely uncivilized, and both the men and the women went almost naked. The others were clothed in the skins of divers kinds of beasts. Such was the first condition of the human race, at the beginning being rough and barbarous, until through the passage of time necessity forced men to invent a number of things for the preservation and maintenance of their lives." (153v–154r)

(Say, for instance, "arquebus and clothing . . . a firesteel . . . biscuit, cider, linen, iron tools"?)

Early "cultural anthropology," or ethnology, has a tendency to look rather like—and sometimes even to be—a kind of literary history, a collection of revealing parallels to the historical and cultural conditions of (variously) Homer's Greece, Herodotus' Scythia and Arabia, Livy's Rome, Tacitus' Germania, the Biblical Mesopotamia, Palestine, and (especially) Ophir.[30]

---

30. This is still thriving in Joseph Lafitau's *Moeurs* (1724). See Chapter 9.

Difference is glossed over, and the *context* of any given practice or craft ignored in deference to a parallel or analogous morphology. As Mary Louise Pratt's devastating 1985 account of early-nineteenth-century observation makes clear, the absence of a historical dimension in the represented world of the Other would eventually serve as the genre's characterizing feature. The historical dimension has not yet disappeared, though. It is kept trivially alive in the anecdote, which is the narratological manifestation of the tendency, already observed in spatial terms, to package the newly emerging "world" of America in small boxes.

As I have discussed elsewhere (in *Witness and the Other World*), especially with regard to the highly narrativized and anecdotal quality of Walter Ralegh's *Guiana*, first-person narrative and in particular the miniaturized narrativity of anecdote require characters with whom the narrator can interact; thus they tend to humanize (to nefarious ends, often enough) by particularizing the alien people Europeans encountered in the New World. Although Thevet's most commonly articulated message about locations in the "new lands" is "it . . . could be easily settled and a fine fortress placed there to control the entire coast" ("Grand insulaire," 150^v), his anecdotes nonetheless tend to portray native Americans as sympathetic and often as oppressed by Europeans. In fact, as can be seen more systematically in Hariot's text, the connection between urging European settlement and representing Amerindians as "benign" (even civilized) is rhetorically causal. After a chapter, particularly urgent in its settlement fever, in the "Grand insulaire" on the Isle of St. Julien (*S. Juhan*) and describing other habitable islands around Newfoundland, Thevet concludes by narrating the following poignant moment:

> The people there are benign and gracious who seek only the friendship of the foreigner and I remember when we had set foot on land, which was on my first voyage returning from the Southern Lands, we were reluctant to accost these barbarians. A kinglet [*un Roytelet*] of the country all clad in skins of wild beasts accompanied by some others, believing that we were angry and that we feared them, said to us in a quite friendly way in his language, Cazigno, Cazigno, Casnouy danga addagrin: which means, Let us go, let us go, on shore [disembark] my brothers and friends; Caoaquoca Ame Couascon, Kazacomy: come drink and eat of what we have; Arca Somioppah, Quenchia dangua ysmay assomaha: We swear to you by the sky, the earth, the moon and the stars that you will have no more harm than our very selves. Seeing the good will of this old man we were with him an entire day. [On] the following day we took the route for the gulf of Canada. (150^v)

There is no mention of this incident in Thevet's description of the same area (the Gulf of St. Lawrence) in the earlier *Cosmographie universelle* or *Singularitez*, where much more vaguely located inhabitants are said to be, respectively, "friendly, easy to handle, and pleasant" (*Cosmographie*, 1010^v) and "of quite large stature, very tricky, who usually wear masks and are disguised by lineaments of red and blue" (*Singularitez*, 149ᵛ). The incident is in fact a fiction, or at least untrue, according to the statement in *Singularitez* that Thevet himself had only come "very close to Canada" on his return, via the Caribbean and Florida, from Brazil in 1556. It may have happened, though, to someone else. Several of the native words match closely the words in the two word lists from Cartier included elsewhere in the "Grand insulaire" (though some do not, or are mistranslated). At any rate, it is *not* here as a result of personal memory, "because I . . . set foot on it," but presumably because Thevet has been talking about how good it would be for other Frenchmen to set foot on it.

It is instructively heartbreaking to look at what Thevet felt would favor his case for French invasion and theft of "this old man['s]" lands. We seem to see Paradise in the very moment of its loss as the "old man" repeats what had become the classic trope of encounter narratives, from the "Letter" of Columbus to Fracastoro's epyllion, *De syphilis*, the invitation to consume "what we have." Buried in this self-justifying tale—and not buried deeply—is an alternative desire fulfilled, the desire to be loved by an Other who is fully real, human, safe, transparent in language and "intentions," *as well as* alien, exotic ("clad in skins"), and "new." This anecdote is fiction also, then, in the sense that it represents desire as well as or in place of any identifiable actuality. It is the perfect, and perfectly impossible, encounter. For the two desires that motivate it conflict—as it turned out the French could not have both the love and the land of the native Americans. The only way to represent both fulfillments is through the freeze-frame, the almost-tableau. The anecdote has no narrative future, it is the vessel of ethnographic time: a fully imaginary, scientific time which lasts only long enough for a social event—a barbecue, a dance—that has no consequence but knowledge. The explorer, the ethnographer, goes ashore and examines the large / tricky / pleasant / disguised / gracious native and then walks out of the picture; the picture stays behind in the book-museum like a diorama, gathering dust and encouraging potential colonists to leave home. Thevet, of course, did *not* go ashore. In his book the anecdote simply represents the structure of desire and fulfillment that readers of New World cosmography and voyage literature might have expected—a structure perfectly mimicked, at no risk to life or sanity, by reading itself. Thus (in part) the obsessive banking on personal experience of the mendacious but serviceable narrator; it is an *I*, after all, who wants, in imagination, to meet Others.

Thevet's books of experience-based but comprehensive cosmography are old-fashioned for the second half of the sixteenth century, which had already seen many voyage accounts that restricted their presentation of data to what had been verifiably seen or experienced by the writer-protagonist. But we have been looking at Thevet as an exemplar of something other than cutting-edge ethnographic intelligence. Thevet was writing (compiling) picture books, for a relatively wide audience—indeed the illustrations for *Singularitez* included, as Bernadette Bucher points out in her important book about the de Bry illustrations, "the first visual portrayals of New World inhabitants" (*Icon and Conquest*, 16).[31] Hariot and White's seminal collaboration became, in de Bry's hands, another such commodity, in which pictures are, however, combined with a verbal text of greater scientific pretension and sophistication. But for now let us consider the enthusiastic plagiarist, the self-aggrandizing purveyor of islands, wonders, engravings, anecdotes, distances, lies and imaginary mines, as a norm—a norm for whose product the markets in several European countries were eager. Norms and bestsellers are always old-fashioned but at least as significant as original work for insight into the general structure of such popular sensations as wonder and pleasure—even in their relation to a scientific "revolution."

People have come to use the word "ethnographic" to refer to just about any eyewitness representation of non-Western people at home. But the word also refers to a genre that did not exist in the sixteenth century. What Thevet offered was not data related to the unasked question, What is (a) culture? but data related to the sensation of the frisson. He was not even offering the more pragmatic service of rendering Indian behavior "predictable," as James Axtell puts it when defining the work of the seventeenth century's proto-ethnographers/missionaries ("The Invasion Within," 42–43). Thevet is rarely precise enough for the reader or prospective colonist to be able securely to distinguish, geographically or linguistically, between large, tricky Indians and pleasant, easy-to-handle ones.

He does offer knowledge (ethnohistorians credit Thevet, for instance, with helping them save "the Huron-Wyandot language . . . from oblivion"), but only to those equipped to seek such knowledge out by modern systematic study of the kinds of cultures Thevet represents.[32] What then were Thevet's books doing in 1557, 1575, 1588?

A thousand things, but for the purposes of this discussion, mainly one: tak-

---

31. This is not *quite* true: the 1535 edition of Fernández de Oviedo's *Historia general* included a woodcut of a man paddling a dugout canoe, and there were woodcuts (not book illustrations though) before that. See Sturtevant, "First Visual Images of America."

32. See Marius Barbeau, "How the Huron-Wyandot Language Was Saved from Oblivion."

33. I am not invoking Greenblatt's usage in *Marvelous Possessions*: his book seems concerned with a different aspect of historical development. A fascinating use of the term, with reference

ing possession.³³ On the largest scale, his plagiarizing polyglossia represents an entire civilization's taking possession—imaginative possession, I mean. I have concentrated here on Thevet's *Canadian* representations partly because of Rabelais's refractions of them and, later, Cyrano de Bergerac's (in his *Voyage dans la lune* [1650]) and Lafitau's (in his exhaustive ethnography [1724] of the same region), but also because, like Rabelais and Cyrano, Thevet had not actually experienced Canada firsthand, though he had visited, at least in body, the New World. Desire, then, has the upper hand over historicity in the construction of these parts of Thevet's works. From what he offers we can tell most directly what he and his readers sought. One can observe in Thevet's nuggets of American strangeness the preparation of "America" as in some sense edible, or at least collectible—the first flash and fragrance, perhaps, of the comestibility of foreign, particularly "primitive," cultures that we take so much for granted, under late capitalism, in the "Travel" section of the Sunday paper or the map stores of cosmopolitan cities.³⁴ Hariot's later text would appear as designed to promote and lubricate a more literal appropriation, of land and regional power in Virginia. Thevet's uncovers for readers that complementary *literary* generation of an Other World where fear and desire blossom materially as visitable (but unvisited), imaginable, representable tableaux.

One's readerly nonvisit to Canada is made through the medium of the putative eyewitness, a self-authorized vessel of knowledge with whom the process of "identification" is relatively easy, since we readers experience *ourselves* as narcissistically omniscient. The creation of a first-person narrator for prose narrative was a project that required many experiments—the first-person framing of third-person fictions by their characterized tellers in Boccaccio's *Decameron* (1353) was one, the absorption of narrative matter into the sensibility of the protagonist-narrator in the heroic voyage accounts of Columbus comes to my mind as another. Thevet's contribution took advantage of both these narrative possibilities, but in appropriating the words and

---

to information and representation, can be found in Brackette Williams's study of the historical tradition preserved through spirit possession in Guyana, where certain descendants of slaves are routinely "possessed" by the spirits of colonists dispossessed after the slave rebellion of 1763 ("Dutchman Ghosts"). See also Gesa Mackenthun, *Metaphors of Dispossession: American Beginnings and the Translation of Empire*, and, for a different inflection on the metaphor, Findlen, *Possessing Nature*.

34. The Levantine cultures encountered during the Crusades were literally comestible to the Europeans who brought home spices and sugar, but the political contacts of Islam and Europe were too mutually anxious, the military balance too even, for such a profoundly aestheticized response. That would come later, during the period of European ascendancy in Asia.

experience of other writers, who were eyewitnesses to the events and situations described, he makes something new: a fictional narrator who stays "in character" and in the first-person singular continuously, a *personified* medium (as opposed to *li conte*) for the transmission of narrative knowledge.[35] We take this for granted in modern prose fiction, but it is a much newer phenomenon than the character, which is after all just another narrative object, similar to buildings or battles or miracles. The *subject* of narrative, and in particular the subject of fiction, is a recent construct.

It makes sense according to psychoanalytic and especially Lacanian models of subjectivity that this capacity of written language for the production of an apparently coherent (if secretly patchworked and stolen) subject of narrative would emerge in a period of shocking contact with strangers, and partly through the medium of an "homme nouveau," a man of humble origin whose own status and visibility had to be invented. Although Peter Hulme rightly warns us to steer clear of discussing the psychodynamics of groups in terms designed to analyze individual experience, it is in fact individuals who produce and consume books, however collective the sources and contexts of the texts they contain (and individuals too who encounter large, tricky, pleasant, easy-to-handle Amerindians in the pursuit of their own social and economic aspirations).[36] If the novel will eventually emerge as a kind of tool for individual readers' reenactments of the mirror-stage, anthropology cannot emerge to satisfy the *same* requirements, and its function as a state-supported academic discipline was and is obviously institutional and collective. But here at the vanishing point where Thevet stands, self-centered, aspiring, state-supported, previous to the novel *and* to the science first codified (also in eastern Canada) by his countryman Lafitau, we see the distinctions waver. It is hard to say which history his books belong to, and that fact has implications for both.

---

35. Thomas Nashe's *Unfortunate Traveler* (1594) will take recognizably fictional advantage of this narrator-character, stolen partly from the Spanish picaresque: "I, Jack Wilton, (a Gentleman at least) was a certaine kinde of an appendix or page." Nashe's association of such a first-person character with travel forms a bridge between the grungy milieu of the picaresque and the later moon voyages of Godwin ("I, Domingo Gonsales") and Cyrano de Bergerac, whose narrators, like Thevet, are notable for pretensions that are lacking in the eponymous story of Lazarillo de Tormes, the abandoned dwarf bastard of a miller and a whore.

36. For discussion of our problematic lack of a "vocabulary in which to talk about collective subjects," see Hulme, *Colonial Encounters*, 82–83. A quick historical sketch of the perceived value of such a "vocabulary" is available in chapter 3 of Stanley Tambiah's account of modern anthropology, *Magic, Science, Religion, and the Scope of Rationality*.

THOMAS HARIOT, JOHN WHITE, AND THEODOR DE BRY: THE *BRIEF REPORTE* AND *AMERICA, PART I*

It is easier at first glance to categorize Thomas Hariot and John White's ethnographic collaboration, as orchestrated by Theodor de Bry, the *Brief and true reporte of the new founde land of Virginia* (1590), often cited by modern ethnologists as a sole "source material" for the North Carolina Algonquins (see note 42). The work is strictly speaking a "voyage," printed in its classic form with de Bry's engravings of White's watercolors as the first volume (*America, Part I*) of what became the *Grands voyages*. But it is a voyage of quite another kind than those recounted or manufactured by the lone and self-serving sensibility of a Columbus, a Ralegh, or a Thevet.[37] In his history of the emergence of ethnology (the study of peoples & cultures) in the sixteenth and seventeenth centuries, Anthony Pagden does not mention Hariot, White, or even de Bry, but their (largely inadvertent) contribution to the history of ethno*graphy* (field-work-based description of cultures and communities) matters quite a bit — not so much as an influence on future science but as a revealing illustration of how, in part, this science came into recognizable being. The relationship of ethnography to imperialism is generally viewed as the coincidental (if lamentable) consequence of ethnography's usefulness to the imperial state and/or of ethnology's early dependence on the mobilizations of colonial conquest. These are indeed functional coincidences; a close look at de Bry's edition of Hariot and White will show other coincidences operating as well, including the commercial and aesthetic motives of an artist, a master engraver, and a printer. Without the conflicting and complementary aims of its multiple authors, we would not have this artifact so often called ethnographic today. Nor would we have it, paradoxically, if a scientific ethnology had already existed in Hariot's day. For it is in the relations between the verbal and visual texts of de Bry's book that we will find the most hauntingly "modern" of its implications, and the pictures were motivated, in part, by the absence of an ethnographic language.[38]

---

37. See Julie R. Solomon, "To Know, to Fly, to Conjure," for an analysis of class-related differences in scientific representation. Her examples are Francis Bacon, Hariot, and White, and her division is between courtly modes of reading exemplified best by Phillip Sidney and Edmund Spenser and an "objective," self-distanced mode engendered by the interests and circumstances of the "commercial class" and proselytized by Bacon. She classifies Hariot among the pre-Baconian, courtly scientific sensibilities. I disagree, but the article is suggestive in many ways.

38. William of Rubruck offers a thirteenth-century version of the predicament when he writes back to King Louis in France from the Levant that "the married [Mongol] women make for

If Thevet's huge works as Royal Cosmographer seem motivated by the former monk's desire to impress and satisfy a clientele of detached and narcissistic connoisseurs, the brief, collectively produced work on "Virginia" that constitutes *Part 1* of de Bry's *America* series might seem the inversion of Thevet's *Cosmographie universelle*. Though both works combine the perceptions and representations of several people into illustrated books designed to produce pleasure in powerful consumers, Thevet's is structured to please and succeed through the illusion of a single author-protagonist in whom the reader can contain and possess the imagined world. The texts braided in de Bry's book, which are aimed at understanding Roanoke rather than imagining the New (or the whole) World, will succeed if they provoke participation, and lead (with equal fantasy) in the direction of a cultural rather than personal self-image. Pleasure is largely repressed in the verbal text, reappearing, in mute company with terror, among the book's exotic illustrations of naked Algonquins and tattooed Picts.

The mathematician and colonial investor Thomas Hariot also saw himself as a missionary, and did some proselytizing while he was at Roanoke in 1585 mapping and taking notes on the doomed colony. Describing the Algonquin reaction to the Bible, he suggests at once their readiness for Christianity and their savage misapprehension of its transcendental Book:

> Manie times and in every towne where I came, according as I was able, I made declaration of the contents of the Bible, that contayned the true doctrine of salvation through Christ, with manie particularities of Miracles and chiefe poyntes of religion, as I was able then to utter, and thoughte fitte for the time. And although I told them the book materially and of it self was not of anie such vertue as I thought they did conceive, but only the doctrine therein contained; yet would many be glad to touch it, to embrace it, to kisse it, to hold it to their breasts and heades, and stroke over all their bodie with it.[39]

To Hariot what mattered was various points of doctrine; to the Algonquins of his *reporte*, the book was a whole and single object with a magical meaning

---

themselves really beautiful carts which I would not know how to describe for you except by a picture: in fact I would have done you paintings of everything if I only knew how to paint" ("The Journey of William of Rubruck," 95). The inadequacy of ordinary language to an experience so visual and fraught with confused or missing communication as distant travel helped give rise, in the sixteenth and seventeenth centuries, to both a rebirth of scientific illustration and to the illustrated travel book. Thevet's and Hariot's are among the first to include eyewitness illustrations.

39. Quotations are from the text of *A brief and true reporte* printed in Theodor de Bry, *America, Part I* (1590).

and efficacy all its own. We are starting to come around to the Algonquin point of view (which has always been at least partly our own as well—after all, we swear on Bibles in court, which soldiers in the American Civil War warded off bullets by wearing). Much contemporary theory and criticism concerns itself with the legal, religious, and economic systems in which books participate as illegal, holy, cheap, or expensive objects; cultural materialism tends to see these systems as largely determining the significance and application of such an object's "doctrinal" contents.

I would not disagree that Hariot's work *is* important for its doctrinal content: it has been rightly viewed as one of the first recognizably scientific accounts of the New World, and White's watercolors, even as engraved by de Bry, are an ethnographic document of lasting usefulness.[40] For art historians, ethnologists, and literary critics, the book published by de Bry as *Part 1* of *America*, the beginning of the *Great Voyages*, is a mine of information, of doctrine relevant to their fields. But the doctrines alter when the book is treated *as* a book: a whole which not only contains several types of data, overt and covert, but in which the particular *combination* of information systems and the *mixture* of aims and functions constitutes a single (if complex) object of attention. In particular, it is as a material object—a purchasable commodity—that the book has been overlooked.

Hariot's text had been published by itself in 1588, two years before de Bry's illustrated publication. Only six copies of that pamphlet survive. White's original watercolors (lent to de Bry to be engraved) were not published until 1964—indeed, they were not even located and identified until 1706, and were lost again thereafter. But as the first publication of the *Great Voyages*, the illustrated *Brief reporte* was a publishing coup, and a confident one at that. It came out in its very first edition in four languages simultaneously, in Latin, French, German and English. It is still in print today.[41]

There was, then, a market for this artifact, and a market much wider than the scientific community (for whom the Latin text alone would have sufficed) or Hariot's original audience of prospective English backers and

---

40. See, for instance, Paul Hulton's introduction to his *America 1585: The Complete Drawings of John White*: "Hariot's method is analytical and White matches it by the precision of his drawings. Just as their maps set a new standard of topographical accuracy, White's drawings of Indians and Hariot's notes achieve a new level of ethnological recording" (12). (For a richer sense of their "precision" and "accuracy" and disagreement about their "match," compare Solomon, "To Know, to Fly, to Conjure.")

41. Part 1 of the *America* series was printed again three times in the seventeenth century: the Latin in 1608, the German in 1620 and, as part of an abridged reissue of Parts 1–9, in 1617. To confuse matters further, the *Briefe report* is often referred to as Hariot's *Virginia*.

Fig. 3. Adam and Eve, Theodor de Bry's frontispiece to the final section of his *America, Part I* (Frankfurt, 1590). By permission of the John Carter Brown Library at Brown University.

colonists (for whom the English text would have sufficed). The market seems to have been literate Europe, or more specifically, literate bourgeois Europe. *America, Part I* was a coffee-table book, at a time when the owners of coffee tables (and the books that adorned them) were as likely as not to be thinking themselves of investing in a venture like the Roanoke colony. The book combines four very different texts: a frontispiece by de Bry depicting Adam and Eve beneath the Tree of Knowledge (see figure 3), Hariot's *Brief reporte* on the colony, de Bry's engravings of the watercolors White made on the same expedition with captions by Hariot (figures 7 and 8), and a separate collection of engravings of ancient Picts and Britons (see figure 9). Each of the texts constitutes part of the context of the others, and all of them function in more than one representational arena. Hariot explains the motives of his *Brief reporte* quite openly. It is divided "for your more readie view and easier understanding" into three parts, each of which gives account of a different kind of Virginian "commoditie": merchantable commodities, commodities for sustenance, and commodities for building. The ethnographic remarks of the third part (which occupy about six out of the work's total of twenty-nine folio pages) are set down "that it may appeare unto you that there is good hope [the inhabitants] may be brought through discreet dealing and governement to the imbracing of the trueth, and *consequently* to honour, obey, feare and love us" (italics mine). This echo of the marriage service is not fortuitous; we will return to it later.

White's portfolio from Virginia included many watercolors of plants, animals, birds, fish, and insects, and they are quite informative (see figures 4 and 5). De Bry used none of them in the book, however, though they might have been imagined to illustrate appropriately the information on edible and "marchantable" flora and fauna in Hariot's text. In fact, since de Bry chose only ethnographic pictures to engrave, the whole of Hariot's verbal text comes to seem ethnographic. Its concern is not, as a peek at the index might at first suggest, with natural history, but with culture and its raw materials.

The scientific study of colonized territories has generally absorbed natural history into the more immediate plot of commercial exchange, of nature as transformable into product. In this plot, it is no surprise that the Algonquins are mainly described in terms of their potential for transformation: into Christians, into British subjects, into laborers on future plantations and in future mines. But it is notable that the resulting ethnography is *structured* by a concern with raw materials, that the watercolors are used in de Bry's book to illustrate a verbal discussion of "commodities."

Since the book is perceivable under the category of ethnography largely as a result of de Bry's choices for its illustration and his unusually suggestive opening and closing illustrations (which have to do only theoretically with the situation depicted by the rest of the book, and to which we will return), it is of some interest to consider the dividing line between what we now call art and science in the Algonquin illustrations themselves, as a sign of what this particularly commercial and pragmatic ethnography brought into focus. When we call an illustration "scientific," we do not necessarily mean that it

Figs. 4 and 5. "The Pyne fruite" and "Scorpions." Watercolors by John White (copyright the Trustees of the British Museum, British Museum Press).

is representationally mimetic, only that it presents or demonstrates data — in particular, data that will be applicable to and "true" of other objects in the category represented. What such an illustration signifies (at least by intention) is limited, unambiguous, and theoretically discoverable and demonstrable by any other trained observer. It aims to be empty of ideological or expressive content.

This kind of data is present in White's drawings wherever an artifact is represented. The most reliable kind of detail they represent is detail concerning tools, weapons, jewelry, clothing, agriculture, and cuisine.[42] But the bodies of his Algonquins are rather European, and their poses and gestures are traceable to the iconography of Mannerism (see figure 6).[43] De Bry's engravings exaggerate these features: first, by eliminating the skin color indicated in White's watercolors and second, by altering one of the few depictions of physical gesture recorded by White which are clearly informational. The watercolor of an American man and woman squatting at dinner has been changed in the engraving to show the couple sitting more like Europeans at a picnic (see figures 6 and 7).

It is a peculiar move, but we can see its logic in the context of a discourse one might call "coloniology" and its transformational bias — clearly these Indians will soon *learn* to sit like Europeans. Already they are playing with European dolls (see figure 8) and rubbing their bodies with European Bibles. What is perhaps more peculiar is the general lack of interest evident in the physiognomy of a different ethnos. It is not that early European explorers and colonizers did not see the differences unrepresented here — travel accounts routinely mention them. But they were not a compelling challenge to either artist, and apparently commercial motives led de Bry to suppress what information White did provide.

The result of these distortions and omissions is exoticism. The familiar features and poses of the Indian bodies set off their clothing, jewelry, weapons, and so forth as "strange." We know that White's predecessor Bevin (whose pictures were lost at sea along with the artist) had explicit instructions to

---

42. Such modern ethnologists as John R. Swanton seem content to rely on White as a sole source, along with Hariot, for information on the Algonquins of North Carolina (see *Indians of the Southeastern United States*, 831). Swanton includes fourteen of White's drawings, along with three of de Bry's engravings, among the illustrations to his own monograph. According to Hulton, "what documentary and archaeological evidence there is tends to support their accuracy to the extent that where there is no such evidence we can expect their written and visual information to be soundly based" (*America 1585*, 12).

43. There are also some startling similarities between White's drawing of the Algonquin dancing and the contemporary prints of morris dancing assembled by Frederick Kiefer in "The Dance of the Madmen in *The Duchess of Malfi*."

Figs. 6 and 7. "Theire sitting at meate." Watercolor by John White (copyright the Trustees of the British Museum, British Museum Press); Theodor de Bry's engraving from White's watercolor, in *America, Part I* (Frankfurt, 1590). By permission of The John Carter Brown Library at Brown University.

Fig. 8. Algonquin woman and child, with European doll. Theodor de Bry's engraving from John White, in *America, Part I* (Frankfurt, 1590). By permission of The John Carter Brown Library at Brown University.

"drawe to lief one of each kinde of thing that is *strange to us in England*" (Hulton, *American 1585*, 9; emphasis mine). On one hand, the effect of strangeness is diffused if a represented object is too fully contextualized. And, on the other, the potentially erotic pleasure of the reader in these naked and half-naked representations might be lessened by too *much* strangeness. The balance struck by de Bry's engravings gives us what are among the earliest examples of the long and rich tradition of erotic exoticism in European art. That they should also take pride of place as the first scientifically respectable visual document of ethnography brings us back to Hariot's encomium in the verbal text, in which he eroticizes the colonial relationship by promising that the Algonquins will conform to the demands of the bridal vow. Commercial motives and eroticized representation seem to go hand in hand, and in the case of preinstitutional ethnology, commercial or at least economic interests always underwrote and enveloped fieldwork. Perhaps redundantly, I point out the sexualized structure of historical relations between Europe and the nations it dominated because I am concerned with that structure in the context of science, with its place in the originating documents of the social science that still brings most middle-class Americans their earliest access to dirty pictures, in the pages of the *National Geographic*.

Of course Hariot's text does exhibit the kind of scientizing rhetoric that would be institutionalized soon (as antirhetorical) in the Royal Society's *Philosophical Transactions*, and it is anything but sexy. It comes equipped with a coherent set of headings and subheadings, an index, a habit of quantifying, a plenitude of native terms for indigenous flora, and it is dependent on close interrogation of the Algonquins in their own language, which Hariot in an unusual step had learned before his voyage and which he had invented a graphic system for reproducing phonetically.[44] Stephen Greenblatt even identifies through the report a threefold mode of operation that sounds like a compressed definition of ethnographic fieldwork — "testing, recording and explaining" — and what he sees as being tested by Hariot is the Machiavellian anthropological concept of religion as a universally effective tool of political oppression.[45] The universal (and prerequisite) instinct to religious awe and its cultural expression was to be precisely the motivating hypothesis for Lafitau's seminal work of comparative ethnology a century or so later; even in its cynical Machiavellian application here, it remains a genuinely theoretical idea, testable in the field and demonstrably tested by Hariot. The Algonquin tendency he records of kissing and stroking the Bible is adduced in support of their "hunger," as he calls it, to be transformed by Christianity — and thus made malleable to Christian purposes, subject to Christian law.[46] Like all the other anecdotal material in the section on Algonquin mores, it is recorded in the service of a hypothesis about the local culture. Anecdote is not, for Hariot, what it was for Thevet — the fictional expression of desire or fear, the trace of the Author's subjective density. It is another form of illustration, the empirical cousin of the *exemplum* of sermons.

Even anecdote is not common in Hariot's report. His avoidance of narrative is more pointed than Thevet's (and more notable, too, in the account of

---

44. See John Shirley, *Thomas Hariot*, 106–12. Shirley and an associate take credit for first recognizing the samples of cipher writing among Hariot's papers as, not the cabalistic code identified by Ethel Seaton in 1956, but in fact a phonetic alphabet (and one much more sophisticated than that recorded in the *Orthographie* [1569] of John Hart, sometimes considered the first modern phonetician). The impulse to see sorcery and heterodoxy in Hariot's sophistication has persisted from his day to our own. According to John Aubrey's *Brief Lives*, the mathematician "Dr. [John] Pell . . . tells that he finds amongst [Hariot's] papers (which are now, 1684, in Dr. Busby's hands) an alphabet that he had contrived for the American language, like Devills'" (quoted in Shirley, 107).

45. See Greenblatt, "Invisible Bullets." It is difficult to see such missionary manipulation of the Word of God as "heterodox," as Greenblatt does in this article. Renaissance Christianity in the field was normally militant and coercive. Indeed, it is almost as easy to see Christian imperatives motivating colonial imperialism as to see that imperialism manipulating the Christian message "to serve its own purpose" — conquest *was* a Christian purpose, divinely suggested ("go ye and teach all nations").

46. Cf. E. B. Tylor, about 300 years later: "It is a harsher . . . office of ethnography to expose the remains of crude old cultures which have passed into harmful superstition, and to mark these out for destruction. . . . Active at once in aiding progress and in removing hindrance, the science of culture is essentially a reformer's science" (*Religion in Primitive Culture*, 539).

a man who had *really* "been there," and seen terrible things). Where Algonquin bodies are effaced in de Bry and White's visual images, story is abandoned in the verbal text. Any such object of representation as an event is avoided generically: the only kind of violence or mortality that can have a presence in the present tenses of this text is the customary or repeated kind. Although the political situation created by such an event as the razing of the town of Asquacoqoc (represented as still standing on de Bry's frontispiece map of the region) might be important to a prospective colonizer, it is not significant to a scientist on the prowl for pattern, design, or regularity in nature. The reader is left with an image of violence very different in its import:

> There was no towne where we had any subtile deuise practised against us, we leauing it vnpunished or not reuenged (because we sought by all meanes possible to win them by gentlenes) but that within a few dayes after our departure from euerie such towne, the people began to die very fast, and many in short space; in some townes about twentie, in some fourtie, in some sixtie, and in one sixe score, which in trueth was very manie in respect of their numbers. This happened in no place that wee coulde learne but where wee had bene, where they vsed some practise against us, and after such time. . . . [The inhabitants] . . . were perswaded that it was the work of our God through our meanes, and that wee by him might kil and slai whom we would without weapons and not come neere them.[47] (28)

This native vulnerability to disease is not mentioned in Hariot's captions for the engravings. Here he emphasizes the wholesomeness of the environment, the moderation of Algonquin eating habits, the people's resulting longevity and freedom from sickness. He points out as well in his main text how, despite privation, the colonists were unusually healthy during their stay, crediting the physical environment of Virginia for the European invulnerability elsewhere hinted at as their distinguishing feature in God's providential scourging of the Algonquins with plague. That this environment be free from illness and death is important to its marketable Edenic image, but as there *was* illness and death there, along with unusually good health, they are represented inside a different paradigm—the providential postlapsarian

---

47. The natives of Virginia agreed with the English courts and modern critics in seeing Hariot as a magician. Solomon's article offers some insight into Hariot's vexed relation to sorcery—for her it stems from the courtly (rather than commercial) attitude toward scientific interpretation and practice: active, manipulative, oriented toward transformation and production, and therefore difficult to distinguish at all points from such illicit sciences as alchemy and astrology ("To Know, to Fly, the Conjure," 533–45).

paradigm in which death is the wages of sin and sin a lack of faith in the True God.[48]

We will consider more fully the spatializing of natural philosophy's discourse in the next chapter, its gradual abandonment of narrative and reliance on the depiction of isolated "objects" and the visually oriented schemata of classification and taxonomy. (It is in part this tendency, as it appears in ethnography, that encourages de Certeau to see ethnography in dialectical opposition to history.) Other motives concern us here. This absence of narrative has, for instance, a plain commercial motive — the story behind the text was not a happy one, as had been "bruited abroade by those who returned" (5). Shipwreck, deceit, betrayal, murder, warfare, storms, the destruction of most of Hariot and White's notes and maps in the final flight, the plague of smallpox the English brought upon their prospective allies and eventual enemies — this is the stuff of best-selling travel books, but it does not encourage colonial investment. Hariot was himself a member of the colony's advisory council.

There is a positive as well as a negative motive for replacing such a story with quantifiable facts; the reduction of experience to facts (a word just beginning to distance itself from its etymological meaning of "deeds") is a form of imaginary — and potentially practical — control. Francis Bacon was soon to make the equation famously explicit between empire and empirical knowledge in his *New Atlantis*, when a governor of his utopian technocracy explains a prototype of the Royal Society to the traveler/narrator: "The end of our foundation is the knowledge of causes, and secret motions of things; and the enlarging of human empire, to the effecting of all things possible" (Spedding 3:156).[49] Hariot's scientific reduction hands down its major legacy in the piece of rhetorical decorum James Clifford has labeled the "ethnographic present."[50] Not only does Hariot forbear to narrate the adventures of the

---

48. This conflict of paradigms — pre- and postlapsarian, you might say — bumps into itself for one poignantly funny moment in the caption to the idyllic illustration of fishing methods: "Dowbtless yt is a pleasant sighte to see the people, sometymes wadinge, and sometimes sailinge in those Riuers, which are shallow and not deepe, free from all care of heapinge opp Riches for their posterite, content with their state, and liuinge frendlye together of those thinges which god of his bountye hath giuen unto them, yet withoute giuinge hym any thankes according to his desarte. So sauage is this people, and depriued of the true knowledge of god" (sig. B5).

49. This and all subsequent quotations from Bacon, in Latin or English (with the exception of quotations from the *Novum Organum*), will be taken from the fourteen-volume edition of *The Works of Francis Bacon*, ed. and trans. James Spedding, Robert Ellis, and Douglas Heath (1861–79). For more on versions of the Spedding edition, see chapter 3, n. 5.

50. See his influential article, "On Ethnographic Allegory," in Clifford and Marcus, *Writing Culture*, especially 110–11. In *Time and the Other*, the anthropologist Johannes Fabian coined the term "allochronic" for such representation. His agonized stance with regard to the business of

colony at Roanoke, but his representation of the native inhabitants is delivered in the same habitual present as his description of the area's other "commodities." They exist in the same thoughtless, passive, ahistorical moment as do the plants and minerals of parts 1 and 2, and where they perform in anecdote, the performances are always emblematic, signs of their essential, observable, recordable nature. They are, in this sense, "part of nature" and thus, in coloniologic, destined to become "part of us." Hariot's only suggestion of a historical existence for these people is in the future tense — the colonial future tense rehearsed in the words of the marriage vow.

The text and its visual illustrations reinforce each other in their objectification of Algonquin life. The imaginary Mannerist poses of the engravings freeze for our curious gaze an existence whose longest era appears, in the verbal discourse, to stretch from planting to harvest. That such a radical detemporalization can be useful to science should not obscure the fact that, in more than one way, it can be useful to commerce. But there *is* a narrative in the book, and paradoxically, it is the person to whom the book was most immediately a commercial object, de Bry, who is responsible for it. The frontispiece of the collection of illustrations ("The True Pictures and Fashions of the People in That Part of America . . . ," see figure 3) depicting Adam and Eve (whose myth it will be the business of anthropology to replace) puts the ensuing material into the narrative of sacred or mythic history and has itself a narrative structure. In the foreground our first parents are about to fall; in the background they are fallen, and cultured — engaged in the primitive husbandry and homemaking we are about to see replicated in the illustrations of present-day life in Virginia. The volume closes with a group of engravings taken from watercolors imagining ancient Picts and Britons. De Bry separates this section rigidly from the Algonquin illustrations, but links it to them on its title page as included "to showe how that the inhabitants of the great Bretannie have bin in times past as savage as those of Virginia." These engravings set up a parallel between colonists and colonized which portrays civilization, as Thevet does in the *Singularitez*, as a cultural maturing process — a matter of historical development rather than a sign of absolute European difference.

Like the engraving of pre- and postlapsarian Adam and Eve, the picture reproduced in figure 9 tells a story. The decapitated heads in the picture, trophies of savage victory (and one of them an unsettling echo of the European

---

ethnographic writing forces him, however, to object to narrative as well as to the "allochronic." Both in time and out of time, "native" characters are rendered imaginary by ethnography in Fabian's account of his profession.

Fig. 9. Pict carrying European head. Theodor de Bry's engraving from John White, in *America, Part I* (Frankfurt, 1590). By permission of the John Carter Brown Library at Brown University.

doll held by the Algonquin child in figure 8), are unmistakably the heads of contemporary Europeans. It is the fantasy of an impossible parricide, according to the direction of de Bry's narrative understanding, in which the Picts represent the childhood of modern Britain. In the other direction, it is infanticide; the ancestors have killed—and mutilated—their descendants. The fantastic suggestion here of violent confrontation between the present and the past is easy to read as prophetic of *colonial* violence in the context of the documents this engraving follows, despite the erasure of such inevitable violence from Hariot's verbal text. It is also a very early instance of that radical destabilization of temporal direction characteristic of science fiction, especially in its mode of social commentary.

Read either way, the historicizing message of de Bry's pictorial brackets is far less cheerful than the static display of Virginian raw materials they bracket. Silently, but all the more emphatically for that, it lays out a program for ethnology that is only recently coming into question. The frozen present tense of Lévi-Strauss's "cool cultures," so exaggerated in the objectifications of "scientific" description and illustration, has always been for Europeans a window into their own past, a past they define themselves as having escaped, sublimated, transcended. But this window rarely becomes a mirror; if the proper study of mankind is Man, its most comfortable focus is Other Man. The "ethnographic present" serves not only to express the sense of gazing into one's own fixed and finished past but to differentiate the reader from the object of that gaze. The European reader exists in history, in a subliming metamorphosis; "they" exist outside it—until Europeans arrive and teach them how to read. Thus "primitive" non-European peoples are placed *taxonomically* in the structure of European history, as illustrations of a certain type occurring at earlier moments in a European narrative.

The notion that there are many histories, many qualitatively nonparallel channels of historical time, has liberatory potential, as we can dimly detect in de Bry's eerie management of White's portfolios. But it could also lead to a sense among ethnologists that their researches were being conducted outside of history, in inconsequential innocence.[51] In a different key, twentieth-century ethnologists have frequently mourned the transformative effect of their fieldwork, its inevitable intervention in what it is difficult not to see as a fantastic, Edenic present tense, made believable by the grammatical tenses

---

51. Some thoughtful (and wonder-positive) essays building on this notion of historical nonlinearity are collected in the anthropologists and cultural historians Florike Egmond and Peter Mason's *Mammoth and the Mouse*. Their declared allegiances are to the work of Carlo Ginzburg, E. P. Thompson, and Eric Hobsbawm.

of their writing.[52] This characteristically modern emotion is opposite to Hariot's missionary-commercial enthusiasm, but related to it. The territory of the present tense is the distant land of all we have left behind to become ourselves: adult, civilized, scientific, nostalgic, and fallen. Ethnography, as de Bry's book signals, will find its occupation maintaining the very distinctions between reader and represented that it threatens to collapse. Its rhetorical decorum subtly alienates the intended reader from its subject matter, although that matter is "our fellow man" and the goal of anthropology is an explanatory description of a species to which we all belong. De Bry's engravings balance exotic data with familiar bodies to present this paradox as an experience of erotic looking; the objects of our wondering and titillated gaze are visibly "one flesh" with us. One stops short at suggesting that European readers embraced or kissed this book, or "stroke all over their bodie with it."

Anthropology has been for some time now undergoing a critique led largely by the ethnographers, who must face most squarely the moral ambiguities of their surveillance and its public uses. Most of the historical examination of the field has been directed at the nineteenth century's climax of bad faith; the mutual aid offered each other by academic anthropology and the imperial state has by now been amply documented and lamented.[53] To lament, even to document, the social configurations constraining the earlier literature of the field is in some ways more difficult. Comparison of the texts examined in this chapter reminds us how varied are the aims and uses of a discourse before it is codified and regulated as an institution—a science, an academic "discipline," a genre. To speak about Thevet, Hariot, White, and de Bry in their relation to ethnography is an anachronism that, while it may shed some light on shared features of the later, institutional texts, also obscures much that was salient about their books in their time. The invocation of wonder and pleasure and the function of the aesthetic in "proto-ethnographic" books also participate in the pan-European embodiment of the sensational in the mass-produced. People of all classes, if they lived in a city or near a market

---

52. Classic examples of this include Bronislaw Malinowski's *Argonauts of the Western Pacific* (1922) and Margaret Mead's *Coming of Age in Samoa* (1923). For an antidote see James Boon's essay "Cosmopolitan Moments," which is in part a narrative of his first trip to his first field site, in Java.

53. The literature is vast. See notes in the Introduction, especially n. 18, for lists of classics. Although they hardly need citing, it would be ritually proper to mention here at least Edward Said's *Orientalism* (1979), and George Stocking's *Victorian Anthropology* (1987). But see also Richard G. Fox, ed., *Recapturing Anthropology* (1991).

fair, could usually own at least a broadside trumpeting news of cultural or physical oddities and miracles.[54] The thrills and tingles that had belonged to the oral world of the preacher and the minstrel or to the sacred arena of church architecture and decoration were being absorbed by a print culture that included broadsides and chapbooks alongside grandly illustrated folios, and the sensational pleasures potential in representation were becoming more and more private in their consumption.

It is probable that major innovations in the technology of representation always invoke widespread anxiety about the content and effect of representations. A culture's temporarily unregulated free play, as it learns the powers and limits of a new medium, must necessarily result in the expression, invocation, and dissemination of experiences previously controlled or previously inarticulate. In the next chapter we will attend to the process of rationalizing taste and information. Wonder comes under attack in a number of major seventeenth-century attempts to purify the codification and transmission of knowledge, as well as representation of the scene in which knowledge is acquired. This conscious alienation of the various branches of natural philosophy from pleasurable and sensational modes of illustration eventually produced as oxymoron the phrase "science fiction." But Bacon, Browne, Ray, Plot, and others like them wrote their manifestoes in a culture still capable of producing Kepler's *Somnium*, a fiction and an allegory that carefully and for the first time described and explained the actual topography of the moon. As is evident from a look at almost any well-produced scientific text of the period, the compulsion to make certain kinds of information into beautiful possessions was thriving, and it easily survived the "scientific revolution."[55] But there is a penitential air about modern ethnographic photographs, compared to the overtly seductive Indian princesses of de Bry, and an emptiness where the Royal Cosmographer expanded to fill his texts with an adventuring Self. An exorcism has taken place.

54. See Natalie Zemon Davis, "Printing and the People"; Roger Chartier, *Cultural Uses of Print in Early Modern France*; Jermone Friedman, *Battle of the Frogs and Fairford's Flies*.

55. Consider, for instance, Georg Matthias Bose, who in the early eighteenth century increased his income from lecture-demonstrations of electricity with the *Venus electrificata*, "an insulated electrified young lady whose kisses would be felt long after they were enjoyed" (Heilbron, "Experimental Natural Philosophy," 371). Barbara Stafford's *Body Criticism* (esp. chap. 5) is rich with examples of scientific or at least technological high spirits in the eighteenth century.

III • THE NATURE OF THINGS AND THE VEXATIONS OF ART

Those . . . who aspire not to guess and divine, but to discover and know, who propose not to devise mimic and fabulous worlds of their own, but to examine and dissect the nature of this very world itself, must go to facts themselves for everything.

—Francis Bacon, *Novum Organum* (1620)

But what do I see other than the hats and clothing? Could not robots be concealed under these things?

—René Descartes, *Meditations on First Philosophy* (1641)

WE HAVE SEEN that in the sixteenth century one of the major discourses of wonder, ethnography, could go both ways, though ethnography has come to refer, normally, to only one way—the "objective," scientized, eventually institutionalized way, open to verification by other members of the institution, and conceived as a process of producing and controlling knowledge rather than (readerly and writerly) experience. Like the practitioners of natural magic, André Thevet failed to produce reliable results, at least from the points of view of the state and capital: one could not depend on his books as aids to material acquisition. From the point of view of the consumer of narrative, on the other hand, he produced or at least provoked Jean de Léry, an enraged and articulate Huguenot writer who had spent much more than ten weeks in Brazil and who "in order to refute those falsehoods of Thevet, [was] compelled to set forth a complete report of [his] voyage . . . to prove that everything he says is so much nonsense" (Janet Whatley in Léry, *History of a Voyage*, xlvi–xlviii). Readers have much for which to thank Thevet in the text of the resulting *Histoire d'un voyage fait en la terre du Brésil* (1578), a work that, to complement the origins of Thevet's in literary burglary, was twice stolen or destroyed in manuscript before Léry's fury at Thevet energized a surviving third version. And whatever the depth of furious opposition behind the work, it shares with Thevet the orientation to the pleasurably exotic and a grounding in the authority of witness. In its subtitle it claims to contain "the Navigation and the *Remarkable* Things

Seen on the Sea by the Author; the Behavior of Villegagnon [in another complementary inversion of Thevet, this authentic witness makes someone else the protagonist of his narrative]; the Customs and *Strange* Ways of Life of the American Savages; Together with the Description of Various Animals, Trees, Plants, and Other *Singular* Things *Completely Unknown over Here*" (my emphases).

Léry is not trying to sell Brazil to investors, as Hariot was Virginia. Like Thevet, he was trying to sell it to (intelligent, well-to-do, cultivated) readers. Male readers, mostly, who were not constructed as would-be pilots or imaginary settlers, but who nonetheless appreciated a good frisson. Words and phrases like "different," "strange," "instead of," "monstrous," "exquisite," and "savage" turn up in the headings of most of the chapters not devoted to the tempestuous ocean crossings that bracketed Léry's American experience.[1]

This chapter will eventually arrive at a work very distant from Léry's, or Thevet's — though perhaps as close as the other side of the coin. The British observer of nature and culture, Robert Plot, the first curator of the Ashmolean Museum and its cabinet of ethnological, biological, and mineral treasures, wrote his major work a century later, and in England, and about a single county — the one he lived and worked in. It has roughly the same number of chapters as Léry's book about Brazil, but only three are devoted to cultural productions (as opposed to ten of Léry's) and though it is in fact full of wonders they are not often announced as such. This is a *natural* history of Oxfordshire, and the extranatural when it forces itself upon the author is admitted with pain: "I was prevailed on at last to make the *relation* publick (though I must confess I have no esteem for such kind of *stories*, many of them no question being performed by combination . . .)" (Plot, *Natural History*, ch. 8, par. 37). One gets the sense that the word "natural" has already acquired the positive charge once held by "wonderful" in titles and characterizations.

But before arriving at Plot's exemplary record of his experience, mostly reduced to data and absent the gathering "I," it is necessary to examine some of his century's theorists on the topics of wonder and nature, reading and knowledge. The texts to be looked at closely in this chapter are either classifiable as manifestoes in the case against naive learning and "popular" knowledge, or exemplary documents from the new world of nonacademic science. The concentration here is on the process of this exclusion in the

---

1. For thoughtful readings of Léry's book see de Certeau, "Ethno-Graphy," Janet Whatley's introduction to her translation of Léry's *History of a Voyage*, and Greenblatt's introduction to *Marvelous Possessions*, 14–19.

mentality of a literary culture generally. The exclusion contributes to an eventual class distinction — contemporary antagonism to "science" and preference for the weird or "inexplicable" is centrally located among the disenfranchised. But since the evidence of intentional change comes from traces left in print by those with access to publication, one can find out very little about shifts in the appetite for wonder among the illiterate poor. Until literacy is more general and print cheaper, that appetite will have slight effect on the divisions of literary and scientific labor.

FRANCIS BACON AND THE *NOVUM ORGANUM*: REPRESSION

The Romanian historian Ioan Couliano's characterization of the Reformation, which is derived less from economic than from psychoanalytic theory, is nonetheless nasty, brutish, and short:

> On the practical level, it results in the advent of modern institutions.
> On the psychosocial level, it results in all our chronic neuroses, which are due to the entirely unilateral orientation of Reformation culture and its rejection of the *imaginary* on grounds of principle. (*Eros and Magic*, 222)

In accordance with the principle of conservation of both matter and energy, one might assume that such a rejection can only amount (at most) to a dislocation. Seen from up close it may not even amount to that (the mathematician Hariot, as we have just seen, contributes to an ethnographic stroke book, in an early Return of the Repressed). But it is undeniable that during the period of Reformation and Counter Reformation the work of repression is undertaken on a large scale: from the philosophical treatise to the dinner-talk exchange of pieties among the educated classes, lip service begins to be paid to the visible, the normal, the plain, and to the transparent, desensationalized, "masculine" representation of all that.[2] The simultaneous efflorescence of wonder and sensation in popular culture, occult *scientia*, art

---

2. The effort seemed most ably summarized to all its enthusiasts by the term "masculine." For instance, Bishop Thomas Sprat says, in his *History of the Royal Society* (1667), "the *Wit* that is founded on the *Arts* of men's hands is masculine and durable" and Oldenburg speaks of the Royal Society as planning "to raise a Masculine Philosophy." For these sources see Brian Easlea, *Witch Hunting, Magic, and the New Philosophy*. Other scholars who have isolated this metaphorically gendered development for analysis include Carolyn Merchant in her classic *Death of Nature*, Evelyn Fox Keller in *Reflections on Gender and Science*, Ludmilla Jordanova in *Sexual Visions*, and Londa Schiebinger in her work generally, especially "Feminine Icons."

(in the modern sense), and cult are, as Couliano himself admits, a function of that repression.³ The process is lengthy and it stutters, it has stages. Bacon may have been a Rosicrucian sympathizer—certainly Kepler was a Pythagorean, Newton an alchemist. Still, despite the complex sympathies of his transitional century, Bacon speaks in a forceful, institutionalizing voice for the repressive energies of the emerging cultural formation. Let us have a look at his manifesto, the *Novum Organum* (1620), its invocations and exorcisms.⁴

"Now what the sciences stand in need of [*opus est ad scientias*] is a form of induction which shall analyze experience and take it to pieces, and by a due process of exclusion and rejection lead to an inevitable conclusion . . . extracted not merely out of the depths of the mind but out of the very bowels of nature" ("Great Instauration," 20–21 [Spedding 1:216–17]). Not just any experience can be analyzed: "For the testimony and information of the sense has reference always to man, and not to the universe; and it is a great error to assert that the sense is the measure of things" (besides, "sometimes [the sense] gives no information, sometimes it gives false information," [21]). The natural history Bacon proposes, then, is "to be a history not only of nature free and at large . . . but much more of nature under constraint and vexed; that is to say, when by art and the hand of man she is forced out of her

---

3. "Those who contend that people of the Renaissance felt, thought and acted like us are greatly mistaken. On the contrary, we have the time-honored custom of seeking within ourselves the world image of the Renaissance person, to such an extent that he is confused with our own 'unconscious,' with what we have learned to uproot and mutilate within ourselves. [He] . . . collects all our most infantile and absurd traits. . . . For we have lost that which he had and he lacks what we have mastered" (Couliano, *Eros and Magic*, 184). Couliano seems to be bearing witness to the fulfillment of Bacon's figure of the Renaissance as the "childhood" of natural philosophy as Bacon understood it. For more on this imaginary relation of present to past, see Latour, *We Have Never Been Modern*.

4. The work now usually called the *Novum Organum* was first published under the title *Magna Instauratio*, along with a prefatory piece headed "Instauratio Magna" ("Great Instauration") and a piece at the end, "Parasceve ad historiam naturalem" ("Preparative towards a Natural and Experimental History"), that was to form the preface of the third part in Bacon's opus, the *Phenomena universi*. I will cite these separate parts, normally published together, by their separate titles, as the differences among their functions and emphases are of relevance to the discussion. This means I will cite from the *Organum* proper as "Aphorisms," either book 1 or 2 (giving the aphorism number rather than page number, unless the aphorism is long—likewise with the aphoristic "Parasceve"). I have used Fulton Anderson's edition of Spedding et al. Parenthetical Latin is from Spedding's fourteen-volume edition of the *Works of Francis Bacon* (London, 1861–1879), which I cite simply as Spedding. (It is easy to confuse this edition with several printings and editions of the Spedding *Works* published in London, Cambridge, and Boston in the years between 1857 and 1879; not all these editions include the *New Atlantis* and they appear in sets of anywhere from seven to fifteen volumes!) Julie R. Solomon's admirable analysis of Bacon's work and the development of a new "philosophical" ethic, *Objectivity in the Making*, appeared too late, sadly, to be of help to me in the making of this chapter.

natural state, and squeezed and moulded" (25).⁵ For "the nature of things betrays itself more readily under the vexations of art than in its natural freedom" (25). Experiment, then, is to replace experience as the object of analysis, and nature will be *coerced* to speak.⁶

The direct confrontation with and expulsion of what Couliano would categorize as the "imaginary" appears in the following paragraph (which looks forward to the similarly sweeping exorcism of Milton's "Hymn on the Morning of Christ's Nativity"):

> In the selection of the relation and experiments I conceive I have been a more cautious purveyor than those who have hitherto dealt with natural history. For I admit nothing but on the faith of the eyes, or at least of careful and severe examination, so that nothing is exaggerated for wonder's sake, but what I state is sound and without mixture of fables or vanity. All received or current falsehoods also (which by strange negligence have been allowed for many ages to prevail and become established) I proscribe and brand by name, that the sciences may be no more troubled with them. For it has been well observed that the fables and superstitions and follies which nurses instill into children do serious injury to their minds; and the same consideration makes me anxious, having the management of the childhood, as it were, of philosophy in its course of natural history, not to let it accustom itself in the beginning to any vanity. Moreover, whenever I come to a new experiment of any subtlety (though it be in my own opinion certain and approved), I nevertheless subjoin a clear account of the manner in which I made it, that men, knowing exactly how each point was made out, may see whether there be any error connected with it and may arouse themselves to devise proofs more trustworthy and exquisite, if such can be found; and finally, I interpose everywhere admonitions and scruples and cautions, with a religious care to eject, repress, and, as it were, exorcise every kind of phantom. ("Great Instauration," 26)

The tropes that largely embody "The Great Instauration" leave no doubt about where the imaginary is being relocated *to* — as "fables," it will circulate between nurses and children. The erotic attention of adult males will be displaced (generalized?) from female persons (if that is where it normatively "belongs") to a thoroughly feminized Nature, now being squeezed and molded into experiments by lab equipment and the "machines" (27) of method.⁷

---

5. On Bacon's language of sexual domination, see Keller, "Baconian Science," in her *Reflections on Gender and Science*. For a good example of the theme of violence continued in the language of one of his own experiments see "Aphorisms," bk. 2, aphor. 45, p. 224.

6. The phrase is borrowed from Aaron Fogel's *Coercion to Speak*. Consider Bacon's words: "The best demonstration by far is experience, if it go not beyond the actual experiment" ("Aphorisms," bk. 1, aphor. 70, p. 67).

7. See Keller on the switch registered in Bacon from a homoerotic (Platonic) to a heterosexual (or functionally "hermaphroditic") model of the acquisition of knowledge ("Baconian Science," 40).

Sometimes Bacon's tropes are less domestic, emerging not from the relations of the nuclear family but from the arts of war, in particular the war of colonial conquest, with its accompanying language of journey and exploration. Describing the *Advancement of Learning*, the survey of the sciences in their present state which will constitute the first part of his Great Instauration, he says "we will . . . make a coasting voyage along the shores of the arts and sciences, not without importing into them some useful things by the way" ("Great Instauration," 17–18). "For there are found in the intellectual as in the terrestrial globe waste regions as well as cultivated ones" (18). The journey through the Waste Land sounds at times like romance, in which "the way is still to be made by the uncertain light of the sense, sometimes shining out, sometimes clouded over, through the woods of experience and particulars; while those who offer themselves as guides are . . . themselves also puzzled" ("Great Instauration," 12). But more often the figures are military: "[I] do not propose merely to survey these regions in my mind, like an augur taking auspices, but to enter them like a general who means to take possession" (18). In his discussion of the metaphorical "Idols" of human thought, he proposes a systematic and methodological "confutation," so "the human understanding may the more willingly submit to its purgation and dismiss its idols" ("Aphorisms," bk. 1, aphor. 61). It is easy to read the entire work as an allegorical account of colonial exploration and conquest (consider its famous frontispiece, of the ship sailing West through the Pillars of Hercules), even to wonder whether perhaps Cortés did not provide the model for the "Scientific Revolution."[8]

Bacon is a heavily figurative writer, even for his time.[9] While this attribute might seem ironic in the man whose followers (in emulation of Salamon House in his "mimic and fabulous" *New Atlantis*) founded the Royal Society, with its platform abjuration of "these specious *Tropes* and *Figures*," (Sprat, *History of the Royal Society* 2:112), it is clearly functional rather than ornamental in a manifesto aspiring to alter not only the knowledge but, as prerequisites, the values and the mental categories of a culture. Bacon's intentions with regard to the study of "nature" (which includes much of culture as

---

8. "The most important of these idols, and the ones in whom they have the most faith, I had taken from their places and thrown down the steps; and I had those chapels where they were cleaned, for they were full of the blood of sacrifices. . . . I made them understand through the interpreters how deceived they were in placing their trust in those idols which they had made with their hands from unclean things" (Cortés, "Second Letter," in *Letters from Mexico*, 106). This is Cortés's ritual throughout his Mexican campaigns. Compare Tylor's remarks on ethnology as "a reforming science," quoted in chapter 2.

9. See especially Lisa Jardine, *Francis Bacon*.

well—see the list of "Histories of Man" in the "Parasceve") are made to align figuratively with a number of other available intentions and desired relations in his cultural moment. The constriction of women's legal rights and public functions discussed in Joan Kelly's famous article, "Did Women Have a Renaissance?" (no) neatly matches the language of objectification and constriction applied to Bacon's "nature" under the new science; the "coasting voyages," economic colonialism, and conquest activities of Bacon's intellectual "polity" are resonant with those of the polity in which he functions as chancellor; the prying scopophilia of the new scientist with his "instruments" and "machines" seems closely related both to growing interest in techniques of political and military "intelligence" and to a reemergence of erotic art encouraged by the development of printing and engraving. The expulsion of fable, phantasm, and "superstition" to the nursery (outland of the bourgeois home, where children and peasant or working-class women confabulate under the titular domination of an absent mother) helps constitute a "mimic and fabulous world," one as Other to Bacon as is America or the Moon—and one that is soon to be appearing conflated with them in the escapist, utopian, and satiric fictions that preceded the insurgence of the "realistic" novel. And even Jane Austen's novel-world is a mimic and fabulous one dominated—though not ruled—by mostly unmarried women not yet subject to marital/colonial/scientific containment by the "vexations" of rational culture.[10]

Reading Bacon's plan for the future is bracing work at best and can be openly grim even, presumably, for those who may expect to participate and profit in this "royal work" ("Parasceve," aphor. 4). For all the glamour of its largeness and "truth," it is an explicitly disenchanted realm he imagines, where "we shall no longer be kept dancing within little rings, like persons bewitched, but our range and circuit will be as wide as the compass of the world" ("Parasceve," aphor. 4).[11] If our understanding is "to be expanded and opened till it can take in the image of the world as it is in fact" (aphor. 4)

---

10. See Richard Handler and Daniel Segal on Jane Austen the ethnographer, in *Jane Austen and the Fiction of Culture*.

11. "Haven't we shed enough tears over the disenchantment of the world? Haven't we frightened ourselves enough with the poor European who is thrust into a cold soulless cosmos, wandering on an inert planet in a world devoid of meaning?" Latour's mockery is an effective weapon in his battle to force on cultural historians a sense of the continuity and *longue durée* of certain cultural patterns and dispositions. (For an account of the historiography productive of these "tears," see Daston, "History of Science in an Elegiac Mode.") But the "disenchantment of the world" is not a figment of what Latour calls the "modern Constitution" or "postmodern nonsense," as Bacon attests here, and as do many of his contemporaries and successors in the period visited by this book.

then fact must be divested of its "ornaments" and "superfluit[ies]," "though no doubt this kind of chastity and brevity will give less pleasure both to the reader and to the writer" (aphor. 3). But then, "they who shall hereafter take it upon them to write natural history should bear this continually in mind — that *they ought not to consult the pleasure of the reader*" (aphor. 2, emphasis mine).

As Freud reminds us, repression is costly (or in Bacon's words, "a history of this kind . . . is a thing of very great size and cannot be executed without great labor and expense" ["Parasceve," 271]).[12] The difficulty of paying this expense is already evident in the heavy freight of figurative language with which Bacon dismisses the "treasury of eloquence" and the "ornaments of speech" (aphor. 3). In the paragraph that banishes "superstitious stories," earlier attributed to the nursery, the infant and nursing mother reappear as his figures for philosophy and natural history (aphor. 3). The figure allows him to replace the "old wife" of the illiterate classes with the bourgeois nursing mother, but that the scene of "masculine" scientific reform is envisaged in such a figure at all is a sign of how loudly the "phantasms" continue to knock.

Consistency was not to be the hobgoblin of Francis Bacon's mind: author of a masque and a utopian prose fiction, he also planned to be the author of a "superstitious history of marvels," to include "the history of pretergenerations" along with "the history of prodigies which are natural" ("Parasceve," aphor. 4). Such things were worthy, then, of being committed to print, although "in a separate treatise of [their] own"; indeed, they had their place in the construction of a rationalized science.[13] A thing might be wonderful and

---

12. See Freud, "The Resistances to Psychoanalysis": "Nor is [society] sufficiently wealthy or well-organized to compensate the individual for his expenditure in instinctual renunciation. It is consequently left to the individual to decide how he can obtain enough compensation for the sacrifice he has made to enable him to retain his mental balance. On the whole, however, he is obliged to live psychologically beyond his income" (259). The effort to fantasize the necessary wealth and organization to compensate for the increase in renunciation required by Reformation societies strikes me as a major determinant in the explosion of Utopian fiction, which explosion includes Bacon's own *New Atlantis* as a major instance.

13. See Findlen, "Jokes of Nature and Jokes of Knowledge." Fifty years later, the naturalist John Ray's introduction to his translation and "enlarge[ment]" of Francis Willughby's *Ornithology* (1678) eschews "fabulous birds, such as are confessedly so, *viz. Phenixes, Griffins, Harpyes, Ruk*, and the like" ([a]^r) but then, "because I would not rely too much upon my own judgement," he makes an appendix for "descriptions of some of that nature out of Hernandez, which I refer to the Readers censure" ([a]^v). Only five of the birds described seem obviously fabulous to Ray; the others are excluded from the main text because they are "too briefly and unaccurately described to give us a full and sufficient knowledge of them" (Ray, "Appendix, in *Ornithology*," 386). This brevity and vagueness are the chief rhetorical ingredients of a wonder. The French virtuoso M. Auzout is reported in the *Philosophical Transactions* (1:7, 204) as having examined three kinds of "*Shining Worms* in Oysters," of which, for example, the second was "red, and resembling the common *Glowworms*, found at Land, with folds upon their backs, and feet like the

also true. It would have to be "severely examined," and then redescribed in the "chaste" language proper to science — it would have to be translated into the plain vernacular of Truth.[14] But by then it would no longer be a wonder. Wonder was likely to survive longest where it was beyond the reach of the "vexations" of the scrupulous naturalist. Second- and third-hand reports were to be admitted to Baconian natural history, though under a cloud of "qualifying note[s], such as 'it is reported,' 'they relate,' 'I have heard it from a person of credit'" ("Parasceve," aphor. 8). But even in relation to the inevitable mediation of exotic information, "machines" for extracting usable data would be invented in the form of "correspondence instructions" by such virtuosi as Robert Boyle and William Petty.[15]

The energy exerted against wonder came more and more to be exerted against language — against rhetorical means for inducing wonder and "pleasant recreation," such as similitudes, "superfluity," "curious variety" ("Parasceve," aphor. 3), but also against the historical dimensions of natural language itself, with the resultant ambiguity and near redundancy of its lexicons. Universal language schemes, the search for a "real character" and attention to the sharp precision of cipher-writing and the information it was likely to transmit, emerged side by side with historical linguistics and the hermeneutic relativism provoked by discoveries in that field. After Descartes and Newton, mathematics would eventually provide the kind of definiteness (though not the degree of it) for which Bacon yearned: "everything relating both to bodies and virtues in nature [should] be set forth (as far as may be) numbered, weighed, measured, defined. For it is works we are in pursuit of, not speculations; and practical working comes of the due combination of physics and mathematics" ("Parasceve," aphor. 7).[16] But even mathematics

---

former; and with a nose like that of a dog, and one eye in the head." This sounds as if it were right out of *Mandeville's Travels*.

14. Two among many fine articles in a large literature on this "vernacular" rhetoric are Peter Dear, "*Totius in verba*: Rhetoric and Authority in the Early Royal Society," and James Paradis, "Montaigne, Boyle, and the Essay of Experience." John D. O'Banion's *Reorienting Rhetoric: The Dialectic of List and Story* devotes two chapters to the forging of a scientific antirhetoric.

15. For important examples, see Robert Boyle's "General Heads for a Natural History of a Country" (1666) and William Petty, "Quaeries Concerning the Nature of the Natives of Pennsylvania" (endorsed 1686). With Petty, as with Bacon, the machine to dry up wonders is not waterproof: *quaerie* no. 5 under the heading of "Religion and Marriages" asks "Do they believe that God can raise men from Death to Life? Make a man a woman? dry up the sea? make the sunne stand still? remove the greatest mountaines from one place to an other, at any future time prefixt?"

16. On universal languages and their relation to seventeenth-century philosophy, see James R. Knowlson, *Universal Language Schemes in England and France, 1600–1800*, Mary Slaughter, *Universal Languages and Scientific Taxonomy in the Seventeenth Century*, and Lia Formigari, *Language and Experience in Seventeenth-Century British Philosophy*. On their functions in imaginary voy-

would have to undergo a purification, shedding the mystical semiotics of numerology and Pythagorean geometry.[17]

The control of language in the service of "works" and the scientific power to accomplish them finds a significant countertrend, in literary writing, in the increasing prevalence of irony (though in the overdetermined flourishing of that trope, political and religious pressures were clearly more directly causal). This atmospheric change in literary sensibility coincides with what might be called a topographical one—nostalgia, exoticism, and eventually sentimentality increasingly pervade the construction of setting. Both fictional and nonfictional vernacular writing, whether dramatic or page-bound, concerns itself with what is other, what is elsewhere: the "passed world" and "Subterranean world" of Browne's *Urn Burial*, the "New World" of the voyage literature, the foreign worlds displayed (as fashion shows) in ballet and masque, the world in the Moon of the lunar fantasies that we will look at later, the better worlds of *Utopia*, the *New Atlantis*, the *Civitas solis*, the worse worlds of *Lazarillo de Tormes* and *La Picara Justina*, the infinitesimal worlds of Hooke and Leeuwenhoek, the transformed world of the alchemists, the next world of the many fervent Reformation cults.[18] The proliferation of other worlds in the sixteenth and seventeenth centuries is so great that perhaps one can be forgiven for looking at "the" world projected by Bacon's research program as just one more. What is unique about it is its claim to be the "true" and only world; its power to compel submission to that claim is manifest in the "confessed" unreality, or surreality, of most of the others.

CATALOGUES AND TABLES: ITERATION

Wonder . . . is broken knowledge.

—Francis Bacon, *The Advancement of Learning*

The most pungent, if elusive, sign of Bacon's inconsistency with his own renunciation of pleasure and wonder comes at the end of the 1620 volume that includes the prefatory "Magna Instauratio," the "Novum Organum," and the

---

ages written by seventeenth-century virtuosi, see Paul Cornelius, *Languages in Seventeenth- and Early Eighteenth-Century Imaginary Voyages*.

17. See Marjorie Hope Nicolson, *The Breaking of the Circle*, and S. K. Heninger, *Touches of Sweet Harmony*.

18. For microscopic worlds, see Chapter 6; for better worlds the classic work is Frank and Fritzie Manuel's *Utopian Thought in the Western World* or for a more literary treatment Marina Leslie, *Renaissance Utopias*, and Amy Boesky, *Founding Fictions*; for materials on seventeenth-century millenarianism, start with Keith Thomas, *Religion and the Decline of Magic*, 140–44.

"Parasceve," in the form of a list (or "Catalogus") of potential natural histories, subdivided after several meteorological items into histories of "The Greater Masses," "Species" (including mineral "species"), "Man," and "Pure Mathematics": "And now should come the delineation of the particular histories. But I have at present so many other things to do that I can only find time to subjoin a Catalogue of their titles" ("Parasceve," aphor. 10). He goes on to make one final figure, of the scientist as lawyer examining "nature herself and the arts upon interrogatories" (which is his second version of putting nature "to the question," the old phrase for the "vexations" or tortures suggested in the "Great Instauration"). But what we get instead of the dry logic of these interrogatories and their results is one of the most beautiful of all the many suggestive and cornucopian lists of the period, a testimony to the reality of that erotic strain in Bacon's theorizing of the relations between "philosopher" and nature. Properly interrogated, Bacon's phenomena will lose their numina, but until that time it is precisely wonderful to skim this "Catalogue of Particular Histories." Its language is anathema to that required and by now achieved in scientific writing; "Eruptions of Fire from the Earth" would have to be replaced by a *term*, or better, several:[19]

9. History of the Blue Expanse, of Twilight, of Mock-Suns, Mock Moons, Haloes, various colors of the Sun; and of every variety in the aspect of the heavens caused by this medium.
...
18. History of the greater Motions and Perturbations in Earth and Sea; Earthquakes, Tremblings and Yawnings of the Earth, Islands newly appearing; Floating Islands; Breakings off of Land by entrance of the Sea, Encroachments and Inundations and contrariwise Recessions of the Sea; Eruptions of Fire from the Earth; Sudden Eruptions of Waters from the Earth; and the like.
...
71. History of Smell and Smells.
72. History of Taste and Tastes.
73. History of Touch, and the objects of Touch.
74. History of Venus, as species of Touch.

---

19. For instance, here is a brief passage from an article at hand by Gordon B. Bonan, "Effects of land use on the climate of the United States": "The principal changes from the CCM2 are related to the cloud, radiation, convection and boundary layer parameterizations and include: improved diagnosis of cloud optical properties (maritime versus continental effective radius, liquid water path), incorporation of trace gases in the longwave radiation (CH_4, N_2O, CFC11, CFC12); incorporation of radiative properties of ice clouds; incorporation of background aerosol; modifications to the cloud fraction parameterization including a new convective cloud scheme and modified layered cloud scheme; modified moist convection; evaporation of stratiform precipitation; and diagnosis of boundary layer height in the non-local atmospheric boundary layer scheme" (452).

That time never really came for Bacon, who died six years later of pneumonia he had caught while stuffing a dead chicken with snow in one of his experiments with cold. It is almost as if he preferred to leave behind a poetic iteration of the "things of this world": the fullness of this world was never lost on the philosopher who (in Fulton Anderson's unintentional echo of Freud) "was always living beyond . . . [his] means" ("Introduction," *New Organon*, ix).[20]

The list, with its intense referentiality and its inevitable emphasis on the noun rather than its syntactical setting, is (despite its cornucopian possibilities) an ideal medium for linguistic purifiers and demystifiers such as Bacon. Catalogues were a common genre in that period of plans and inventories, cabinets and, increasingly, museums.[21] A catalogue like that of "Tradescant's Ark" (the first public museum in England), published by the younger John Tradescant in 1656, offers a slightly different version of sensational plenitude from that produced by Bacon's invocation of the "phenomena universi."[22] The items in Tradescant's catalogue refer to individual objects rather than "species," and thus the catalogue, though divided into chapters on the basis of some concept of category, is more like a merchant's inventory or a detailed will. It has reference to property rather than, or anyway more than, to knowledge.[23] And of course wonder is more properly a response to the singular and potentially palpable than to the general and categorical.

Obviously catalogues take us a long way from any sort of narrative and thus apparently from some kinds of early modern worldmaking, both epistemological and "imaginative."[24] But such catalogues, like (if also quite unlike)

20. Bacon did write a few of the 130 "particular histories" listed in his catalogue—"History of the Winds," "History of Life and Death," "History of Sulphur, Mercury and Salt," and an "Inquiry Concerning the Loadstone." (For all but the "History of Sulphur," see Spedding, vol. 2.)

21. On museums and cabinets in this period, see Oliver Impey and Arthur MacGregor, eds., *The Origins of Museums*, as well as the catalogue of the exhibit of "wonder cabinets" mounted by Dartmouth's Hood Museum, Joy Kenseth, ed., *The Age of the Marvelous*. See also Daston and Park, *Wonders and the Order of Nature*, chap. 7, and Findlen, *Possessing Nature*.

22. *Musaeum Tradescantianum: Or, A Collection of Rarities Preserved at South-Lambeth neer London* (1656).

23. See R. F. Ovenell, *The Ashmolean Museum*, 65–69 and 162–70, for early major thefts from the museum that Plot curated. The list of objects stolen in 1691 by an unknown "gent." (possibly a Mr. Beverland, "Publisher of prohibited obscene and profane Books" [68, Ovenell is quoting from Wood's *Fasti*]) is itself a lovely little catalogue: "11. A small Agat of various colour in the form of a heart, set in gold . . . 12. An Oval piece of Amber encompassed with a silver hoop having a frog enclosed . . . 17. A small picture of John Aubrey Esq. in water colour done by Cooper set in a square frame of Ebony" (67).

24. On the oppositional relation between narrative and the list (or more generally the schematizing impulse in Western rationalism) see Hayden White's essay "The Value of Narrativity in the Representation of Reality" and John D. O'Banion's *Reorienting Rhetoric*. In the course of

Thevet's plagiarizing, offer an atomized ingredient to the discursive situation in which the novel and the discipline of anthropology were emerging: it is in fact *only* in the novel (and in modern travel writing) that one can still find details of such careless exactitude as "Two feathers of the Phoenix tale" (*Musaeum Tradescantianum*, 2) or "A Trunion of Capt. *Drake's* Ship" (43) or "A Cherry-Stone holding 10 dozen of Tortois-shell combs, made by *Edward Gibbons*" (37). It is a mistake to think of the novel as primarily or above all a *narrative* form; it is also the artistic location of the "gratuitous detail," the isolated frisson, the individual passion or observation. (Chapter 6 will return to the importance of the detail, in a discussion of micrographia.)

What a catalogue like Tradescant's does not have, that the taxonomies of species to be constructed by Linnaeus *will* have, is blanks. That is, there is no paradigm, no signifying grid, in which to register differences between the logic of the paradigm and the "facts" of nature, or between the potentially full paradigm and the incomplete state of the data with which it is being filled. One cannot really glean a collecting program from reading the catalogue of the collection-so-far. The structure is simply additive (as was that of the later Ashmolean catalogue, where Plot's assistants and successors merely wrote in new acquisitions between the lines of the entries on old ones).[25]

A trace of what might be embarrassed awareness of his catalogue's meaninglessness appears in Tradescant's occasional, largely decorative, use of the brackets from the Ramist dichotomized outline tables familiar to all Renaissance schoolboys. (See Robert Burton's *Anatomy of Melancholy* for a complete table.) They do not go beyond a straightforwardly syntactical function, serving as a shorthand device rather like ditto marks. At their most complex (with one exception) they are two-fold:[26]

$$\text{A Match-coat from} \begin{cases} \text{Virginia} \\ \text{Canada} \end{cases} \text{of} \begin{cases} \text{Feathers} \\ \text{Deer-skin} \end{cases}$$

O'Banion's (ironically overschematized) account of the decline of rhetoric and its replacement by a logical, a-narrative mode of thought he calls "List," he often mentions the novel as a kind of byproduct of the rejection of narrative as a form of knowing. For O'Banion (drawing on Kenneth Burke) this rejection is a subset of the abandonment of rhetoric as a field irrelevant to the disembodied forms of transmission in a print culture.

25. See Ovenell, *The Ashmolean Museum*, e.g., 68, 79, 85, 91. Some acquisitions were not entered at all; erasures intermittently marked thefts and losses. There was no regular accessions register until 1757.

26. See Tradescant, chap. 10, "Garments, Vestures, Habits, Ornaments."

If the catalogue of the museum is worth publishing, it is presumably to benefit "such ingenious persons as would become further enquirers into the various modes of Natures admirable workes" ("To the Ingenious Reader," a^r). So Tradescant claims. But such an "enumeration" is more likely to serve his vaguer purpose as "an honour to our Nation"—the catalogue should prove his assertion that his "Rarities [are] more for variety than any one place known in Europe could afford" (A7^v). What it conveys is an open-ended amount of property, not any relations between the properties of things.

Stripped of narrative (that is, of the travel writing that might have narrated the acquisition or described the original contexts of the rarities), the list is certainly "objective" (it has no grammatical subject). But it remains wonderful—in Bacon's equation, broken—as did no doubt the "Ark" itself, or any of the cabinets the Germans so aptly called *Wunderkammern*. As in the Ramist table, the principle behind the cabinet's display is visual and spatial rather than abstractly logical or taxonomic. Walter Ong says of the Ramist dialectic: "[it] represented a drive toward thinking not only of the universe but of thought itself in terms of spatial models apprehended by sight. In this context, the notion of knowledge as word, and the personalist orientation of cognition and of the universe which this notion implies, is due to atrophy. Dialogue itself will drop more than ever out of dialectic. Persons, who alone speak (and in whom alone knowledge and science exist), will be eclipsed insofar as the world is thought of as an assemblage of the sorts of things which vision apprehends—objects or surfaces" (*Ramus, Method, and the Decay of Dialogue*, 9).

Of course, a "thing" is no more direct or self-evident a signifier than a word or number, and for the purposes of science an individual thing is almost never a signified. Bacon, proselytizer of light and prophet of illumination ("all depends on keeping the eye steadily fixed upon the facts of nature and so receiving their images simply as they are"), yearning to write "an apocalypse or true vision" ("Great Instauration," 29), is nonetheless concerned to write the "histories of Species" (along with those of such other generalities as the "Greater Masses" and of "Man"—see headings in the Bacon's closing "Catalogus"). The invention of a workable, biological definition of "species," much and long contested, as even the *Oxford English Dictionary* will tell you under definition 2, no. 10, was not yet, and Bacon feared glib, make-believe categorizations.[27] His new science was to determine immanent kinships

---

27. "When man contemplates nature working freely, he meets with different species of things [*species rerum*], of animals, of plants, of minerals; whence he readily passes into the opinion that there are in nature certain primary forms which nature intends to educe, and that the remain-

among phenomena, not to describe singular and portentous deviations from an indefinite norm. That is, sooner or later the new science was to lead to the systematic application of a grounded notion of what is now understood by the term "species." But in the meantime, while Ramism and the print revolution rendered knowledge increasingly visible and visual, "species", (a very complex word, with a dense history) had another set of meanings and uses, now obsolete but of particular relevance to our concerns with display, tableau, the spectacular and the occult, the visible and the hidden. Definitions 3, 4, and 5 in the *Oxford English Dictionary* offer the following: "the outward appearance or aspect of a thing," "a thing seen, a spectacle: *esp.* an unreal or imaginary object of sight, a phantom or illusion," "a supposed emission or emanation from outward things, forming the direct object of cognition for the senses or . . . understanding."[28] The given first uses of all these meanings date from the first half of the seventeenth century, revealing at the level of the word that which Ong points to at the level of text and system and which cabinets, museums (not to mention zoos), and their published catalogues confirm at the level of what might already be called the institution.[29]

---

ing variety proceeds from hindrances and aberrations of nature . . . or from the collision of different species and the transplanting of one into another. To the first of these speculations we owe our primary qualities of the elements; to the other our occult properties and specific virtues; and both of them belong to those empty compendia of thought wherein the mind rests, and whereby it is diverted from more solid pursuits" ("Aphorisms," bk. 1, aphor. 66). For modern controversies over "species," see the major articles by biologists collected in sections 5 ("Essentialism and Population Thinking"), 6 ("Species"), and 7 ("Systematic Philosophies") in Elliott Sober, *Conceptual Issues in Evolutionary Biology*.

28. Pushing a bit farther into the etymological forest, we find that the cognates and derivatives of "species" under the listing for its Indo-European root in the *American Heritage Dictionary* include most of the thematic concerns of this book: "*spek-*. To observe. . . . I. SPY, . . . ESPIONAGE, . . . SPECIMEN, SPECIOUS, SPECTACLE, SPECULATE, . . . INSPECT, INTROSPECT, PERSPECTIVE, . . . SUSPECT, . . . SPECIES. . . . II. Extended o-grade form *spoko-* metathesized in Greek *skopos*, one who watches, goal, and its denominative *skopein* . . . , to see: SCOPE, . . . HOROSCOPE, SCOPOPHILIA, . . . TELESCOPE." As for "phantom" and its derivatives, Wayne Shumaker tells us (*Natural Magic*, 7) that *phantasia* is "the instrument by which the *anima* knows the external world. . . . *Phantasia* may be translated as 'spirit,' 'fancy,' 'imagination,' and other terms but means, essentially, in its Greek root, [*phan-*], 'what brings to light, makes to appear, reveals, discloses, exhibits.'" The "idol" (from *weid-*) which will be so crucial in Bacon, who uses it etymologically, is linked to many of the same thematics, especially through its Greek and Latin forms, *eidos* (EIDETIC, perhaps HADES ["the invisible"], IDEA), and *videre* (VIEW, VISION, VOYEUR, SUPERVISE, SURVEY). In suffixed form *wid-tor* it gives us STORY and HISTORY.

29. Other published catalogues besides Tradescant's include Pierre Borel, *Bibliotheca chimica. Seu Catalogus librorum philosophicorum hermeticorum* (Paris, 1654); Olaus Worm, *Museum Wormianum, seu Historia rerum rariorum, adornata ab Olao Wormio. . . . Variis et accuratis Iconibus illustrata* (Leyden, 1655); Cornelis de Bie, *Het Gulden Cabinet van de Edel vry Schilderconst* (Antwerp, 1662); Paulo Mario Terzago, *Museo o Galerie Adunata . . . Manfredo Settala* (Tertona, 1664); Lorenzo Legati, *Museo Cospiano Annesso a quello del Famoso Ulisse Aldrovandi* (Bologna, 1677); Nehemiah Grew, *Musaeum Regalis Societatis, or a Catalogue and Description of the Natural and Artificial Rarities*

If the objects of knowledge are most properly visible things, if the categories that make these objects signify are understood not as Platonic ideas, occult and hidden in the mind of God, but as "outward appearances," then certain features of the world will be more salient to educated men than others. The invention of scientific instruments to reduce other aspects of phenomena to visibility (the thermometer and barometer, for instance) will increase the diversity of these features. But still the emphasis objectifies, constructing a "theatrum" of knowledge. To quote Ong on the "traffic in space" again: "Words are believed to be recalcitrant insofar as they derive from a world of sound, voices, cries; the Ramist ambition is to neutralize this connection by processing what is of itself nonspatial in order to reduce it to space in the starkest way possible. . . . Displayed in diagrams, words transmute sounds into manipulable units like 'things'" (Ong, 89–90).

This desperate attempt to wake up from a profoundly self-centered, "subjective," and symbolic dream-world into the clear light of day, to believe one's own eyes and not what one had been told, to *see* (as in "Ah, I see!"), is not radical enough in conception to achieve its deepest objective ("true vision"), and the desperate learned woke into another dream. As with Hariot, the attempt to know (to see) began increasingly to diverge from and even to replace the attempt to narrate (to tell). "Histories" of the New World were largely geographies; ethnology concentrated on artifacts (stowed in cabinets and museums) and spectacles (of the Brazilians at Rouen, for example, or in general early modern "ballet").[30] Bacon's "histories" were finally only subjunctive lists of objects one *might* write histories of, as close to a catalogue as a person without a collection could come. Improvements in engraving were starting to make scientific illustration a serious opportunity (realized, for example, in Ray's and Willughby's *Ornithology*).[31] Cryptographers made language accessible *only* to the eye—and even poets produced emblem books and pattern poems in which the "cry" is "reduce[d] to space." And Nature

---

Belonging to the Royal Society and Preserved at Gresham College (1681); Claude de Molinet, *Le Cabinet de la bibliothèque de Saint-Geneviève* (Paris, 1692).

30. For connections between spectacle, early forms and venues of ballet, and the New World (as well as other non-European places and peoples), see Chapter 7 and its citations, especially Suzanne Boorsch, "America in Festival Presentations."

31. The print revolution itself made scientific illustration possible and thus altered the nature of the scientific text: "Every copy of a printed picture exactly duplicated the original plate made or approved by the author, so that the didactic effectiveness of a text could be safely extended by graphic illustrations; . . . beyond its capacity to extend the author's written statement, the mechanically duplicated image could itself be the statement. In all fields of technology and descriptive science, books could be produced in which the substance was communicated primarily by images" (James Ackerman, "Artists in Renaissance Science," 103).

then, along with her children, the "salvages," is "searched out and brought to light" (Bacon, "Aphorisms," bk. 1, aphor. 50), "dissect[ed] into parts" (bk. 1, aphor. 51), but naturally not listened to. Nature's interior is to be reached and made visible by the violence of dissection and conquest. She is to be "penetrated," not spoken with. Tradescant's catalogue instances concisely the extraction of visible speech from a woman's silent interior, listing "A copper Letter-case an inch long, taken in the *Isle of Ree* with a Letter in it, which was swallowed by a Woman, and found" (*Musaeum Tradescantianum*, 54)].

It is quite ineffective to ask a tree or a storm cloud a scientific question. But "Nature" is a personification that represented from the start quite a bit of culture, including most non-European peoples and (all but explicitly) European women. Natural history museums in Europe and the United States still contain ethnographic collections and tableaux of non-European "life" in the form of dioramas. As late as 1815, Sarah Bartmann, the "Hottentot Venus," was being displayed alive in European anatomy theaters; dead and "dissect[ed] into parts" she is still visible in a jar at the Musée des Hommes in Paris.[32] When Bacon can decorate a lengthy Latin treatise on scientific methodology with a continuous conceit of nature as a fertile woman "unveiled," "exposed," "penetrated," "interrogated," and "vexed" (tortured), readers are also given to understand something of what can be decorously said "between (Latinate) men" of women. The woman in the illuminated initial letter from the *Instauratio Magna*'s dedication to King James (figure 10) is turning into a tree: frontally nude, rooted to the ground, her raised arms sprouting branches and leaves, Nature is a mute Daphne whose dream of metamorphic escape will be dashed by the sublimated but still avid pursuit of a scientist who wants to penetrate trees.

SIR THOMAS BROWNE: "IRONICALL MISTAKES"

As is clear from the more suggestive items in Bacon's list of histories or Tradescant's catalogue, merely reducing marvelous matter to the form of a list did not succeed in censoring the marvelous out of it. Bacon calls for "express proscription," and Thomas Browne among others heeds that call in his *Pseudodoxia Epidemica* (first edition 1646), a fond and fascinated exorcism of "vulgar errors" in a genre that before his work mostly attended to "Errors in Physick."[33] The work is the wicked twin of Bacon's list of worthy objects of

---

32. Sander Gilman, "Black Bodies, White Bodies."
33. "For though not many years past, Dr. Primrose hath made a learned and full Discourse of vulgar Errors in Physick" (Browne, *Pseudodoxia*, 5); Browne also cites the works of physicians

Fig. 10. Nature as Daphne. Initial in first edition of Francis Bacon's *Novum Organum* (London, 1621). By permission of the Houghton Library, Harvard University.

scientific pursuit: it is an anecdotal and copious (but not systematic) register, augmented and reissued five times during the author's lifetime, of what must be rejected by science—redolent with wonder but in the mode of irony and nostalgia. Plot some years later would construct his *Natural History of Oxfordshire* in obedient mimicry of what he perceived as Nature's order; Browne here structures his book in imitation of the wonder books it is meant to refute, which in turn have an associative structure that imitates the stream of consciousness in the wondering mind, and underscores the relations of metonymy by which the monstrous attaches itself to the margins of Nature. As Browne says in his letter to the reader, "Knowledge is made by oblivion; and to purchase a clear and warrantable body of Truth, we must forget and part with much we know" (3). But as Freud might add, "There are no negatives in the Unconscious," and the relaxed reader, then as now, was likely to delectate the curios here rather than learning firmly to eschew them.

Despite its anecdotal amplitude, the work itself is structured as a catalogue—though it contains more lore than description of material artifact or natural object, the doctor was a major collector of artifacts: in his diary John Evelyn called Browne's house "a paradise and cabinet of rarities," and he stored lore in commonplace books for forty-six years.[34] Margaret Hodgen, the historian of anthropology, approvingly called the *Pseudodoxia Epidemica* "a

---

Girolamo Mercurio and Laurence Joubert (5). Citations of Browne are from Keynes's edition (based on Browne's final edition, 1672) of the *Pseudodoxia*, in *Works of Sir Thomas Browne*, vol. 2, usually by book, chapter, and page number. Other works of Browne are also cited from the Keynes edition, by volume and page number.

34. *Diary of John Evelyn* 2:270.

serious work, designed to destroy by logic and ridicule countless errors and false theories still stubbornly held by the commonalty of men" (*Early Anthropology*, 130–31). A temporal chauvinist, she may have been assuming withering irony even in passages where it is missing. Would she be equally quick to imagine Browne "destroying" his commonplace books or his collection? Probably not. She would probably consider an object or private note innocent of "wonder" in itself, as not designed to stimulate or deceive another person—as having no rhetorical dimension. One suspects nevertheless that Browne enjoyed his collection (and his Errors) as a "paradise," in Evelyn's term, rather than as the rational torture chamber idealized by Bacon.[35]

Which is not to say that Browne does not idealize Bacon to some degree—his work is confessedly a response to the call, in *Advancement of Learning*, for a "calendar of popular errors" (Spedding 3:365): "Nor can we conceive it may be unwelcome unto those honored Worthies, who endeavour the advancement of Learning: as being likely to find a clearer progression, when so many rubbes are levelled, and many untruths taken off. . . . And wise men cannot but know, that Arts and Learning want this expurgation" (*Pseudodoxia*, 5). Like Bacon too, Browne refers to vision (here and elsewhere) as the final arbiter of truth: "We are not Magisteriall in opinions . . . [but] have only proposed them unto more ocular discerners" (6).[36]

Thinkers and writers working in what later thinkers and writers will call a "transitional" period are disadvantaged by a rather murky sense of what they are part of a transition to. It is easy, if erroneous, to imagine an assembly line of seventeenth-century writers, especially in England, some like Bacon decrying the unexamined lore of the *Auctores* and the folk, while others like Browne collect and store it for the later use of poets and fabulists. It is certain that Melville made loving use of Browne's spermaceti whale (*Pseudodoxia*, third edition) but unlikely that Browne had him in mind or that our fictitious assembly line of gentlemen virtuosi and academics in fact placed

---

35. Browne is also the author of a hilarious and beautiful parody of collection catalogues, the *Musaeum Clausum, or Bibliotheca Abscondita*, which includes, among other rarities, "A *Sub Marine* Herbal," "A fair English Lady drawn *Al Negro*, or in the Aethiopian hue excelling the original White and Red Beauty," and finally "A Glass of Spirits made of Aethereal Salt, Hermetically sealed up, kept continually in Quick-silver; of so volatile a nature that it will scarce endure the Light, and therefore onely to be shown in Winter, or by the light of a Carbuncle, or Bononian Stone" (Keynes 3:109–20).

36. See also the reference in chapter 5 of Browne's essay "The Garden of Cyrus" to "delightful Truths, confirmable by sense and ocular Observation, which seems to me the surest path, to trace the Labyrinth of Truth" (Keynes 1:226). Bacon also relies on the conceit of the Labyrinth, an extremely awkward figure for an environment to be negotiated by sight, as anyone can attest who has been to the maze at Hampton Court—surely both Browne and Bacon had. Perhaps it was the memory that led Browne to refer also to the "maze of Error" (*Pseudodoxia*, 69).

much emphasis on conservation of lore. There were distinguished antiquarians, and Browne was among them, but I do not want to suggest that nostalgia had yet won the day (see, for example, Browne's "Urne Buriall"). Ambivalence is the emotional posture we must eventually attend to, which mirrors—or presages—a cultural splitting that will lead to nostalgia *later* (among educated middle-class adults, that is).

The analysis of Error as a social phenomenon in Book 1 of the *Pseudodoxia* is many pronged and polemically addressed to a number of moral causes—Credulity, Supinity, "prostration unto Antiquity" and, in general, "the common infirmity of humane nature." To that extent, it does not interest us here. Where Browne approaches the matter of the imaginary and its rhetoric in this opening summary is in chapter 4, which treats especially of "Misapprehension." Here he makes the point, not for the first time in recent history (see Bacon's "Idols of the Marketplace" in "Aphorisms," bk. 1, aphors. 59–60), that mistakes about reality have verbal sources, of which two are "worthy our notation; and unto which the rest may be referred: that is the fallacies of Equivocation and Amphibologie, which conclude from the ambiguity of some one word, or the ambiguous Syntaxis of many put together. . . . By this way many Errors crept in and perverted the Doctrin of Pythagoras, whilst men received his Precepts in a different sense from his intention; converting Metaphors into proprieties, and receiving as literal expressions, obscure and involved truths" (*Pseudodoxia*, bk. 1, chap. 4, 32).

This literal apprehension of metaphorical discourse is a root ingredient of the enjoyment of prose fiction as well as of, in a very different posture, the fundamentalist beliefs then current among millenarian cultists. It is also of course a major ingredient in the construction and reception of reported monsters and wonders. The contemporary keenness for universal language schemes or "real characters" was partly energized by the aversion to ambiguity Browne here articulates; it bespeaks a desire for a two-dimensional instrument of reference which, should it (or could it) have been successfully invented and put into use, would have fatally restricted the theoretical development of the sciences by naming only those phenomena visible to current theory or to common sense.[37]

"The circle of this fallacy [of literalism] is very large, and herein may be

---

37. For discussion of the language that George Dalgarno did, in fact, manage to construct in his *Ars signorum, vulgo character universalis et lingua philosophica* (1661), along with a reproduction of fifteen pages from the 1834 edition, see Wayne Shumaker's "George Dalgarno's Universal Language." It is interesting to consider the invention of such "thin" and secret codes as emerging from an impulse precisely complementary to the ethnographer's proclivity for cracking "thick," public ones—the indigenous languages that are only occult to the dangerous foreigner.

comprised all Ironical mistakes for intended expressions receiving inverted significations; all deductions from Metaphors, Parables, Allegories unto real and rigid interpretations" (*Pseudodoxia*, bk. 1, chap. 4, 34). Irony can come in other forms than mistakes: utopian and fantastic fiction are ironic forms in their divergence from the rules and regulations of the "world" we "know," and so is travel writing: the charm of all such narrative lies in the postulated difference of its world-setting. Much depends upon whether one is to believe, disbelieve, or suspend disbelief in a divergent world; these states of belief are productive of very different ironies.

Browne's work is a major canonical text with which the reader is probably at least glancingly familiar. I will examine his method, then, in the invocation and rejection of only two errors especially pertinent in our context of attention to an emergent anthropology: that "the causes [of Blackness in Negroes] . . . are . . .: [1] The heat and scorch of the Sunne; or [2] the curse of God on Cham and his posterity," which he discusses in two chapters on "The Blackness of Negroes" (bk. 6, chaps. 10 and 11). In these errors one can watch Browne working with the most historically ticklish of differences, attempting to replace the literalist ironies allied to alienation with more openly literary ones. Such errors reveal clearly the relation Browne understood, more complexly than Bacon, to bind together words and the "things" they so often create.[38]

The literalism or misapprehension of the two wondrous errors about "the Blackness of Negroes" comes in two kinds: in the first case, darker African "complexions" are seen as environmentally caused—as a generalization to whole populations and to inheritability of the familiar phenomenon (to Europeans) of tanning. There is no seventeenth-century genetic biology to render implausible the notion of acquired traits,[39] and any northern European who had been to North Africa or the western coast would be rational to appreciate the convenience of dark skin, harder to burn in a sunnier climate. The absurdity a modern person sees at first in this error is not the same as that confronted by the subtle Dr. Browne. Indeed, Browne can offer no self-convincing counterexplanation ("how, and when this tincture first began is

---

38. Since this chapter was written, interesting work on "Blackness" in the early modern English imagination has appeared, notably Kim Hall, *Things of Darkness: Economies of Race and Gender in Early Modern England*; Edward Washington, "'At the Door of Truth': The Hollowness of Signs in Othello"; and Mary Floyd-Wilson, "Temperature, Temperance, and Racial Difference in Ben Jonson's 'The Masque of Blackness.'" (as well as her forthcoming work on Browne and Shakespeare). On skin color and race in the scientific and fictional works of Margaret Cavendish see Rosemary Kegl, "This World I Have Made"; unpublished work on that topic by Sujata Iyengar should be forthcoming soon.

39. Browne discusses the possibility of such a natural process (*Pseudodoxia*, 466–67).

yet a Riddle" [466]); he can only "level rubbes" here by reminding his readers of a fact made salient by the new institution of slavery.[40] The vulgar "mistake" is, ignoring the physical dislocations occasioned by the slave trade, to imagine that all "Africans" are, *ipso facto*, in Africa, participating universally in a convenient anatomical adjustment to African climates (of which, as Browne also points out, there are several).[41]

Browne dispenses with the notion that "the Sunne is . . . the Author of this Blackness" (466) by pointing out that Africans "transplanted, although into cold and flegmatick habitations [i.e, enslaved and sent to Europe or North America] continue their hue both in themselves, and also their generations; except they mix with different complexions [i.e., are raped by their masters]. . . . And so likewise fair or white people translated into hotter Countries receive not impressions amounting to this complexion . . . as Edvardus Lopes testifieth of the Spanish plantations, that they retained their native complexions unto his days" (i.e., no Black or Indian men impregnated colonist women, producing racially mixed later generations) (463).[42]

That narrative-suppressing quality of Hariot's might be pointed out here, which permits a guiltily dynamic situation to become inaudible behind a screen of "data." But it is precisely the absence of historical awareness in "the commonalty of men" that Browne is here "expurgating." In 1646, only a picture of the world that literalized an ethnographically ornamented map like Willem Blaeu's (1630) would put Black Africans exclusively in Africa.[43] And Browne's comparison of Black resistance to climate-based change with White resistance reduces the fetishizing division between reading subject and textual object.

40. "If the fervour of the Sun were the sole cause hereof in Ethiopia or any land of Negroes; it were reasonable that Inhabitants of the same latitude . . . should also partake of the same hue and complexion, which notwithstanding they do not. . . . This defect is even more remarkable in America; which although subjected unto both the Tropicks, yet are not the Inhabitants black. . . . And although in many parts thereof there be at present swarms of Negroes serving under the Spaniard, yet were they all transported from Africa, since the discovery of Columbus; and are not indigenous or proper natives of America" (463–64).

41. By 1650 the numbers of Africans "transplanted" annually averaged about 30,000 (Manning, *Slavery and African Life*, fig. 1.1). Many of these, though a minority, were bound for Europe or British North America and the Caribbean—almost 23,000 in the first half of the seventeenth century (see Phillips, *Slavery from Roman Times to the Early Transatlantic Trade*, table 9.2).

42. Female slaves (accounting for only at most a third of the New World's enslaved population) tended to be house servants. As such they were sitting ducks for the "sexual liaisons" (!) that, as Phillips puts it, "often developed between the masters and their female household slaves" (200)—especially in the "cold and phlegmatick habitations" where there were no sugar plantations and a higher percentage of slaves were employed as domestic servants. (See also Phillips, chap. 8, "African Slaves in Europe," esp. 162.)

43. See the facsimile edition of the Blaeus' ultimate collection, the *Atlas Major* (1662–63).

I am concerned, not to demonstrate Browne's relative moral attractiveness, but to articulate the process by which he transforms popular knowledge into marginal lore (a process fraught with class conflict, to which we will return later). I do not know if it is possible for a person to exceed her own horizons of expectation, and Browne's was a racist and colonialist society. Still, I have been looking at passages in which he supported a cognitive taste for the nonimaginary in considering human difference. The ensuing discussion of "the generation and sperm of Negroes," however, is another story:

> Now although we conceive this blackness to be seminal, yet are we not of Herodotus' conceit, that their seed is black. An opinion long ago rejected by Aristotle, and since by sense and enquiry. His assertion against the Historian was probable, that all seed was white; that is without great controversie in viviparous Animals, and such as have Testicles, or preparing vessels wherein it receives a manifest dealbation. And not only in them, but (for ought I know) in Fishes, not abating the seed of Plants; whereof at least in most though the skin and covering be black, yet is the seed and fructifying part not so: as may be observed in the seed of Onyons, Pyonie and Basil. Most controvertible it seems in the spawn of Frogs, and Lobsters, whereof notwithstanding at the very first the spawn is white, contracting by degrees a blackness, answerable in the one unto the colour of the shell, in the other unto the Porwigle or Tadpole; that is, that Animall which first proceedeth from it. And thus may it also be in the generation and sperm of Negroes; that being first and in its naturals white, but upon separation of parts, accidents before invisible become apparent; there arising a shadow or dark efflorescence in the outside; whereby not only their legitimate and timely births, but their abortions are dusky, before they have felt the scorch and fervor of the Sun. (469–70)

What makes the passage so distressing is the scene assumed by its dispassionate analysis of "the sperm of Negroes" in light of the historical relations possible at the time between European doctors and African bodies. Where the doctor has seen or heard of the internally white seed of "Onyons, Pyonie, and Basil," or "the spawn of Frogs, and Lobsters" to which he calmly compares "the sperm of Negroes" is not so problematic, but there have been few innocent "enquiries" in Browne's day into the reproductive physiology of African men. The image of the Black patient quickly becomes the image ("species") of a specimen ("species") in the lab where Bacon wants to "dissect it into parts"; this is the ultimate image of the overdetermined passivity and reification enforced by slavery and subsequently sustained by eugenics.[44]

---

44. This is not to say that biology itself is racist. See Gilman, *Difference and Pathology*, and Nicole Hahn Rafter, *Making Born Criminals*, for critical surveys of racist and sexist research programs in biology, comparative anatomy, and physiology.

The epistemology of visibility is pushed in the passage on sperm to an extreme that provides a rhetorical/aesthetic balance for the passage on skin color. Browne's enquiry into "Blackness," which he sees as a color, is in part a demonstration of the obscurity and "subtilty" of even so apparently transparent a sensible datum as an impression of color: "even in proper and appropriate Objects, wherein we affirm the sense cannot err, the faculties of reason most often fail us. Thus of colours in general . . . few or none have yet beheld the true nature" (460). To fully problematize his chief example of color puzzles (the list of which begins "Why Grasse is green?" [461]), Browne moves from the exterior surface to surfaces more and more interior, or anterior, from scorched skin to "inward use of certain waters" (466) to seminal properties. When such a paradoxologist ends a discussion of black skin by speaking of the color (white) of sperm, he is certainly parodying the positivist faith in the truth of what we see. More concretely, he is deconstructing the "complexion" as a coherent surface which "reflects" its real world, in this case the hot climate to which vulgar error attributes the ultimately elusive "Blackness of Negroes."

This error brings us to Browne's second chapter on the topic ("Of the same") and the second kind of literalism, a parochial misapprehension of what might be signified by "the curse on Cham," the notion that "Blackness" could be construed by Blacks as a curse—the notion, then, that values are absolute and norms universal. The whole long comi-tragedy of the "monstrous races" was underwritten by such a notion, under the terms of which "monstrous race" is not a paradox.[45] (To be fair, "monster" did not initially specify aberrance but simply portentousness, but soon enough portentousness came to be associated with what looked mistaken or at best "playful" in nature—"men whose heads do grow beneath their shoulders.")

The first half of the chapter is a philological argument amplified by considerable geographical detail on the subject of the origins of a number of scriptural characters. The upshot of this argument is to substantiate its rather simple initial claim: "The curse mentioned in Scripture was not denounced upon Cham, but Canaan, his youngest son" (bk. 6, chap. 11, 470). This curse was supposed to take the form of Blackness, but there do not seem to have been any black populations in Canaan. Ergo, the curse was what "is plainly specified in the Text . . . Cursed be Canaan, a servant of servants shall he be unto his brethren, which was after fulfilled in the conquest of Canaan, subdued by the Israelites, the posterity of Sem" (bk. 6, chap. 11, 472).

---

45. See John Friedman, *Monstrous Races in Medieval Art and Thought*, and (on early Greek paradoxography) James Romm, "Belief and Other Worlds" and *Edges of the Earth in Ancient Thought*.

Philological victory does not however exhaust Browne's response (which is notably ample—given three successive chapters, this Blackness is the single most scrutinized site of Error in the *Pseudodoxia*). Montaigne-like, he wants to go on to make a pitch for aesthetic, perhaps erotic, relativism: "Lastly, Whereas men affirm this colour was a Curse, I cannot make out the propriety of that name, it neither seeming so to them [Africans], nor reasonably unto us" (472). He consults pagan and Jewish authors, both Platonic and Aristotelian, on "the definitions of beauty" (472) and finds it is ultimately "determined by opinion"—not a matter for scientific certainty, it "seems to have no essence that holds one notion with all" (474).

The final objection in the chapter is to the un-Philosophical methodological habit, in "points of obscurity," of "fall[ing] upon a present refuge unto Miracles" (474). This is said to be even worse than "Antipathies, Sympathies, or occult qualities" (the kind of Neoplatonic science of resemblances and analogies that Foucault took as paradigmatic for the Renaissance in *The Order of Things*)[46]—it is an abridgment of the very process of knowledge, substituting wonder for empathy when it "lay[s] the last and particular effects upon the first and general cause of all things" (475). Every miracle redounds to the power of a narrowly imagined God, but blots out the quiddity of whatever natural "effect" it is perceived in place of. Thus the tendency to "affirm" Blackness as a biblical curse is seen in the *Pseudodoxia* as part of a tendency to wondering prostration, and it links fundamentalist religion to the preference for wonder over analysis and definition. Of course, the notion that "Blackness" refers to some one essence in the world (and that essence, simply a color pertaining to certain "other people") has played, as well, a part in what one might call fundamentalist anthropology as an object of both fascination and "scientific" predication. Browne does not reach so far in his expurgations as to undo the symbolic tangles of this reified category.

The third chapter in this sequence, "A Digression concerning Blackness," comes as a storm of analysis and definition, as if consideration of miraculous explanation had unleashed in Browne a scientific tantrum. Here he addresses the general referent of "Blackness," separating it from the putatively paradoxical relation to human skin. He attends, not to the source of the phenomenon in "the receptions, refraction, or modification of Light," but to "certain materialls" which may give rise to the color—it is a pointedly, even hyperbolically materialist examination of a general phenomenon whose

---

46. See chapter 2 of that work, "The Prose of the World": "The universe was folded in upon itself: the earth echoing the sky, faces seeing themselves reflected in the stars, and plants holding within their stems the secrets that were of use to man" (17).

spiritual explanation in a specific instance he has just trounced (bk. 6, chap. 12, 475). "Things become black by a sooty and fuliginous matter proceeding from the Sulphur of bodies torrified; not taking *fuligo* strictly, but in opposition unto [*atmis*] that is any kind of vaporous or madefying excretion," et cetera, et cetera, for six pages.

If analytic materialism as figured by Bacon in the *Novum Organum* seemed a dark prospect, Browne's use of it here to beat a fundamentalist and racist wonder into the ground is an effective antidote. Natural knowledge as imposition and domination was not the only vision, and though one can see it functioning in Browne's chilling discussion of sperm, one can also read in his text an example of the new science as an intimate dismissal of surfaces, an unraveling of the customary knowledge which forms precisely the material of ideology. It is notable that "Blackness" seems to have provided this epistemological warrior the inspiration for his most concentrated *tour de force*. Black Africans (by the 1640s, people from "Senegambia" to Angola on the West Coast and, on the East Coast and the Horn, people from the Sudan as far south as Mozambique), becoming both politically and figuratively an Other of a sort most easily assimilated to the refiguring of the ends and means of European science, are a logical, even magnetic focus of interest for a work like Browne's. Perceived as a single human group both like and unlike Europeans, they provoke the possibility of a more finely articulated knowledge of the human species; as a group whose historical relation to Europeans (Browne would say "Christians") is already provocative of guilt and thus of fear, "Negroes" offers a strong example of the kind of imaginary Browne calls Pseudodoxia. And as a collective term for certain subjugated persons, "Negroes" means something about which one could speak and think clinically without offense or surprise to one's (white, European) readers.[47] There are many reasons, not all of them sinister, why the "normal" (relative to the investigator or his/her institution) tends to be the last category rather than the first to be studied in the case of the human species. One reason is surely that doctors and scientists find it more "natural" to study what is not themselves, that is, "nature." And social circumstances in modern European nations render those in the categories of Other more vulnerable to examination. Indeed, it is definitive for early modern and modern culture: the Other is she who is examined, dissected, *articulated* by the Self. (Medical experi-

---

47. That guilt was "thinkable" in the 1640s we know from, for example, Las Casas's (posthumous) regretful retraction of his suggestion that African slaves be imported to the New World to replace the labor of frailer Amerindian populations. It was not thought of, consciously, very often (until, as Eric Williams explained it, slavery was no longer useful to capitalism—see his *Capitalism and Slavery*).

mentation in the Nazi camps is the extreme example of this doubly "normal" tendency.)[48]

What is more interesting though, because less predictable and less easy to dominate with explanation, is the overdetermined quality of this little chain of chapters, with its many, many reasons why people should think differently about Blackness and its climactic hurricane of materialist chemistry. The chemical "digression" ends with a near apology expressed in an appropriate figure from the kind of "adventure capitalism" that had first brought Blackness to the forefront of European common knowledge with the sudden expansion of the Portuguese slave trade in the 1440s (a trade which had subsequently increased *1,600 percent*).[49] Browne calls his "Digression on Blackness" an "adventure in knowledge," and adds that "although on this long journey we miss the intended end, yet are there many things of truth disclosed by the way; and the collaterall verity may unto reasonable speculations some what requite the capital indiscovery" (481).

The economic figure reminds us of the frequent cajoling of the sixteenth-century explorers who, like Columbus, had so often to bring home parrots, gum mastic, and display-natives in place of the gold they hadn't found yet, and in most places never did find. Finally slave traders managed to find reliable sources of gold in America by trading slaves for it with the Spanish and later with English and French landowners.[50] Like the adventure capitalists, Browne hopes this digressive Blackness may "some what requite the capital indiscovery." In the spirit of emergent European racism (and dominant academic Aristotelianism) he makes up a substance called Blackness, associates it with sheer Matter, and plays punningly with negative senses of "darkness": "Why [should] some men, yea and they a mighty and considerable part of mankind, . . . first acquire and then retain the gloss and tincture of blackness? Which whoever strictly enquires, shall find no less of darkness in the cause, than blackness in the effect it self" (bk. 6, chap 10, 461).[51] How is it that, at

---

48. Examples closer to home include the black men used as "controls" in the famous Tuskegee syphilis experiments and the retarded children subjected to radioactivity without their knowledge (or their families') at the Fernald Institute in Massachusetts.

49. The total Atlantic slave trade in 1451 came to 15,000 persons; in 1650 it was over 242,000. See Phillips, *Slavery from Roman Times*, table 9.2. For "adventure capitalism" see Michael Nerlich's Marxist account of literary change in a changing Europe, *Ideology of Adventure*, vol. 1.

50. For a fascinating account of the conjunction of goldlust and the slave trade, and its consequences for ethnological theorizing, see William Pietz's series of articles in *Res*, "The Problem of the Fetish," particularly the one subtitled "Bosman's Guinea and the Enlightenment Theory of Fetishism." Pietz is concerned with the West African rather than the American end of the route.

51. It is the objectifying habit of school-text Aristotelian dualism that makes it so easy to conceive and objectify something like Browne's "Blackness"; this does not mean that such ob-

the same time, Browne is so benevolently, plurally anxious to include, to make amends, to connect, and to equalize, so anxious and interested as to stay on the topic for three chapters in a row (four, if you count the chapter on the also persecuted Gypsies, introduced as "counterfeit Moors" [bk. 6, chap. 13, 481])?

It is as if anthropology were the secret love of Browne's scientific mind, and a guilty love at that—though there is no institutional anthropology yet, he already sees what the material conditions of it will be and so is already speaking out of both sides of his mouth, like many enlightened liberal anthropologists to come.[52] Browne's desire to undermine racist projection and rejection does not extend to an ability to separate himself from the social dynamic that stimulates and perpetuates racism. His tropes and figures, and his cold eye, give him away. Even as he cannot silence his wonders, but only iterate and condemn them, he cannot abdicate the seat of power from which he surveys the *phenomena universi* and hands down interpretations.

KNOWLEDGE IS POWER

> "And so those twin objects, human Knowledge and human Power, do really meet in one, and it is from ignorance of causes that operation fails."
>
> —Francis Bacon, "Instauratio Magna"

Knowledge *is* power, and knowledge of human beings is power over human beings, as Sir Francis Walsingham and his employers had known on a practical level, and as Sahagún and Olmos had known more altruistically but in the same interventionist mode. "We are not Magisteriall in our opinions," says Browne in his preface "To the Reader," and Bacon likewise denounces "those who . . . render sciences dogmatic and magisterial" ("Aphorisms," bk. 1, aphor. 67); in both Browne and Bacon the word "magisterial" may be aversive because of its connection with *magia*, a fruitless science and a failed power. For in fact they are both quite Magisteriall. They are masters, and masters must have not only their bachelors and apprentices but their illiter-

---

jectification can't be called racist, only that racism, like all cultural phenomena, is composed of the ingredients at hand and will exploit or blossom out of philosophy as well as out of economics. For a succinct and erudite account of the textbook reading of such seventeenth-century Europeans as Browne and the cast of mind it helped to (re)produce, see Rief, "The Textbook Tradition in Natural Philosophy, 1600–1650."

52. See, e.g., Andrew Apter, "Que Faire? Reconsidering Inventions of Africa," or Fabian, *Time and the Other*. For critique, of Clifford Geertz in particular, see Vincent Pecora, "The Limits of Local Knowledge," and Peter Dale Scott's relevant annotations in his poem *Coming to Jakarta*.

ates; they must have Error, in fact, as an object for their talent at correction, or at least as a "darkness" in which their illumination glows.

Hierarchy, then, is not absent from the *Pseudodoxia*'s theoretical grounding in book 1, where Browne has a chapter on "the erroneous disposition of the People." Here he assigns much of the blame for the pitiful condition of human knowledge in the aggregate to the uneducated lower classes, whom he saw as naive positivists and literalists from their lack of training: "For the assured truth of things is derived from the principles of knowledge, and causes which determine their verities. Whereof their uncultivated understandings, scarce holding any theory, they are but bad discerners of verity" (bk. 1, chap. 3, 26). Not only are they bad judges but brutal, dominated by appetite and therefore "swarming" with Errors and vices (27). They can understand "the wisdom of our Saviour" (28) no better than "the point and unity" of secular truth (26) and therefore, welcome heresies and "oblick Idolatries" (29). Mountebanks fool them easily for profit: "Fortune-tellers, Juglers, Geomancers, and the like incantatory imposters, though commonly men of Inferiour rank, and from whom without Illumination they can expect no more than from themselves, do daily and professedly delude them" (30).[53] Politicians of course delude them also — and here at last comes Browne's redeeming irony: according to Pliny, "the proper and secret name" of Rome was concealed from its people, "lest the name thereof being discovered unto their enemies, their Penates and Patronal Gods might be called forth by charms and incantations" (31). Thus in the penultimate paragraph of the chapter he begins to let up his harsh attack, or at least to democratize it — the powerful of Rome who fear the name-magic alluded to here are no closer to verity than the people they keep in ignorance. And so, in the end, "whoever shall resign their reasons . . . although their condition and fortunes may place them many Spheres above the multitude, yet are they still within the line of Vulgarity" (31). The sphere of the "vulgar" has been expanded at the last moment into a metaphorical catchment, "vulgarity," that anticipates what will later be a common, even dominant usage.

It is a place in the text that requires a pirouette of irony — either that or a utopian proposal for social modification. For in a vernacular book dedicated to the "expurgation" of human knowledge, Browne has here consigned most of his countrymen to a permanent error and ignorance determined by the ef-

---

53. Note the opposition here between the "incantatory" and "Illumination," precisely the opposition of values discussed by Ong as emerging in the culture of print and here used for class markers: the oral is popular and low, the visual associated with text, print, and the "higher" learning inaccessible to "the People."

fect on "the fallible nature of man" of their class circumstances. "Nor have we addressed our Pen or Stile unto the people (whom Books do not redress, and are in this way incapable of reduction)" (5), he announces categorically in his preface after explaining that he wrote the book in the vernacular (sort of) as a service to the English gentry. But already the personnel of natural philosophy reached further down than the gentry.[54] As can be seen in Plot's *Oxfordshire*, craftsmen and farmers were at least as likely to understand the concrete operations of nature as were university-trained scholars, though they might have no "theory" (which, after all, the Royal Society and the French Académie des Sciences proscribed as expressly as Bacon did wonders). The Royal Society would be open, theoretically, to English males of all classes and occupations.[55] Like the "Statists and Politicians" of Rome, however, Browne wants to conceal the proper name of the city from its unqualified, incompetent, uncultivated, illiterate, confined, gross, farraginous, deluded, irrational rabble of heretics and idolaters (bk. 1, chap. 3, 25–28).

The allegory of Pliny's Rome has some interesting implications for the social structure of knowledge in an increasingly imperial (and politically restless) Britain. Once again it represents knowledge—"the proper and secret name of Rome"—as something only the powerful have, and which they have inexplicably, because they are powerful. (How was the secret name of Rome first "learned"?) This is the opposite of Bacon's condensed narrative, in which knowledge *leads* to power. If the People gain this knowledge, it will not make them powerful instead of their leaders; it will be circulated into the hands of "enemies"—the powerful of some other nation. Knowledge inevitably rises, it would seem, to the level of the social class that appropriately possesses it. The people can only betray it; they cannot keep it. And yet, they are sensed in this drama of political occultism as a kind of power themselves—a force necessarily subversive and dangerous unless kept in ignorance of themselves, of their own name. And so Browne shows us two kinds of knowledge, each hermetic: "verity," available only to those with "theory" (the gentry and aristocracy with their Latin educations), and the cheap

---

54. On membership and class status see Lotte and Glenn Mulligan, "Reconstructing Restoration Science" and Steven Shapin, "'A Scholar and a Gentleman.'" Tradescant's list of benefactors includes merchants and sea captains; the range of class status among the first members of Royal Society can be seen in the membership list printed in Sprat's *History of the Royal Society* 2:431–33; see also 1:62–76 ("The qualifications of the Membership of the Royal Society").

55. This was not the case in the Académie des Sciences, though it had been part of the intentions of its immediate prototype, the unrealized Compagnie des Sciences et des Arts, "probably elaborated in 1664 or early in 1665 by Thevenot and the astronomers Auzout and Petit." See Roger Hahn, *The Anatomy of a Scientific Institution*.

magic of "Geomancers," with which "men of inferior rank" delude, presumably, men of inferior rank.

But popes and kings retained "Fortune-tellers," and natural magic of various kinds was a vocation of highly educated men well equipped with "theory." It is of interest, then, that Browne locates the deluded and delusive knowledge of *magia* in the realms of the popular, "incantatory" and socially marginal. It is as if one way to expurgate knowledge was to push error down the social scale. Thus Vulgar Error lives on, and the People, for whom "an Apologue of Esop [is] beyond a Syllogism in *Barbara*" (bk. 1, chap. 3, 26), will persist in all manner of *phantasia* and "Fable" (as Browne's laconic marginal note has it). "Unsufficient for higher speculations, they will always betake themselves unto sensible representations" (27) and unto the Idolatry induced by a rapt fascination with visible wonders and display. Never mind that collectors like the Tradescants or Browne himself share the same fascination, and Bacon had made himself rather a prophet of Light than of the occult properties of things and "second intentions of . . . words" suddenly elevated by Browne as the finer preoccupations of the "knowing."[56] The distinction aimed at is not between the verbal and the visual as much as between the spectacular (or incantatory) and the lucid. "Sensible representations," which seems to mean wonders, stun the mind, slowing the consumption and production of ideas. Even the catalogue represents some kind of productive response to the wonders it lists. What the intellectuals fear is rapture.

Once the class war surfaces, it is hard to take sides between popular error (that is, popular culture) and progressive (that is, elite) science. The lack of "theory," that Browne complains of, and which supports the construction of the singular and inexplicable wonder, leads also to a political ductility that can be exploited to terrible ends. But then, the *curiositas* of highly trained twentieth-century atomic physicists helped to staff Los Alamos and Hanford. An intellectual passion can be exploited as easily as any. Is the class conflict manifested in Browne's text just the trace of a free-floating prejudice in the author, or is the conflict here specific to the business of an epistemological revolution? It is difficult to say. Eventually, whole disciplines, anthropology and sociology, will be constructed to rationalize the objectification of human groups to which their practitioners do not as a rule belong.[57] By 1646 those disciplines show few signs of existence, but Browne's

---

56. See Keith Hutchinson, "What Happened to Occult Qualities in the Scientific Revolution?"

57. On the early days of fieldwork-based sociology in this country, see Carla Cappetti, *Writing Chicago*.

chapters on "the People," "Negroes," "Gypsies," and "Pygmies" are among the signs.

The contribution to vernacular literature of this (anti-)pseudodoxical genre and particularly Browne's work is the Error itself, the "Ironical mistake" he catalogues so fully and faithfully in all its varieties. Not only does he, in proscribing them, preserve for history the *species*, *icons*, and *phantasia* of "the People" and the European past, but he even provides models for the construction of new Errors and wonders. The capacity for willed literalism is a key ingredient in the formation and enjoyment of fiction in any medium (and Browne pays ambivalently corrective attention to a number of painterly fictions in book 5, "Of many things questionable as they are described in Pictures"). If culture cannot absolutely discard, we are looking at the waste management techniques of the seventeenth century with regard to "species," or images.[58] Eventually, some Europeans will be able unambivalently to enjoy false propositions and exaggerated, sensational representations. The false, the unreal, the surreal, the perverse, and the impossible will all be sites of permitted or at least recognized pleasure. That kind of pleasure, second nature in modern cultures at all social levels and in most countries, was an epistemological achievement as much as was the restructuring of truth conditions. Though it had precursors in romance and epic, both forms were attached with more or less intensity to believable genealogical or pseudo-historical narrative axioms. Their narrative worlds were not different in kind from that of historiography (though such historiography was quite different in kind from our own which, like our fiction, has been dialectically reconfigured by "science").[59]

But we are looking at something muddier than the direct contribution of formal elements by one genre or author to the *technics* of another. In a print climate still dominated by the educated and the elite, we are looking at the repositioning of the reading experience (verbal or not) of wonder and the

---

58. See Marshall Sahlins on "structural transformation" in *Historical Metaphors and Mythical Realities*: "The complex of exchanges that developed between Hawaiians and Europeans, the structure of the conjuncture, brought the former into uncharacteristic conditions of internal conflict and contradiction. Their differential connections with Europeans thereby endowed their own relationships to each other with novel functional content. This is structural transformation. The values acquired in practice return to structure as new relationships between its categories" (50). Sahlins's ethnographic example is not isomorphic with the situation of cultural reproduction within a single culture split along class lines—but then it is European contact, cultural and commercial, with such places as Hawaii that helped to initiate the transformation of European "knowledge" and the redistribution of its types and branches.

59. Consider Geoffrey of Monmouth's *History of the Kings of Britain*, a foundational text for the growth of romance in England and France but written in faithful accordance with conventions of historiography and, in a sense lost to modern culture, believed.

redefinition of the imaginary, which now includes elaborate Error. These changes in appetite and production, along with others yet to be discussed, will alter the cultural field within which "worlds" are constructed, projected, described, imagined, invented, oppressed, collected, and consumed.

ROBERT PLOT: CYBERNETICS

The suggestively named Robert Plot, LL.D., F.R.S., was the first curator of the Ashmolean Museum (the alchemist Elias Ashmole's gift to Oxford of a collection based on the contents of "Tradescant's Ark"). A disciplined Baconian collector of natural histories, he also practiced alchemy for profit and claimed to have discovered the secret of the "first matter."[60] This loyal follower of "the Lord Bacon" offers a smaller world than we have seen so far — the political and biogeographical island of Oxfordshire, as represented in his *Natural History of Oxfordshire* (1677).

Nothing could be less like the *Cosmographie universelle* of the French Royal Cosmographer than Plot's *Natural History*. As Plot points out himself in the dedicatory letter to King Charles, after comparing himself to Aristotle and Pliny (both objects of imperial patronage): "This attempt seems more justly to belong to Your Majesty, than any of their Histories to their respective Patrons, it being so far from exceeding Your Majesties Dominions, that it contains but an Enquiry into one of the smallest parts of them; *viz.* Your alwaies Loial County and University of Oxford, whereas their Volumes are bounded only with the Universe." Here instead of a world there is precisely an island,[61] and, although it is Plot's home and base of operations, his book of it, far from creating an "Isle of Plot," is what Clifford Geertz has termed "author-evacuated" (*Works and Lives*, 9). There is no instinct to appropriate the points of view, identities, or information of anyone else in the construction of an authorial character, and Plot's citation system produces quite modern-looking footnotes on almost every page.[62] He represents his work

---

60. See Taylor, "Alchemical Papers of Dr. Robert Plot," 69. On Plot's career, especially in relation to the Ashmolean, see chap. 3 of Ovenell, *Ashmolean Museum*.
61. It is tightly enough boundaried that Plot is a little unsure of himself in relating a dream that was not dreamed in Oxfordshire, though it was *about* events in Oxfordshire and indeed helped the police of Oxford solve a case of burglary: "The *dream*, 'tis true, of which I am now writing, was had at *Bocton* in *Kent*, but the most important concern of it relating to *Oxford*, I thought fit rather of the *two* to place it here" (46). (Plot hoped to make a book about each county in England but in the end only managed one other, Staffordshire.)
62. His contemporary John Ray was similarly scrupulous. He digressed in the introduction to his translation (and augmentation, every instance of which was clearly marked in brackets in the main text) of his friend Willughby's *Ornithology*, to file this complaint about some-

from the start as belonging to the progressive program sketched out for the scientific community by Francis Bacon half a century earlier.

The book is scrupulously structured and comprehensive ("it being my purpose in this *History of Nature* to observe the most natural method that may be" [chap. 1, par. 1]), and succinctly self-similar, as the chaos theorists say. Its ten long chapters (divided into numbered paragraphs) cover "the Heavens and Air" (1), "Water" (2), "Earths" (3), "Stones" (4) and "Formed Stones" (5), "Plants" (6), "Brutes" (7), "Men and Women" (8), then "Arts" (9), and "Antiquities" (10). The final two chapters on "Artificial Operations" repeat somewhat figuratively the order of attention to "natural Bodies" ("Arts," for example, starts with instruments for Celestiall Observation and moves on to Water-works, and so forth). Throughout, the treatment of each category is divided into three by kinds (or rather degrees?) of relation to "Nature": "I shall consider, first, Natural Things, such as either she hath retained the same from the beginning, or freely produces in her ordinary Course. . . . Secondly, her *extravagancies* and *defects* . . . as in *Monsters*. And then lastly, as she is restrained, forced, fashioned, or determined, by Artificial Operations. . . . Things of Art (as the Lord *Bacon* well observeth) not differing from those of Nature in *form* and *essence*, but in the efficient only" (par. 2).

It is in some ways a peculiar project, for all its mimetically "natural" method, since there is nothing natural about the total object so anatomized. Oxfordshire was not even a particularly homogeneous county geologically or topographically, or therefore agriculturally, and *no* county is an island. The detailed foldout map in the front of the book is framed by representations of the coats of arms of all the county's peers, and much of the preface to the reader is concerned with this frame and Plot's method of representing heraldic colors graphically in a black and white engraving. The Earl of Lichfield (who improved a remarkable Water-works at his seat at Enston, on which I will say more later) is referred to on the map as "Lord of the Soil." The effect of this framing of the natural history of Oxfordshire is to saturate the natural environment with the symbolic reality of those who draw their political and social power from it. The positioning of the chapter "Of Men

---

one more of Thevet's stripe: "Here by the by I cannot but reflect upon the Author of a late *English* book, entitled, *The Gentlemans Recreation*. For having had occasion to examine and compare Books upon these Subjects, I find that all that he hath considerable concerning *Fowling* is taken out of the aforementioned Book of *Markham*, and yet hath he not to my remembrance made any mention of his Author. . . . I do not blame him for epitomizing, but for suppressing his Authors names, and publishing their Works as his own, insomuch that not only the Vulgar, but even learned men have been deceived by him" Willughby, *Ornithology*, ([a]ᵛ). (Note that the Vulgar are here imagined to read books of ornithology, if not as many as Ray does.)

and Women" between the one on "Brutes" and the one on "Arts" seems inevitable at first, as natural as Plot assumes it is, in part because the connections between the land and the people are so nearly chthonic still, in seventeenth-century rural England. Plot is doing his bit to change all this, his project intended for "not only . . . the advancement of a sort of *Learning* so much neglected in *England*, but of *Trade* also" ("To the Reader"). But for the moment, the matter of anthropology is diffused throughout the spheres delineated here. The woman who "was brought to bed . . . of a Son" in the hollow of a huge elm is filed under "Plants," the woman who returned to life after being hanged from one belongs to "Men and Women" (a chapter only briefly ethnographic, in fact, devoted largely "to the unusual *Accidents* which have attended them" [chap. 8, par. 1]).

Despite Plot's Baconian loyalties, wonders seem to spring up at his every footfall. In the opening paragraph of chapter 1, "Of the Heavens and Air," he is already crestfallen. According to his strict method, he can't talk about celestial observations until chapter 9 ("Of Arts"), since most of what is worth describing depends on the telescope, "known here above 300 years ago" when it was invented (probably not) by his other hero, Roger Bacon. The delineation of the method, in fact, unfolds as part of his elaborate regret at being thwarted by it of starting his book with something from the register of the sublime: "Since then the Celestial Bodies are so remote, that little can be known of them without the help of Art, and that all such matters . . . must be referred to the end of this Book, I have nothing of that kind to present the Reader with, that's local, and separate from Art, but the appearance of two *Parahelia* or Mock-Sunnes, one on each side of the true one, at Ensham on the 29th of May, early in the morning in the year 1673" (par. 4). The status of wonders as bargain-basement sublime was ne'er so well expressed.

The book contains its wonders by intention and confessedly. But, delivered in "plain, easie, unartificial Style, studiously avoiding all ornaments of Language" ("To the Reader") and, where required by Francis Bacon's protocol, modified with "qualifying notes, such as . . . 'I have heard it from a person of credit'" (Bacon, "Parasceve," aphor. 8), the wonderful finds itself suspended at times in a medium of almost hilarious defensiveness. The delight Plot clearly takes in waterworks and mechanical contrivances, for instance, was characteristic of the period that produced Descartes across the English Channel.[63] But it is paid and apologized for in a language equally mechani-

---

63. See Shumaker's article "Accounts of Marvelous Machines in the Renaissance" as well as his chapter on the German Jesuit Gaspar Schott (author of *Mechanica hydraulico-pneumatica*, 1657) in *Natural Magic and Modern Science*.

cal—no word picture is painted for the reader whose pleasure Bacon had charged future writers not to consult:

> Amongst the *Water-works* of Pleasure we must not forget an *Engine* contrived by the Right Reverend Father in God, *John Wilkins*, late Lord Bishop of *Chester*, when he was *Warden* of *Wadham College*, though long since taken thence; whereby, of but few gallons of *water* forced through a narrow *Fissure*, he could raise a *mist* in his *Garden*, wherein a person placed at a due distance between the *Sun* and the *mist*, might see an exquisite *Rainbow* in all its proper *colours*: which distance I conceive was the same with that assigned by *Des Cartes*, *viz.* where the Eye of the *Beholder* is placed in an angle of 47 degrees, made by the *decussation* of the line of *Vision*, and the rays of the *Sun*; and the *Fissure* such another as in his *Diagram*. (chap. 9, 48)

The subject of this rainbow-effect—"a person placed at a due distance"—has disappeared into the language of measurement and replicability, and indeed the narrator is disappearing into it as well. Elsewhere in the chapter on "Arts," after descriptions of the Air pump, the Barometer, the Circular Thermometer, and the Hygroscope, Plot brings out the *ne plus ultra* of scientific instrumentation, what amounts to a mechanical scientist, or at any rate a mechanical writer:

> [Christopher Wren's] contrivance to make *Diaries* of *wind* and *weather*, and of the various qualifications of the *air*, as to *heats*, *colds*, *drought*, *moisture*, and *weight*, through the whole year. . . .
> Now that a constant observation of these qualities of the air both by night and day might not be insuperable; he contrived a thermometer to be its own *Register* and a *Clock* to be annexed to a weather-cock, which moves a *Rundle* covered with White Paper; upon which the Clock moving a black-lead *pensil*, the *observer* [!], by the traces of the *pensil* on the paper, may certainly know what winds have blown, during his sleep or absence, for 12 hours together. (chap. 9, pars. 30–31)

As Peter Dear's article ("*Totius in verba*") on the rhetoric of the *Philosophical Transactions* in their early years makes clear, the evacuation of the subject from the experiment report was by no means an obvious desideratum or immediate achievement of scientific writing: the etymological bond between "authority" and "author" is strong and resilient. What is seen here, as if in a figure from a modernist poem, is an adumbration of the cybernetic ideal realized at last in our own century, and answered in the 1980s with such chilling panache by the break-dancers' mimicry of Descartes's animated robots. That "sleep or absence" during which a robot makes observations of the world and writes them down is not so far from Descartes's *noir* version of Genesis, his six-nights' attempt, in the *Meditations*, to create a belief in what his senses tell him of the created world: "I will regard the heavens, the air, the earth, colors, shapes, sounds, and all external things as nothing but the deceptive games of my dreams, with which [God] lays snares for my credulity. I will regard myself as having no hands, no eyes, no flesh, no blood, no senses, but as nevertheless falsely believing that I possess all these things. I will remain absolutely fixed in this meditation" (*Meditations on First Philosophy*, 60).

"The reader's pleasure" repudiated by Francis Bacon is related to the writer's pleasure, and both to the possibility of pleasure in the human subject (a possibility Plot still honors, who after all "undertook [this *History*] at first for my own *pleasure*, the subject of it being so pleasant" ["To the Reader"]). But it is a strange pleasure, bespeaking the satisfaction of a strange desire, that this prosthetic subject seeks:

> But of all [the Garden walks] that I ever met with, there is a walk at the Worshipful Mr. *Fermors* of *Tusmore*, the most wonderfully pleasant, not only that it is placed in the middle of a *Fish-pond*, but so contrived, that standing in the middle no Eye can perceive but it is perfectly *streight*, whereas when removed to either end, it appears on the contrary so strangely crooked, that the Eye does not reach much above half the way. Which deception of sight most certainly arises from a *bow* in the middle, which seems only an *ornament*, and the incapacity of the *Beholder* of seeing both parts of the *Walk* at one time; which that it may be better apprehended, see the manner of it, Tab. 13 Fig. 3. where the letter *a* shews the *walk* from the *garden* tending toward that in the *Fish-pond*. (chap. 9, pars. 116–17)

This is one of the most enthusiastic descriptions in Plot's book. It would seem that the best pleasure imaginable for the New Scientist is the pleasure of being deceived, literally unable to trust his own senses, like Descartes in the opening fantasy of the *Meditations* that leads him to his imaginary disembodiment. The causes of the deception are (1) an "ornament" and (2) the limitations of the beholder's senses—allegorically, beauty and the human body, the lures Plot has been taught by his masters to resist, who can only now enjoy them in their role as open deceptions.[64]

The set piece of the chapter on "Arts" suggests the distance traveled by the curious since the days of Thevet, Münster, and Agricola. Paragraphs 50–54 of chapter 9 concern an artificial island in the midst of the figurative island of Oxfordshire, an elaborate "Water-works" at Enston into whose absolute interiors Plot penetrates with ease, revealing everything in a pair of elaborately detailed plates, to which the written text is alphabetically keyed. Thomas Bushell, Esq., constructed a grotto and banquet house in 1636 after

---

64. The first two chapters of Svetlana Alpers's *Art of Describing* contain a rich and lucid treatment of the "breve confinium artis et falsi" and the relations understood in the world of Constantijn Huygens, Kepler, and Vermeer among deception, perception, art, and scientific equipment. Plot belongs to that world, but does not share the subtle sensibility of these visual experts.

cleaning a spring called Goldwell on his estate. While undertaking the task, he "met with a *Rock* so wonderfully contrived by *Nature* her self, that he thought it worthy of all advancement by *Art*." It was presented in a wonderful show to Queen Henrietta, who "commanded the *Rock* to be called after her own *Princely* Name," then neglected during "the late unhappy *Wars*," and finally restored by the Earl of Lichfield (Lord of the Soil) who "made a fair addition to it, in a small island situate in the passage of a rivulet" (pars. 50–51). Its chief features are iterated in a charming list keyed to Table 11 (see figure 11):

9. The streams of water from about 30 Pipes set round the Rock, that water the whole Island, and sportively wet any persons within it; which most people striving to avoid, get behind the Man that turns the Cocks, whom he wets with
10. a spout of water that he lets fly over his head; or else if they endeavor to run out of the Island over the bridge with
11. 12. which are two other Spouts, whereof that represented at 11, strikes the legs and that at 12 the reins of the back.
. . .
16. A Cistern of stone, with five spouts of water issuing out of a ball of brass, in which a small Spaniel hunts a Duck, both diving after one another, and having their motion from the water.

In the original Grotto, a system of hidden pipes rising from "a very small *Cistern* of *Water* behind a *stone* of the *rock*" intermittently causes a Nightingale to sing (par. 54; see figure 12).

Artificial nature (different from the rest of nature, as Plot and Bacon tell us, "in the efficient only") is admirably suited to scientific investigation and description (indeed the modern experiment requires just such artifice). For one thing its occulted interior can be penetrated successfully (unlike "Blackness") and explained: "having a *mouth* and a *Languet* just above [the Cistern's] surface, the *air* being forced into it by the approaches of the *water*, a noise is made near resembling the *notes* of a *Nightingale*" (par. 54). For another thing, the wonder can be retained without the inaccurate, obscuring rhetoric of hyperbole and similitude that is necessary to produce it in relation to organic nature (normally plural, law-abiding, familiar). Here the "Facts themselves" are fabulous, the carefully investigated song of the Nightingale a concrete fiction which can be nonetheless described in the plain language of fact. Because the singing Nightingale and hunting Spaniel are designed to repeat their actions without variation, an encounter with them can be "iterated" without a subject. There need be no narrative context, no experiential origin of Plot's knowledge: it is the mechanical design that hunts, dives, sings.

Fig. 11. Waterworks of the "artificial island" at Enston. Table 11 of Robert Plot's *Natural History of Oxfordshire* (Oxford, 1677). By permission of the Houghton Library, Harvard University.

Fig. 12. Grotto at Enston. Table 12 of Robert Plot's *Natural History of Oxfordshire* (Oxford, 1677). By permission of the Houghton Library, Harvard University.

The only narrative that counts is that of the queen, the only subject whose experience of the wonderful Rock is significantly individual, whose encounter is contrived as a masque, never again to be performed: "The then *Queens* most excellent *Majesty*, . . . in company with the *King* himself, was graciously pleased to honor the Rock not only with her *Royal* Presence, but commanded the same to be called after *her* own *Princely* Name, HENRIETTA: At which time as they were entring it, there arose a *Hermitt* out of the ground, and entertained them with a *Speech*; returning again in the close down to his peaceful *Urn*. Then was the *Rock* presented in a *Song* answer'd by an *Echo*, and after that a *banquet* presented also in a *Sonnet*, within the Pillar of the Table; with some other Songs, all set by *Simon Ive*" (par. 51).⁶⁵

This artificial environment, as a fictional setting for a show, is then inhabited by a subject — who, subject-like, names it after herself (like Thevet, like the Painter's Wife) — or who, like the monarch she also is, appropriates it. As far as it is hers, her subjective narrative experience of it is significant. For the rest of us it exists in the numbered table and the scientific passive: "when that *pipe* is filled there is no more singing" (par. 54).

The successful explanatory description of a man-made natural environment and its man-made wildlife is a step toward the general mechanical description of a law-bound nature in technical terminology and reported in an "author-evacuated" passive. We are close to the idea of the model. The animate tableau of the Spaniel and the Duck suggests the further possibility of animal and even human behavior isolated, island-like, in a "culture," for a law-bound interpretation: ethnology and ethology proper.⁶⁶ The irony, of course, is that this most perfect object for the methods and rhetorical forms of the new science is in fact a fiction — and a wonder. Freud told us that the Repressed returns in the uncanny — as a doll, a puppet, a mechanical Duck, a "robot concealed beneath [these hats and clothing]." Or perhaps as the "strangeness" and "charm" that bind the quarks of which it is now supposed "this very world itself" is made.

---

65. Note that the presentation of the artificial wonder is to the queen, not the also present king — a foreshadowing, perhaps, of the habit of later times of considering the audience of the novel as constitutively female, even though about as many men as women read novels.

66. The notion of a scientifically conducted ethnographic field experience is anticipated in a figure of Bacon's comparing the new and the old philosophy: "The one just glances at experiment and particulars in passing [like the traveler], the other dwells duly and orderly among them" ("Aphorisms," bk. 1, aphor. 22).

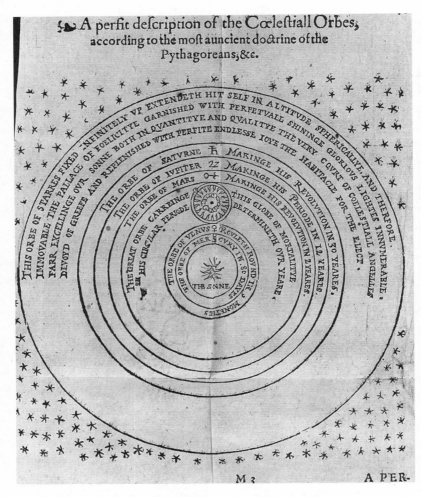

Fig. 13. The Copernican universe. From Leonard Digges's *Prognostication everlasting*, rev. Thomas Digges (London, 1576). By permission of the Houghton Library, Harvard University.

Part II
ALTERNATIVE WORLDS

PART 2 offers readings of a diverse group of seventeenth-century texts that take up the controversial issue referred to, then as now, as the "plurality of worlds." Most of these texts enter the fray by representing planetary other worlds with various degrees of fictionality, but Chapter 4 begins with a philosophical work by Giordano Bruno that offers theoretical or ideological support to the concept of the plurality of worlds; the chapter moves on to Galileo's dazzling announcement in the *Sidereus nuncius* (1610) of telescopic discoveries that bear Bruno out (Jupiter's four moons, the Earth-like surface of our own Moon). Chapter 6 includes treatment of a text that represents another kind of new world witnessable only through the prosthetic eye of an instrument, the microscopic world of Robert Hooke's *Micrographia* (1665).

The peculiar ontological status of these inaccessible new worlds—evidence for whose existence depends on interpretation of technologically derived images—makes for a significant opportunity to consider the possibilities and the historical development of the fictional as a mode of inquiry and representation. The works considered in these three chapters cover a wide spectrum of fictionality, with wonder and the sublime and sensational marking the ultraviolet limit. Texts, such as *Micrographia* or the *Sidereus nuncius*, which make no effort to report imaginary events, even as epic similes, may nonetheless belong to the family of imaginative literature, even of fiction, as a consequence of their efforts at stimulation and the peculiar intensity of their vividness with regard to the "invisible."

More commonly understood as fictions or quasi-fictions are works that take speculation to the point of invention, such as Francis Godwin's *Man in the Moon* (1638), or start out in invention with no debt to the astronomical "real" at all, as does Margaret Cavendish in the case of her *New Blazing-World* (1666). All the texts of that kind examined here bear some relation, however, to astronomy and the theoretical models of the new mechanical philosophy, and some of them are dense with information. The information-bearing and epistemological work of fictional texts is material to this book's analysis of the changing management of knowledge during the information explosion that focuses my attention. It is also material to an understanding of fiction's formal geography, its territorial extent, and of European fiction's *history*: Did science fiction (as we tend to label these texts) precede the realistic novel? Did its concerns inform the conventions of what came to be an apparently more domestic and sentimental tradition?

I think so, and Chapter 6 in particular discusses some of novelistic fiction's patrimony from the scintillant work of the natural philosophers. It shows as well the scene of division, the emerging and gendered opposition from within the camp of the imagination to what Blake would eventually denounce as the "Corporeal Eye" of thinkers such as Newton (or painters such as Reynolds). According to Newton, "in bodies we see only their figures and their colours, we hear only the sounds, we touch only their outward surfaces, we smell only the smells, and taste the savours; . . . their inward substances are not to be known either by our senses, or by any reflex act of our minds."[1] But Margaret Cavendish knew their inward substances, and if she didn't, she made them up.

1. Quoted in Steven Shapin, *Scientific Revolution*, 157.

IV • ON THE INFINITE UNIVERSE AND THE INNUMERABLE WORLDS

You would say that there is no need to posit a spiritual body beyond the eighth or ninth sphere; but that just as this same air surroundeth and containeth earth, moon and sun, so also it is extended infinitely to contain other infinitely numerous stars and great animals; and this air becometh thus the common and universal space, the infinitely spacious bosom which holdeth and embraceth the whole universe, no less than that part which is perceptible to us owing to the innumerable lamps thereof.

— Giordano Bruno, *De l'infinito universo et mondi* (1584)

There are no longer any absolute directions in space. The universe has lost its core. It no longer has a heart, but a thousand hearts.

—Arthur Koestler, *The Sleepwalkers* (1959)

THE WORD "WORLD" has always been one of enormous force, and its plural form still provides a frisson even to those of us who grew up in the space age, watching men land on the Moon before we'd learned to drive a car. "Worlds" has many valences, and the paradoxical structure of the word, its fusion of the Many and the One, is irresistible to intellectuals from pluralist societies who trace their political modernity back to the very century in which this cosmological "plurality" gained acceptance.[1]

The story of its rise to power has been told dramatically many times, usually beginning with the "martyrdom" of the wandering apostate monk, Giordano Bruno (1548–1600). After fifteen years of exile among European courts and centers of learning (he had fled his convent in Naples under suspicion of heresy), and after eight years of incarceration and interrogation by the Venetian Inquisition, he was burned (muzzled) at the stake in Rome in

---

1. In *Imagined Communities*, Benedict Anderson makes a number of suggestive remarks about the concept of the pluralization of worlds in relation to the origins of modern nationalism (a decidedly negative outcome) and modern print genres, such as newspapers and novels (see chaps. 2–5).

the first weeks of the century, for reasons thought by many to have included his philosophical teachings on the plurality of worlds.[2] The indictment has never been found, in fact, and we have only the *Summario* of the Venetian Inquisitor's case and the few words of a hostile witness, Gaspar Schopp, who narrated the event in a letter, to help us guess what any of the "eight points of doctrine" might have been for which Bruno was deemed too dangerous to live.[3] The historian Frances Yates thinks they were probably theological points, to some degree separable from his philosophical ideas; her drama of Bruno does not celebrate European emancipation from the Vatican stranglehold, but rather the hermetic enthusiast's importance as a counter to the usual heroes of the period's scientific history: "surely he would have written violent dialogues against Kepler, the 'pedant'" (*Giordano Bruno*, 444). The English translator of *De l'infinito universo et mondi*, Dorothy Singer, clearly believes Bruno was executed for his relativism and in general his heterodox attitude. The missing documents permit innumerable imaginative investments.

Attention to individual texts should show that despite their differences of style and position, Bruno and Kepler have something profoundly in common in their cosmological animism and the imaginative density of their response to the idea of "innumerable worlds." Though it may be far-fetched to imagine that the plurality of worlds sparked much in the way of political pluralism during the signally intolerant Counter Reformation, it is clear enough that the period saw, in Susan Bordo's words, "the cultural reawakening to the multiplicity of possible human perspectives" (Bordo, *Flight to Objectivity*, 115) or indeed to the salience of perspective itself, in several senses of the word. When the idea of other worlds became current, not only as a metaphor (the "New World" and the "Old World") but as an astronomical actuality (four new satellites orbiting Jupiter, planets as other "earths" made of matter the

---

2. The most relevant writings are *De l'infinito universo et mondi*, written and published in England (though the title page says "in Venetia") in 1584, after Bruno spent time at Oxford and in London with various intellectuals connected with Ralegh's "School of Night," including Hariot, and the later Latin treatise, his last work, *De innumerabilis immenso et infigurabili*, published in one volume with *De monade* in Frankfurt, 1591. Recent books on the issue of the plurality of worlds and its textual history include Steven J. Dick, *Plurality of Worlds*; Michael J. Crowe, *The Extraterrestrial Life Debate*; and, from the point of view of literary history, Karl S. Guthke, *The Last Frontier*.

3. The *Sommario del processo di Giordano Bruno* was discovered among the personal archives of Pope Pious IX and published in 1942. It is not the official report but a summary of the Venetian *processo* drawn up for use by the Roman Inquisition after Bruno was finally sent on to Rome. According to Frances Yates it contains a reply by Bruno "to censures on propositions drawn from his works." She goes on to remark that "I find this document very confused and confusing, but it touches on . . . the earth as animated by a sensitive and rational soul, on their being many worlds" (*Giordano Bruno*, 354). The letter of Schopp to Conrad Rittershausen appeared in full for the first time in the *Acta litteraria* (Jena, 1707).

same as the matter of our planet, a conceivably habitable Moon), it offered an irresistible object to the fictional imagination.[4] We will look at some full-fledged other-worldly fictions in later chapters. The present chapter is chiefly concerned with that "infinitely spacious bosom," the space itself of multiplicity, within which these celestial islands would be floated, and with the texts that announced or elaborated it.

Space is not a void, at least not for the purposes of this discussion. It is better imagined as a lattice, even a syntax. It is a set of relations and dimensions: relations between objects and between locations, and also between objects and observers (or possible observers). Heliocentrism is a concept of space, as is Ptolemy's multiply bounded and concentric system of spheres. The relativist and pluralist space of Bruno's infinite universe has as many centers as it has worlds. Kepler's *Somnium* positions readers on another world in order to observe the heavens (and Earth!) from an alternative "center." Galileo's *Sidereus nuncius* announces an empirical confirmation of a multi-centered space by discovering telescopically another planetary system in orbit around Jupiter.

The classic popularizing text, Bernard le Bovier de Fontenelle's *Entretiens sur la pluralité des mondes*, revised and republished in several editions by the author between 1686 and 1742 and frequently translated, is a charming fiction with a female protagonist, a work compared by the author to Madame de Lafayette's *La Princesse de Clèves* and translated into English early on (1688) by one of England's first woman novelists and its first professional woman playwright, Aphra Behn. What happened to the idea between its appearance in the embattled Bruno's philosophical dialogue of 1584, *De l'infinito*, and this exquisite Enlightenment bauble? Between its emergence in the male homosocial intellectual world of the AWOL Dominican and its canonization in a text inviting women to the banquet of science? What uses were found for it? And how did it end up so close to such an exclusively aesthetic notion as an idea for popular consumption?

What may seem surprising choices of text here — Galileo's *Starry Messenger* instead of his *Dialogue Concerning the Two Chief World Systems*, and Kepler's *Dream* instead of the more comprehensively cosmological *Mysterium cosmigraphicum* or the *Harmonice mundi* — have to do with the empirical emphasis

---

4. For bibliographies of mainly imaginative works which took up the challenge (as well as challenges posed by the New World and the explorations of Oceania), see Phillip Gove, *Imaginary Voyage in Prose Fiction*, Geoffroy Atkinson, *Extraordinary Voyage in French Literature*, and especially Marjorie Hope Nicolson, *Voyages to the Moon*; Guthke's literary history, *Last Frontier*, brings what we now consider a branch of science fiction from Lucian and Iamblichus through the riches of the seventeenth and eighteenth centuries into the twentieth century.

and narrative qualities of these two texts. All four texts considered here border very closely on the fictional and can be seen as outliers in the field of science fiction. In every case the space they structure is conceived as crucially habitable or inhabited—lived in, like the "islands" of the earth's changing space and of the "human sciences" under construction.

SUBLIMITY:
BRUNO'S *DE L'INFINITO UNIVERSO ET MONDI*

Cosmology is a sublime science, and its quotient of accessible sublimity was suddenly enlarged when Bruno invoked the immensity of the universe, an intuition on his part unrelated to telescopic experience or the empirical discovery of new "stars." His late medieval forerunner, Nicolaus of Cusa (1401–1464), had argued both infinite universe and innumerable worlds on the theological basis of a distaste for limiting the expression of God's creative power; Bruno was powerfully influenced by Cusanus. He was additionally excited by the apparently hermetic diagram of the solar system in Copernicus's *De revolutionibus* (1543), a diagram in which he read, as Yates put it, "a hieroglyph of divine mysteries" (*Giordano Bruno*, 155).[5] It is easy enough to see Bruno as simply capable of imaginative response. In a Neoplatonic environment, where hieroglyphs were all the rage and sun-worship the public secret of many occult philosophers, one might expect more than a single Bruno. And in fact it is easy to find textual passages suggesting that his warm response to the signifying power of the heliocentric system was one of many.

This particular sublimity, however — of the infinitude of the universe — involved a really new conception. Heliocentrism had been thought up before, by respectable Greek philosophers, and it fit the monarchism and nascent state building of the age. The cautious (and tactful) Copernicus was not hounded by the Inquisition and in fact *De revolutionibus* was dedicated to Pope Paul III. The philosophical poem of Lucretius (99–55 B.C.E.), *De rerum natura*, which makes fun of the concept of a bounded universe, was only rediscovered and put into circulation in Cusanus's youth, and attracted mainly literary attention. The implications of these works teased out by Bruno (with Thomas Digges's help—see figure 13 for his map of the heavens) are also vis-

---

5. As Stillman Drake has pointed out in several places, Bruno was probably inspired by a particular version of this diagram, that published in England by the Copernican Thomas Digges in his updated version of his father Leonard's astronomy textbook, *A Prognostication everlasting of right good effect* (1576). It shows, as Copernicus's own diagram does not, an unbounded area beyond the sphere of Saturn for the fixed stars represented as theoretically extending to infinity. See figure 13.

ibly homologous to new developments in political geography (for example, the potential for expansion of colonial empires in an unmapped world), but they require a drastically renovated concept of space. One might intervene here to say that the painters who introduced geometrical perspective also changed the idea (the felt structure) of space and were left alone by the authorities (or were paid by them to paint their palaces), but painting was understood to be representation, while cosmology was not.[6]

Or rather, cosmology was not usually understood by mainstream ecclesiasts as representation. In the magus Bruno's hieroglyphic world, the distinction was not so clear between signifier and signified. It would have to become so for scientific language to be rendered efficiently transparent, but for Bruno — and a little later for Kepler (the "pedant") — poetic and quantitative meanings could coexist and interpenetrate.[7] The world had, to borrow from the title of Fernand Hallyn's recent book, a "poetic structure."[8] In Foucault's partial account of Renaissance knowledge in *The Order of Things*, the doctrine of "resemblances" postulates a universe in which the two terms of a metaphor are equally real and significant and their resemblance not heuristic but essential, substantive — a potential channel of influence or communication. Although this is one of Foucault's airier totalizations, certainly Ficino's natural magic bears him out, and Bruno learned much from Ficino (even getting in trouble during his lectures at Oxford for having stolen long passages verbatim from the master).[9] So a revolutionized space described in the philo-

---

6. Copernicus advanced his heliocentric hypothesis only as a mathematical model capable of "saving the appearances," not as a description of reality; his cosmology was explicitly then a representation and nothing else. But his important readers read the work as in fact a description of reality; this is why it was eventually (1616) put on the Index of prohibited books, "pending correction." On the alteration of pictorial space, see William Ivins, *On the Rationalization of Sight*, and Samuel Edgerton, *Heritage of Giotto's Geometry*.

7. I am aware that Kepler was not a numerologist, that his "mathesis" was quantitative and modern. But no modern astronomer could have written the *Mysterium cosmographicum* (1597), in which the distances of the planetary orbits from the sun turn out to be a function of the sequence of the five nested polygons; there are, in this intuition, notions of neatness and homology that are aesthetic only, though certainly "rational."

8. All Hallyn initially claims is that cosmological hypotheses are, as we have been assuming, man-made, and originate in the imagination. By the end of his book it is clear, however, that not just any cosmology would attract his aesthetic attention; Kepler's universe, for example, has "depth," is "hypercodified" — its mathematical features "take on meaning with the introduction of a multiplicity of codes grafted onto the global order" (285). It is the metaphysical resonance built into these complex hypotheses of the Renaissance and the Baroque that enables Hallyn to justify a "poetics" of cosmology. See Hallyn, *Poetic Structure of the World*.

9. On Bruno and Ficino see Ioan Couliano, *Eros and Magic*, 68–72 and Yates, *Giordano Bruno*, 155 and 208–11. The story of Bruno's lifting from Ficino in a debate at Oxford is told by (the expostulating Puritan) George Abbott in a pamphlet called *The Reasons Which Doctour Hill Hath Brought, for the Upholding of the Papistry* (1604).

sophical texts of a magus was poetic in the archaic sense of the word (where *poesis* means making, shaping), rather than in the vague, sentimental Enlightenment usage more appropriate to Fontenelle's aesthetic.[10]

In his only work devoted solely to this matter, the dialogue (or "quatrologue") *De l'infinito universo e mondi*, Bruno laughs at an older poetic structure, or Elpino laughs for him (addressing Bruno's stand-in Theofilo): "You would deny that they [the fixed stars] are as it were embedded in a single cupola, a ridiculous notion which children might conceive, imagining perhaps that if they were not attached to the celestial tribune and surface by a good glue, or nailed with stoutest nails, they would fall on us like hail from the air immediately above us" (Singer, 299).[11] This heaven, the Ptolemaic (and in fact the Copernican, as far as the cupola and the fixed stars go), is architectural. Its different areas have very different natures, and different levels of prestige. Earth is the scurrilous basement of the universe; the sphere of the fixed stars is in closest contact with the Primum Mobile, the fastest-spinning sphere, which forms the interface between God's surrounding will and the physical universe. The relationship of the areas in this universe to each other is concentric, on an axis of rising spiritual value: the universe is very emphatically unified and single. It is, whether geo- or helio-, *centric*. Bruno's universe has many centers ("a thousand hearts"). "Thus the earth no more than any other world is at the centre; and no points constitute definite determined poles of space for our earth, just as she herself is not a definite and determined pole to any other point of the ether, or of the world space; and the same is true of all other bodies. From various points of view these may all be regarded as centres, or as points on the circumference, as poles, or zeniths and so forth" (280).

One might consider such a space anything but radically new—it sounds like a functional description of feudalism. But feudalism, no matter how multicentric in historical practice, was a structure theoretically unified by hierarchy, with a king, however weak, at the top and center. The model (perhaps residual) of normal, fully developed political organization was empire, even when reality was the France of the Hundred Years' War, or the strife of the Italian city-states.

---

10. "The threat of heterocosmos or alternative worlds [is] mounted against the hero-God, Jehovah, whose claim on representation altogether, whose authority (which the Bible attests), is his one creative act, the making of the one world there is, which is of course subverted by any second making, such as the Poet . . . intends" (Allen Grossman, *Poetry: A Basic Course*, lect. 6).

11. The English is quoted from Singer, *Giordano Bruno*, which includes a complete and annotated translation of *De l'infinito*. Where it seems necessary to provide the Italian, I quote from *Opere Italiane*, 3d ed., ed., Bruno Gentile.

The older structure is not only centralized, but bounded, as had been the *oikumene* (the Old World's "circle of lands") before the cosmographers realized that Columbus had found a new and separate continent. It is not however bounded by a huge expanse of palpable emptiness, such as the "Ocean Sea": it is bounded by finiteness in the absolute, imaginable only as a cupola to which stars are glued like ornaments. In a familiar trope, Bruno is imagining the old "world" as a theater, a definable place — a place of illusion and nearness, where we can see all there is to see and even so are fooled by glue.[12] The sublime cannot operate in such a universe — should the stars fall they will annoy, like hail or confetti.

The chief, though not the only, theological problem with the new plural system of worlds has to do with salvation history, and recalls the problem suggested by the discovery of people in the New World: who fell when Adam did, and who was saved by Jesus? If the other worlds are inhabited (and one must assume they are, for "Omnipotence doth not grudge being" [Singer, 262]), then their inhabitants must be accounted for in the dramatic narrative of salvation. But how? As Fontenelle would later put it, "The descendants of Adam have not spread to the Moon, nor put colonies there" (*Entretiens*, 6). Kepler will cleverly evade the issue (see below); Fontenelle will offer a sophistic solution (the inhabitants are not "men"). Bruno more grandly ignores it.

Bruno's only remarks on the "innumerable and excellent individuals" who must inhabit his innumerable worlds come towards the end of the fifth and last of the dialogues that largely make up *De l'infinito*, in response to Albertino's objection that "civil intercourse" is a great virtue and that "the gods who had created diverse worlds would have done ill, in that they had not contrived that the citizens thereof should have commerce one with another" (359). Surprisingly (or not), given his position as a permanently displaced wanderer outside his home country, Bruno says "we learn by experience, that it is best for the living creatures of this world that nature hath distributed their diverse kinds throughout seas and mountains. And if by human

---

12. "During periods in which long-established images of symbiosis and cosmic unity *break down* (as they did during the period of the 'scientific revolution'), may we not expect an increase in self-consciousness, and anxiety over the distance between self and world — a constant concern, to paraphrase [Margaret] Mahler, over the 'whereabouts of the world'?" (Bordo, 58). "PHIL [Bruno's stand-in]: . . . If the world is finite and if nothing lies beyond, I ask you WHERE is the world? Aristotle replieth, it is in itself. The convex surface of the primal heaven is universal space, which being the primal container is by naught contained. For position in space is no other than the surfaces and limit of the containing body, so that he who hath no containing body has no position in space. What then dost thou mean, O Aristotle, by this phrase . . . ? What will be thy conclusion concerning that which is beyond the world?" (Singer, 251–52).

artifice there hath befallen traffic among them, good is not thereby so much added to them as removed, since communication tendeth rather to redouble vices than to augment virtues" (375). More significantly, Bruno represents Albertino's "courteous exchange of intercourse" as a kind of fatal blurring of individual boundaries, saying there is no more need for it "than that all men should be one man or all animals one animal" (375). Whatever this may say about Bruno's private psyche (and perhaps it helps explain his notorious difficulty with the niceties of civil intercourse), it suggests that his philosophical system privileges difference and heterogeneity over unity and harmony. The model of empire tends to homogenize. Bruno loves the plural moment, the moment he lives in, geopolitically — the moment the astronomers are about to enter empirically, when in 1608 a Dutchman shows off the first known telescope to some official visitors from the King of Siam.[13] Thomas Hariot (who must have met Bruno while he was in England writing *De l'infinito*) would look at the moon through another telescope and draw the first pictures of it the following summer, and a few months after that, Galileo would make the observations that revealed the moons of Jupiter.[14]

Lest Bruno be absorbed entirely into the Foucauldian homologies, we should keep our attention on his peculiar dismissal of the "commerce" that characterized the earthbound discovery of plural human or cultural worlds. Perhaps it is partly the clarifying exaggeration of a polemical work, a work that wants to introduce both multiplicity and *difference*, by a man sophisticated enough to realize that colonial empire will destroy difference as far as possible. What interest then does Bruno have in the actuality of different and unvisitable worlds, worlds *effectively* singular?

Of course, I have no idea. But the "infinitely spacious bosom" enfolding all these infinitely different worlds, that "common air" filled with splendid, mutually undiscoverable "animals," is a figure of hyperbolic possibility. It is also a figure of fear, or at least awe. The hypothetical and rarely mentioned "excellent individuals" inhabiting this larger sky are strong images of "the other" as absolutely other — not to be despised as inferior, like the "natural man" of the pagan Americas (if anything they will be, like Arthur C. Clarke's satanic angels, greater than we are), but not to be touched or communicated with either. Whatever potential sterility this set of permanent blanks might have for a nascent ethnology, it is extraordinarily fertile for those who would imagine the alternative (or parallel) "worlds" of fiction and utopia. Its most powerful

---

13. See *Ambassades du Roy de Siam envoyé à la Haye le 10. Septemb. 1608*, 11.
14. For close chronologies of these events see Roche, "Harriot, Galileo, and Jupiter's Satellites," and Bloom, "Borrowed Perceptions: Harriot's Maps of the Moon."

feature might be the tension between the actuality and the unknowability of these celestial islands. On earth the exotic islands of Columbus and Thevet are quickly understood as European fortresses supplying necessary goods and materials to people at home, but the islands of space are in and to themselves. They are alternative being, rather than supplementary. They escape the dynamics of power and in so doing, suggest the possibility that power is escapable. In Bruno's case at least, this was not true.

Like many people who have contemplated alternative worlds, Bruno was eager to create a real one and tried to interest the Holy Roman Emperor (Rudolf II), among others, in his synthesis of Copernicanism, hermeticism, and anti-Calvinist theology as the basis for a "tolerant realm and benevolent government" (Drake, "Copernicus," 29). He was not successful, but Frances Yates at least (in perhaps her least respected book) thinks he was an important influence on the Rosicrucian subculture that, in her narrative, underlies the brief, utopian reign of the Bohemian "Winter King."[15] Bruno's fears of the chaos intolerance could provoke were fully realized, at the personal level as well as the international. The Pope and the normally antagonistic secular authorities of Venice and Rome combined forces to eliminate him. It is worth noting then that his "tetralogue" on the infinite universe and the innumerable worlds—a work that lays out a metaphysical and cosmological basis for his ideas about terrestrial reform—includes an interlocutor whose resistance to Bruno's thought is so persistent that he is dramatically ejected from the circle of speakers, eliminated from the text.

Burchio represents the voice of conventional deference to authority, especially the authority of Aristotle: "Would you then render vain all efforts, study and labours on such work as *De physico auditu* [sic] and *De coelo et mondo* [sic] wherein so many great commentators, paraphrasers, glossers, compilers, epitomizers, scholiasts, translators, questioners and logicians have puzzled their brains? Whereon profound doctors, subtle, golden, exalted, inexpugnable, irrefragable, angelic, seraphic, cherubic and divine, have established their foundation?" (324). He is dismissed at the end of the third dialogue by Elpino, a more docile student of Bruno's stand-in Theofilo: "I pray you, magnificent sir, do not trouble yourself to return to us, but await our coming to you" (326). But after an intervening fourth dialogue, Bruno

---

15. Yates's *Rosicrucian Enlightenment* is a grandly erudite conspiracy theory, but it is difficult to shake the sense it provides of a European subculture anxious to steer clear of the dogmatism and violence submerging "civil intercourse" among and within the nations of the West. Such horrors as the Thirty Years' War generally do spawn movements of resistance and oppositional ideologies, at least in cultures where communication is relatively efficient and quick, as in a literate print culture like that of Western Europe during the Reformation and Counter Reformation.

feels the need for an additional and resistant interlocutor, so he brings on Albertino, a more malleable Aristotelian, as it turns out, but one who begins in Burchio's position: "What, I who am a doctor, approved by a thousand academies, I who have publicly professed philosophy in the first academies of the world, am I now to deny Aristotle, and crave to be taught philosophy by such fellows [as Theofilo]?" (351). He succumbs to Bruno's eloquence by the end, after a long philosophical duet, and Philotheo/Theofilo is moved to ask Elpino the yes-man "What meaneth it . . . that Doctor Burchio hath not so speedily nor indeed ever consented with us?" (378). Elpino has no real explanation; Theofilo is simply complimenting the agreeable new convert for his "vigilant wit." And we are left to wonder ourselves, why the creator of this multicentered and infinitely spacious new universe created a character whose resistance to it was so intolerable that, although his function remained necessary, so that he had to be replaced, his presence could not be sustained.

It might seem that Bruno was unable to hear his own lessons, or to attend to the ethic hidden in his metaphysics. I think, however, that Burchio is a character with, as modern novelists say, "a life of his own." Where a medieval allegorist would simply have *begun* with a character malleable enough to function both as resistance and later as acquiescence, Bruno—inventor of a plurality of worlds without "intercourse"—lets his character become intolerable and then expels him, rather than preemptively revising him out of existence. The character escapes the author's power, "[does not] ever consent." His outline is sharp, his fate his own. He does not merge with Bruno's personae—"all men [are not] one man" in this representation of civil intercourse. Difference is honored, including, perhaps, "internal difference / where the meanings, are" (Emily Dickinson, *Poems*, no. 258).[16]

If we are looking for what the plurality of worlds had to offer the art of fiction as it would be increasingly practiced in the new age of print, I think we can begin by finding that it provided the possibility, at least in Bruno's influential imagination, of the separate, the distant, the alternative, the other, the "irrefragable [*irrefragabili*]," to use Burchio's own word. Bruno insisted over and over that what he didn't like about the Ptolemaic system and Aristotelean physics was the *order*, the mutual enclosing of sphere by sphere, the nested quality of the finite world. ("The famous and received order of the el-

---

16. John Bossy's *Giordano Bruno and the Embassy Affair* makes a case for Bruno's having maintained at least two identities for a while (the chief other identity was Henry Fagot, the spy for Walsingham in Castelnau's house) and five or six distinct handwritings (several of which are on record as Bruno's own, two or three of which are Fagot's)—an interesting possibility, although the extraordinary cynicism of Bossy's case against Bruno renders his data suspect.

ements and of the heavenly bodies is a dream and vainest fantasy" [322].) He traveled incessantly, abandoning every nest, whether intellectual, religious, or geographical, and finally abandoned travel itself. "Break and hurl to earth with the resounding whirlwind of lively reasoning those fantasies of the blind and vulgar herd, the adamantine walls of the *primum mobile* and the ultimate sphere" (378). People have speculated on and researched with great effort and ingenuity Bruno's motives for returning in the end to an Italy where he would surely be arrested and tried. My own speculation is that even abroad he felt enclosed by fear, that the safety of exile made his apostasy seem, not an escape, but simply the inner surface of the identically adamantine sphere of Catholic hegemony. And that his eight years of opposition, while incarcerated, to the very grounds of the Inquisition's process was a (theatrical) production of distance and separateness, grand evidence that the alternative world could indeed be "at hand."

THE EMPIRICAL: GALILEO AND THE "ENCHANTED GLASS"

> O telescope, instrument of much knowledge, more precious than any sceptre! Is not he who holds thee in his hand made king and lord of the works of God?
>
> —Johannes Kepler, *Dioptrice* (1611)

Events in a more empirical *theatrum* followed Bruno's death with what was often taken as evidence for the plurality he had intuited philosophically. With a telescope, as Roger Bacon had put it, "we might cause the sun, moon, and stars in appearance to descend here below, and similarly to appear above the heads of our enemies" (Van Helden, *Invention*, 28): Bruno's much-scorned cupola emerges again in Bacon's sentence as the desideratum of the technicians, from which stars can indeed be made to fall like hail, for military purposes. Or at least to seem to, and that was enough for such a generally theatrical age. Hariot was one of the first to use the telescope: his earliest mention of it may be in *A briefe and true report*, where he provides a list of technical wonders with which his party overwhelmed the Virginians psychologically: "Moste thinges they sawe with us, as Mathematicall instruments, sea compasses, the vertue of the loadstone in drawing yron, a perspective glasse whereby was shewed manie strange sightes . . . were so straunge unto them, and so farre exceeded their capacities to comprehend . . . that they thought they were rather the works of gods than of men" (27). The theater of war indeed.

Van Helden discounts all reports of "perspective glasses" before 1608, including this one, explaining them as either vaguely theoretical or, if actual, instances of magnifying glasses only. His chief argument is based, not on the sometimes contradictory textual evidence, but on the idea that "if Digges or Bourne had actually been in possession of a combination of mirrors and/or lenses which constituted what we would call a reflecting telescope, surely the instrument would quickly have been adopted for military use."[17] Although in fact it was Hariot's custom to sit on just about any discovery he made (for example, the 1609 drawings of the moon's surface), the remarks of Bruno's fellow magus John Dee on the potential usefulness of perspective glasses support Van Helden:

> The Herald, Pursevant, Sergeant Royall, Capitaine, or who soever is carefull to come nere the truth herein, besides the Iudgement of his expert eye, his skill in ordering *Tacticall*, the helpe of his Geometricall instrument: Ring, or Staffe Astronomicall: (commodiously framed for cariage and use) He may wonderfully helpe him selfe, by perspective Glasses. In which, (I trust) our posterity will prove more skillful and expert, and to greater purposes, then in these dayes, can (almost) be credited to be possible. (Dee, "Mathematical preface," aiiii^v–bi^r).

As Dee's badly punctuated optimism suggests, the introduction of gunpowder into European armories in the fifteenth century had spurred an arms race that provoked considerable technological development. Galileo had participated in this, designing instruments for aiming cannon accurately, publishing a handbook on the use of his "Geometric and Military Compass," and showing himself instantly aware of the telescope's aptitude for observing at sea, for example, "sails and vessels so far away that, coming under full sail to port, two hours and more were required before they could be seen without my spyglass [*ochiale*]." As soon as he had heard about the invention, he set about making one and, "having known how useful this would be for maritime as well as land affairs, and seeing it desired by the Venetian government, I resolved on the 25th of this month to appear in the College and make a free gift of it to his Lordship [the Doge]."[18]

Galileo's interests tended to be in physics and engineering, and less metaphysically extended than the cosmology that obsessed Bruno and the man who admired them both, Johannes Kepler. But Galileo was, in the face of what he saw through the telescope, a poet like the rest of them. *Sidereus nuncius* (1610), the "Starry Messenger" (or "Starry Message"), is a gorgeous

---

17. Van Helden, *Invention of the Telescope*, 14.
18. Letter to Benedetto Landucci, 29 August 1609, in Favaro, ed., *Opere* 10:253–54. (Translation is from Drake, *Galileo at Work*, 141.) On Galileo's relation to patronage and related "self-fashioning," see Mario Biagioli, "Galileo the Emblem Maker," and more generally his subsequent book, *Galileo, Courtier*.

little book, and its major gesture, the naming of the four newly discovered moons of Jupiter after the Medici family, is the act not only of a canny courtier but of someone with a sense of the sky's allegorical depth. It may be hard to imagine Bruno naming a "world" after a man who could give him a job, but it is not hard to imagine him wanting a look at those worlds through the perspective glass (unlike a number of academics who actually refused to look).[19] Along with Galileo's even more specifically Copernican discovery later that year (1610) of the phases of Venus, the new moons offered solid empirical evidence of an important tenet of Copernican cosmology: other bodies than the earth could be at the center of planetary orbits. Instead of fixing his attention on the celestial space opened up for doubters by this phenomenon, Galileo first attended to the other, earthly, term of the resemblance or correspondence — the relationship of nobles to overlord, children to father, a relationship invoked in sociable or erotic terms which are afterwards applied to the activities and sensations of telescopic observations.

The dedicatory letter—in which not only the book but the moons it announces are dedicated to Cosimo de Medici, echoing the mimetic ambiguity of the book's title (Starry Message or Starry Messenger?)—describes Galileo's relationship to his lord, the Duke of Florence, this way: "Since I was evidently influenced by divine inspiration to serve Your Highness and to receive from so close the rays of your incredible clemency and kindness, is it any wonder that my soul was so inflamed that day and night it reflected on almost nothing else than how I, most desirous of Your glory (since I am not only by desire but also by origin and nature under Your dominion), might show how very grateful I am toward You" (Van Helden, 32).[20] This delicate appropriation of the myth of Danae (impregnated by Jupiter's oddly solar "rays" in the famous "golden shower") appears and reappears in the astronomical text where Galileo writes, for example, of "the Sun who penetrates the Moon's vast mass with his rays" (54, "qui radiis suis profundam Lunae solidatem permet" [Favaro, 3.1:73]) or of "the lunar body . . . bathed by light from the Earth" (55, "a Terra ipsum lunare corpus . . . lumine perfundi" [Favaro, 74]) or of "the Earth's surface shin[ing] far and wide, perfused by lunar splendor" (56, "lunari splendore perfusa" [Favaro, 74]). As Margaret Carroll has pointed out in an analysis of rape imagery in the palace decoration of Renaissance absolutists, Cosimo de Medici's palace was one of several

---

19. See index subheading, "Telescopes, refusals to look through," in Drake, *Galileo at Work*.
20. The *Sidereus nuncius* (following a facsimile of Galileo's working manuscript, 15–51) is in Favaro, ed., *Opere* 3.1:51–96. I will occasionally provide parenthetical or footnoted Latin from this edition.

to emphasize the power of Jupiter as an emblem of his sovereignty, and to do so through mythological rape scenes—which send dual messages, both threatening subordinates and also sharing with them as "between men" a fraternal identification through sexual fantasies of domination and "possession."[21] Danae was a popular Renaissance theme: one of Titian's was in Rome in the Farnese Palace, and another, commissioned by Phillip II, hung in his palace in Madrid with the "Rape of Europa."[22]

The incorporation of the pederastic language of courtly "intercourse" into the text of a hugely important scientific announcement can be partially understood as a Renaissance response to the same patronage issue that now causes grant applications to the NSF to begin by invoking politico-magical formulas such as "widespread concern about possible future effects of global warming" to reassure potential sponsors of political payback for their support. But the *Starry Messenger* is a text in an anomalous situation: some historians date the beginning of modern science with its appearance.[23] Like Bruno's dialogue, it must confront a changed universe. Unlike *De l'infinito* it must describe something like a physical experience of that universe: an experience of enhanced gazing—spyglasses, then as now, seem to have touched an erotic nerve.[24] (Kepler again: "Shall we make it a Cupid's arrow, which, entering by our eyes, has pierced our innermost mind, and fired us with a love of Venus?")[25]

The Starry Messenger begins with an account of lunar observations using the telescope (illustrated with engravings from Galileo's drawings) and then, after a brief discussion of the fixed stars and the Milky Way which

21. See Carroll, "Erotics of Absolutism."
22. The paintings were finished in 1546 and 1554, respectively. A third (1555–60) was sent in 1600 from Cardinal Montalto in Rome to Rudolf II in Prague. See Wethey, *The Paintings of Titian*, vol. 3, cats. 5–7. The second and third versions include the wonderful figure of the (old and ugly) nursemaid trying to catch some of the gold as it falls and indeed, in the third, gold coins have actually landed on the bed near Danae's raised thigh. (It may be worth noting that the Farnese never paid Titian for the first version, nor for his portrait of Pope Paul III.)
23. "*Sidereus Nuncius* was not so much a treatise as an announcement: in a few brief words, and in sober language [!], it told the learned community that a new age had begun and that the universe and the way in which it was studied would never be the same again" (Van Helden, *Starry Messenger*, vii).
24. They did so even before Van Helden admits they were invented: see Robert Greene's 1594 play, *Friar Bacon and Friar Bungay*, in which Roger Bacon offers his glass to a lovelorn Edward I: "Within this glasse perspectiue thou shalt see / This day whats done in merry Fresingfield, / Twist loueley Peggie and the Lincolne earle." The clearest case of seventeenth-century telescopic eroticism is Li Yu's "Tower for the Summer Heat," which narrates the Jesuit introduction of telescopes to China.
25. "An magis sagittam Cupidinis, qua per oculos illapsâ mens intima vulnere accepto in Veneris amorem exardescat?" "Preface" to *Dioptrice*, 22.

reveals their innumerability, shifts to the series of observations in the winter of 1609/10 that resulted in his identification of Jupiter's moons: in other words, the protagonist is neither the Moon nor the "Medicean Stars" nor the Milky Way, but "the instrument with the benefit of which they make themselves visible to our sight" (Van Helden, 35). It is a kind of experience that he wants to illustrate, and he does so with narratives that invite our identification more overtly, if less comprehensively, than does Thevet's cosmography. As we will see in greater detail in Chapter 6, "glasses" promote a scopophiliac experience, more immediate and even sensuous than the glamorous identities granted by Thevet's experiential accounts of terrestrial geography. "It is most beautiful and pleasing to the eye to look upon the lunar body" (Van Helden, 35, "Pulcherrimum atque visu iucundissimum est lunare corpus" [Favaro, 59]).

Galileo shared a value common among the virtuosi, if frowned upon by Bacon, believing that pleasure was at the foundation of scientific pursuits: "it will be pleasing and most glorious to demonstrate that the substance of those stars . . . is very different from what has hitherto been thought" (Van Helden, 36). The magnifying effect of the *ochiale* is "truly wonderful" (37, "admirabilis" [Favaro, 60]), although in this text the method of constructing such a glass is "tasted only with our lips" (39); bright spots in dark areas of the moon are "even more wondrous" (42, "maiorem . . . admirationem" [Favaro, 64]) than the unevenness of the terminator discovered by the *ochiale*, the lunar surface is "decorated with spots like the dark blue eyes in the tail of a peacock" (43), shadows moving in the brighter parts of the moon "happen beautifully" (48).[26]

The beauty Galileo is always careful to attribute to the Moon and stars (and in the final sentence, to the "fair [or shining!] reader" [Van Helden, 86, "candidus Lector" (Favaro, 96)]) is offered not in the way of authoritative judgments or canned flattery but as embedded in narrated experience. The

---

26. In his preface to the first English translation of Euclid's *Elements*, John Dee lists three reasons why common people, able only to read the vernacular, might want to do science: "for sundry purposes in the Common Wealth or *for private pleasure* and for the better maintayninge of their owne estate" (quoted in Debus, *Man and Nature in the Renaissance*, 7). One section of Sprat's *History of the Royal Society* (3:10) is a defense of experimental science against charges from businessmen that it is too pleasurable (535–37). A few pages later, another section (13) offers us a sense of what the businessmen were worried about: "What *raptures* can the most *voluptuous* men fancy to which these [pleasures of experiments] are not equal? Can they relish nothing but the pleasures of their *senses*? They may here enjoy them without remorse" (344). On the importance of rhetoric generally to scientific persuasiveness in the seventeenth century (centrally but not exclusively concerned with Galileo and those who disputed him), see Joan Dietz Moss, "The Interplay of Science and Rhetoric."

impersonal temporality of the experiment report has not yet been invented, and Galileo is certainly not impersonal here; this account of his observations makes clear that the manner of this text is narrative and the subject of its experiences a character who, while he may become the reader—"To whatever region you direct your spyglass an immense number of stars immediately offer themselves to view" (Van Helden, 62, "sese in conspectum profert" [Favaro, 78])—is also a living person, himself: "I waited eagerly for the next night. But I was disappointed in my hope" (65).[27]

The narrative of the *dangereuse* new astronomy involves extremely precise scene-setting—gone are the days of the conventional *topos*: "before the city" or "by a spring in a wood." The function of this precision is precisely to make the experience repeatable, not (or not only) in a readerly fantasy but in actuality, by other scientists. If we want to repeat Galileo's experience of the full expanse of Earthshine, we will

> at first glance . . . [see] only a slender shining circumference . . . on account of the darker parts of the sky bordering it, while, on the contrary, the rest of the surface appears darker because the nearness of the shining horns makes our sight dark. But if one chooses a place for oneself so that those bright horns are concealed by a roof or a chimney or another obstacle between one's sight and the Moon (but positioned far away from the eye), the remaining part of the lunar globe is left exposed to one's view, and then one will discover that this region of the Moon, although deprived of sunlight, also shines with a considerable light, and especially when the chill of the night has already increased through the absence of the sun. (53–54)

The book's precision, its extended erotic imagery, the narrative framework ("guided by I know not what fate, I found" [65]), the engravings of the moon, the diagrams of invisible stars hidden in familiar constellations, the sequence of drawings that record the configurations of Jupiter and its moons, the suggestions for further observation and research: all these are strategies for representing and literally sharing an experience—with sharing or repeating considered as the limit case of representation (constitutive of modern scientific practice). The ideal reader reconstitutes rather than merely imagining the represented experience. Bruno had offered innumerable worlds to the philosophical view, but Galileo was trying to show us space as, in the immortal words of Sun Ra, a "*place*." "Indeed, with the glass you will detect below stars of the sixth magnitude such a crowd of others that

---

27. See Peter Dear, "'*Totius in verba.*'" Note also that "experience" is a term as yet undifferentiated from "experiment." To quote John Dee again, speaking of "archemastrie": "Bycause it procedeth by *Experiences*, and searcheth forth the causes of Conclusions, them selues, in Experience, it is named of some *Scientia Experimentalis*. The *Experimentall Science*" (in Debus, *Man and Nature*, 8).

escape natural sight that it is hardly believable.... But in order that you may see one or two illustrations of the almost inconceivable crowd of them, and from their example form a judgement about the rest of them, I decided to reproduce two star groups" (59).

The space Galileo depicts is a literalization of the plenitude crucial to what Arthur Lovejoy called the "Great Chain of Being"—it is like a child's fantasy of the candy store, or Columbus's of West Indian gold, or Thevet's of islands—but it has here a strangely inconsequential dimension of actuality: "to whatever region you direct your spyglass, an immense number of stars immediately offer themselves to view [as the Indians had been said to "offer themselves" to the Spanish].... Moreover... the stars that have been called 'nebulous' by every single astronomer up to this day are swarms of small stars placed exceedingly close together" (62). This is a space that, in Galileo's repeated phrase, "offers itself," "presents itself to view," "bathes" us, and is bathed in turn with light. It is an enormous Arcadia (masking as always an enormous court), where the little Medicis play hide-and-seek around their father Jupiter and Galileo has the time and attention to watch them and chart their motions at play. It is full to the brim with worlds, fuller even than Galileo can give witness to, for he knows a stronger telescope will show him even more.

Of course, Bruno's universe also was full to the brim. An attractive consequence of his theoretical persuasion about the goodness of creation and existence is the crowdedness of the heavens; he is not too clear about how many worlds might be "enough." But Galileo's is complicated by visibility and potential visibility. The erotic troping of both his politic dedication and his cosmological *descriptio* underscore the potency of looking, for him (no doubt rendered more vivid by the phallic "spyglass"). The particular trope he chooses to ground his rhetorical flowers in is one in which light is fertile, or rather seminal. It is the erotic world of the court and the aristocracy we see in Galileo's heavens, a world in which both sexuality and reproductive potency are crucial arenas for acquiring, displaying, and maintaining dominance.[28] A world, as well, in which looking often takes the place of touching, a world strung together by knowing glances and the sightlines they terminate.

The place reported to us (*repertus*, Columbus's word) by the Starry Messenger is visited in the spirit of the *conquistadores*, who also found a lot of

---

28. For analysis in these terms of the "decorative programs" of Italian palaces of the period, see Carroll, "Erotics of Absolutism." Carroll's essay tries to answer the question "Why are the palaces of absolutist rulers filled with paintings of mythological and historical rape scenes?" The figure of Jupiter is central to the self-presentation of these rulers, including Cosimo de Medici, and the relations of ruler and ruled are expressed in images of erotic violence and domination.

nymphs and fountains, but there is a major difference, the very difference there will one day be (deliteralized) between ethnography and science fiction, between science or travel writing and novel writing: Galileo has not in the usual sense "been there."[29] What is especially revolutionary about this Starry Message is the way it problematizes the question of whether indeed Galileo had traveled to, had experienced, other worlds. He had seen them, he had adventurously gone out to them. And the lights they had bathed him in have a quality of tactility that actually exceeds by quite a bit the erotic tactility evoked or openly represented in most colonial discovery narratives. (I disagree with Susan Bordo about the fundamental body horror of seventeenth-century philosophy—at any rate one could make a case for the philosophers' experience of light and vision as sometimes ravishingly bodily.)[30] But it is still important to make a distinction between "being there" and "seeing."

It is a distinction that will be easier to make after a look at Kepler's astronomical fiction, the *Somnium*, which pretends to a slightly more body-present witnessing. But first let's consider the problem (or opportunity) Kepler addresses with his fiction. Obviously the telescope was not needed to provoke an engagement with the idea of a plurality of worlds; Bruno didn't have one, and neither did Nicolaus of Cusa (who didn't even have the imaginative pressure on him of Columbus or Thevet, who were reporting new and multiple earthly worlds). What we see emerge with Galileo's experience is an impulse—the very impulse repudiated by Bruno, traveler and spy—to participate in the life of another place. To chart it and illustrate it and domesticate it with erotic metaphors, as did the explorers of the New World, so as to make it continuous with the world of home, imaginable. Galileo's starry universe is more beautiful than sublime. That is probably the effect of his personality and his situation. But it is a construction someone would sooner or later have recorded, and it has had a long *Nachleben*: the governments of the imperial nations are still spending money on exorbitant efforts to make space our (various) backyards. Or theaters of war. This essentially sociable impulse (however sinister its manifestations) is balked in the case of planetary other worlds, or has been until very recently. They are unreachable in a precise sense: although we can see them more and more clearly, clearly enough

---

29. In *Discourse of Modernism* (158–59), Timothy Reiss provides an interesting literary analysis of Kepler's handling of this issue in the *Somnium*.

30. Consider the combined interests of the Dutch virtuoso, natural philosopher, and art patron, Christiaan Huygens (1596–1695); for some examples, see Lorraine Daston and Katharine Park, *Wonders and the Order of Nature*, 13 and 303, or my discussions of Plot's rainbow spectacle in Chapter 3 and of Robert Hooke's *Micrographia* in Chapter 6.

to name their topographies like any good Columbus, we cannot become a part of their economies. We cannot connect.

The functional emptiness of the other worlds, combined with their optical "visitability," produces a very strange space indeed, though one with which we have become familiar. It is a space characterized by a paradoxical plenitude, rife with frustration for cultures otherwise experiencing an unprecedented freedom of movement and power of plunder. In Galileo's book it is marked with the language of desire, even more than America had been in the books of Columbus or Peter Martyr — at one remarkable point it is presented also in the language of an impossible mutuality: trying to explain his belief in Earthshine as the source of the moon's *lumen cinereum* he says: "What are we to propose — that the lunar body or some other dark and gloomy body is bathed by light from the Earth? But what is so surprising about that? In an equal and grateful exchange the Earth pays back the Moon with light equal to that which she receives from the Moon almost all the time in the deepest darkness of the night" (Van Helden, 55).[31]

Such tropes are illusory enough in the context of colonial exploration, where there are at least others with whom to contend, actual diplomacy to conduct, a potential (if never realized) mutuality to invoke. They bespeak a mournful lonesomeness in the context of humanly empty space.[32] They also bespeak, even in Galileo's actually quite exuberant text, a need to fill what's empty or to imagine an animate life for it. For Bruno, who calls the worlds "great animals" ["grandi animali"][33] and intuits the grievousness of real social contact between worlds (he knew Hariot and Raleigh, and perhaps Thevet, who worked for the same king as he did), there is something appealing (sublime) about the uncrossable distances, and his crowded universe is satisfactorily animate without his imagining its humanoid cultures. But Galileo's narrative of starry experience projects the dynamics of human relations outward onto the sky he spies on with his "instrument." Without

---

31. "Quid proferendum? nunquid a Terra ipsum lunare corpus, aut quidpiam aliud opacum atque tenebrosum lumine perfundi? quid mirim? maxime: aequa grataque permutatione rependit Tellus parem illuminationem ipsi Lunae, qualem et ipsa a Luna in profundioribus noctis tenebris toto fere tempore recipit" (*Favaro*, 3.1:74). Katharine Park has pointed out that the image in the last sentence captures "the essence of the client/patron relationship," a difficult dynamic of equal but not equalizing gift exchange (personal communication with author).

32. Here is the response of an Englishman to that loneliness, a century later: "Samuel Clarke told an acquaintance that he thought it possible that the souls of brutes would eventually be resurrected and lodged in Mars, Saturn, or some other planet" (W. C[oward], *A Just Scrutiny*, 97).

33. I quote from the epigraph to this section of my chapter (*Opere*, 430–31), but Bruno uses the phrase and discusses the issue many times. In a discussion of the world-ness of the innumerable stars, for example, he calls them "animali con maggior e piú eccelente raggione" (*Opere*, 452).

going so far as to invent a propositional fiction (as Kepler will), he implies one, or fictionalizes the *tone* of the enlarged universe.

One of the most interesting questions Galileo's book raises, in the literary context of this study at least, is the (appropriately sociable) question of audience. As Galileo himself was perfectly well aware, the compelling thing to someone like Cosimo de Medici about his protagonist, the telescope, was its military potential. Although he had tutored Cosimo as a youth in mathematics, the great Medici was now occupied with power politics, and Galileo was always game to supply such interests with technical support. His fellow scientists were not the intended recipients of the text's courtly and erotic language, and Cosimo was not the serious recipient of its astronomical reportage. One could split the text into layers or strands and say that the language is for Cosimo, the data for astronomers and philosophers. Or one could try to imagine what the scientific narrative might *mean* to someone like Cosimo de Medici, and how its strategies might resonate with the perspective of an Italian duke. When Cosimo de Medici read of a vastly more crowded (and probably heliocentric) universe, which could be contacted by sight with an instrument of great military potential, but not conquered or traded with or exploited for raw materials, and when he read of it as the erotically delightful adventure of his eager potential servant, his former teacher, what was that reading like? Was it like reading a novel? Was it a "mirror for princes," a lesson in the limits of human control and aggression?

And what did the romance, the breathless excitement of close pursuit and starry nights, mean to the natural philosophers who shared Galileo's interest in the heavenly bodies and our means of investigating them? Best of all objects for speculation, who were the readers outside these two obvious categories? What could one get out of such a book who had neither professional nor political interest in the instrument whose powers it displayed? This third category of readers would also to some extent subsume the first two: we have here a book without the prefabricated audience made possible under late capitalism by the institutionalization of the sciences and the demographic research of publishing conglomerates. There were as yet no scientific journals in which as a matter of course one would announce important discoveries. There was only beginning to be news at all, a "science" characterized and also mobilized by temporal unfolding, "progress," a plot.[34]

---

34. This is surely too large a generalization—not all the sciences were similarly constructed (though certainly the emergence of print had an enormous effect on the speed of developments). See, e.g., Chiara Crisciani's argument in "History, Novelty and Progress in Scholastic Medicine."

So books like Galileo's appeared in a huge, as yet unrationalized space of literary possibility. This *Starry Messenger* was smaller and cheaper than Thevet's recent cosmographical books, less practically consequential than Columbus's similarly slim (and illustrated) *Epistola de insulis nuper repertis* of a century before.[35] Books were more common in 1610 than they had been in 1493; casual reading was emerging.[36] Picaresque novels had been appearing and getting translated for fifty years; one of the genre's classics, *La vida del Buscòn*, was written only a year or so before the *Messenger* (though not published until 1626). I am trying to imagine here a kind of leisured reader (though not as leisured, perhaps, as Thevet's) who finds a detached pleasure in contemplation of the physical universe and in "keeping up," as it is called now. A reader for whom information is a commodity of both social prestige and aesthetic potential, and whose information, therefore, must arrive in structures recognizable at least from other discourses or disciplines, preferably prestigious ones, with its pleasure-giving functions acknowledged and indulged.

LUNAR ASTRONOMY

What are for us among the main features of the entire universe—the 12 celestial signs, solstices, equinoxes, tropical years, sidereal years, equator, colures, tropics, arctic circles, and celestial poles—are all restricted to the very tiny terrestrial globe, and exist only in the imagination of earth-dwellers. Hence, if we transfer the imagination to another globe, we must conceive of everything as changed.

[Note 146:] "Everybody screams that the motion of the heavenly bodies around the earth and the motionlessness of the earth are manifest.... *To the eyes*

---

35. One edition of *Sidereus nuncius* of about 550 copies (a normal Renaissance print run) was printed at the time of Galileo's discoveries, and there was another printing of the same edition in October of that year in Frankfurt; several copies of the first printing were sent as gifts to courts where they would circulate among many readers, often with a telescope attached. For details of the publication history, see the introduction and list of editions to the thoroughly annotated edition and translation of Isabelle Pantin, *Le Messager céleste*.

36. For popular and casual reading in the period, see Roger Chartier's articles (with bibliography) on the *Bibliothèque bleue* and the "literature of roguery" published therein, collected as chapters 7 and 8 of *Cultural Uses of Print in Early Modern France* and for England, Margaret Spufford, *Small Books and Pleasant Histories*. We should bear in mind that even at the end of the eighteenth century, the inventory of Étienne Garnier's warehouse in Troyes contained only 8.8 percent "novels and comic literature," according to the analysis of Henri-Jean Martin, in "Culture écrite et culture orale, culture savant et culture populaire dans la France de l'Ancien Régime." Since the *Bibliothèque bleue* printed only cheap pamphlets and chapbooks, this suggests how very minor a literature prose fiction was, even if one adds the 8 percent chivalric romances (a percentage which might be high because of the relation of Troyes to the history of that genre).

> *of the lunarians, I reply*, it is manifest that our earth, their Volva, rotates but their moon is motionless." [My emphasis.]
>
> —Kepler, *Somnium* (1634)

Kepler's *Somnium, sive astronomia lunae* takes advantage of the "huge space of literary possibility" mentioned above.[37] In the introduction to his English translation Edward Rosen assumes, with reason, that this narrativized account of the moon and its celestial perspectives was intended for an audience of astronomers ("mathematicians"). But there have been many interlopers in that cozy circle, starting with the disgruntled Leonbergers who seized on a manuscript of an earlier version to substantiate *maleficium* charges against Kepler's mother (the narrator's fictional mother is something of a magician).[38] Because of its narrative frame and visualized land- and skyscapes, it has been claimed as the "first" work of science fiction (but so have other seventeenth-century narrative works),[39] and it was certainly an influence on the writers of moon voyages we will consider in the next chapter, Bishop Godwin and Cyrano de Bergerac. Though it mainly conveys hard data, explained mathematically, it has had more appeal to fiction writers and historians of science than influence on scientific discourse or thinking.[40]

Kepler "shuddered" at Bruno's cosmological ideas (Caspar, *Kepler*, 385), though he was a passionate Copernican, and wrote more than one repudiation of Rosicrucian and Hermetic occultism. He was much fonder of Galileo—publishing an enthusiastic *Conversation with the Starry Messenger* al-

---

37. I will be quoting from Edward Rosen's annotated translation, *Kepler's "Somnium"*; where I have provided original Latin it is from Frisch, *Opera omnia*, vol. 8.

38. The story of Kepler's mother and the *maleficium* charges is best pieced together in the long introduction, by John Lear, to Patricia Frueh Kirkwood's translation, published in Lear, *Kepler's "Dream."* Like so many stories of *maleficium*, it seems to come down to neighborhood bickerings, made serious by the availability of a serious legal charge. The mother of Kepler's autobiographical narrator in the *Somnium* is a wise woman who collects and sells medical herbs and also has regular contact with spirits.

39. See Donald Menzel, "Kepler's Place in Science Fiction," on Kepler's *Somnium* as "first"; Roy Swanson, "The True, the False, and the Truly False," in support of Lucian; Thomas Copeland, "Francis Godwin's *Man in the Moon*," for Godwin; and Julien Hervier, "Cyrano de Bergerac," for Cyrano de Bergerac. Michèle Longino has even plausibly advanced Corneille's *Medée* in her "Staging of Exoticism."

40. Historian Alan Gabbey, for instance, discusses the *Somnium* (and its influence on Galileo's debaters in the *Dialogue concerning the Two Chief World Systems* [1632]) in the context of a corrective history of Newton's lunar libration theory ("Innovation and Continuity," esp. 119 and 121–22); Menzel considers the work in the context of science fiction ("Kepler's Place in Science Fiction"). It has an influential place as well in the didactic genre of such textbook or popularizing "journeys" as Gabriel Daniel's *Voiage du Monde de Descartes* (1690) or John Moorshead's *Scientific Dream* (1845).

most as soon as the *Messenger* itself appeared (though pointing out to a friend that Galileo had failed to give credit either to Bruno or himself where it was due). He had something in common with Bruno, however, easier for us to see than for him, in his deeply metaphysical and aesthetic understanding of cosmological structure.[41] I would suggest he also shared with Bruno a sense of the link between cosmology and intimations of social reform; in any case, one can see in the *Somnium* an extension — or occupation — of *De l'infinito*'s space of difference and tolerance. The chief values organizing this peculiar and posthumous text are alternativity and inclusivity. The moon is held up here, at last, as an Other World; not, or not only, a mirror, but a true Other — locked with us in the impalpable embrace of light imagined by the *Starry Messenger*.

The context for Bruno's execution, Galileo's trial, and Kepler's difficulties with his *Somnium* was, in part, the new salience of alternative worlds *on earth*. The theological challenge the New World shared with Bruno's many worlds could not be avoided: that world had, then, to be contained, or perhaps more accurately to be merged, through colonial, missionary and mercantile appropriation, with the one world Christ saved and the Pope controlled. Kepler reproduces the new cosmographical context everywhere in both the narrative of the *Somnium* and its voluminous notes. The notes are full of specific allusion to voyage literature, data from which is properly cited as if it belonged to the same technical literature to which Kepler's text belongs. A continuous textual plane is thus constructed on which Peru and Levania (as he calls his lunar world) are "islands," both islands, for that matter, shielded from excessive sun by low cloud cover. Lunar cold is colder than "Quivera" (Acosta's Kansas, linguistically distorted and geographically misplaced).[42] The "island" was, as we have seen, the landform that functioned as a kind of master trope of New World topography, and that characterized the focus of classic voyage literature, especially where it spoke most directly to private desire: Columbus finds islands, as do André Thevet and Thomas More. *The Purple Island*, *The Isle of Pines*, the islands of Marguerite de Navarre and Shakespeare and Defoe and Swift: we can follow in these dots a trajectory

---

41. See chapters on Kepler in Hallyn, *Poetic Structure*; also see "Kepler, His *Dream*, and the Analysis and Pattern of Thought," chap. 4 of Timothy Reiss, *Discourse of Modernism* (though Reiss tries as much as possible to separate what he calls "the expression of self" from "the presentation of scientific material" [149]).

42. A significantly *exotic* comparison. The mountains of Kepler's native Bohemia might have served as well, but the Moon was closer in emotional geography to Kansas/Quivera. For Quiveran cold see José d'Acosta, *Historia Natural y Moral* (1590), bk. 2, chap. 10; for the confusion and distortion see Rosen in his translation of the *Somnium*, app. J.

that points to richly imaginary fulfillments. In New World writing this meeting or mutual generation of fulfillment and desire can be diverted to the socially and politically useful. "Adventure capitalism" after all requires reachable countries of the heart's desire.[43]

In fact Kepler did think space travel was plausible and that the moon was reachable, and he assumed that Germans would get there first. (Bishop Wilkins was rather of the opinion that the Moon would lose her political virginity to a British flag.) But it wasn't good business, an island "50,000 German miles up in the ether." And this fact meant no containment or practical diversion of the desires provoked by our readerly contact with the planet of madness. Where Columbus or Thevet could conclude or even foreclose a description of New World topography by remarking that an island or bay would be "easily fortified" to ensure control of nearby resources, Kepler punctuates his narrative with variations on the less profitable refrain of changing places in the mind: "For Levania seems to its inhabitants to remain just as motionless among the moving stars as does our earth to us humans" (17). The business of narrative is perception, not acquisition: a mobilized, alterable, positional kind of perception rather than the steely focus of the "imperial eye."[44]

Moon voyages present alternative worlds that offer most saliently the radical fact of alternativity itself. And they do this vis-à-vis an "island" which, unlike America or even Iceland (the frame-setting of Kepler's *Somnium*), is perfectly visible almost every night, to everyone. Kepler's voyage, ostensibly the least satirical or political of the lot, got him in the most trouble of any of the moon writers, and more than once. These troubles generated the enormous expansion represented by the Notes (amounting to roughly six times the length of the original narrative), in which he augments the scientific data already foregrounded in the main body of the narrative and defends various ludic moments against their ludicrous misreading in the events of his mother's imprisonment and trial. The less fictionally inflected Notes function as an exegesis of the narrative Kepler repeatedly terms an "allegory," suggesting that scientific discourse might have been seen as a kind of mediation, even arbitration, between the suggestive new empirical data and the mainstream cosmology of the educated—or not so educated—public. (Kepler's narrative encourages this notion, with its weird opening reminder of the civil war brewing in Bohemia between the Holy Roman Emperor, the virtuoso

---

43. See Nerlich, *Ideology of Adventure*.
44. The structure and historical context of that focused gaze are scrupulously analyzed in Pratt's *Imperial Eyes*.

Rudolph, and his ambitious and angry brother.)[45] We have forgotten this kind of writing (with the possible exception of Edwin Abbott Abbott's nineteenth-century sport *Flatland*), perhaps because the representational war is over, won hands down by the Puritan supporters of "unornamented" denotation, the demotic of institutionalized science.

The narrative's strange first paragraph opens up two more pairs of alternatives. Who would expect a book with either of the titles mentioned above (*Somnium*, or *Lunar Astronomy*) to begin like this: "In the year 1608 there was a heated quarrel between the Emperor Rudolph and his brother, the Archduke Matthias"? The strangeness of this starting point is emphasized by the absence of any segue between the topic of Bohemian civil strife and the dream vision's pivotal formula, "I went to bed and fell into a very deep sleep." We have been offered the picture of a *mutually exclusive* alternativity between two nearly identical persons, in the royal brothers' struggle for a single throne; Kepler (the frame narrator) then starts reading about the strikingly parallel troubles of the legendary Bohemian Queen Libussa and falls asleep over his book. His dream seems to have been provoked by thoughts of the mutual exclusivity of contenders for power in the political realm. The dream offers the Moon as an alternative to Bohemia—the nighttime world of dream, moon, the perennial, and, as we will see, the uncontentious, replaces the daytime political world of strife and change; the book Kepler dreams he is reading about the young astronomer Durocotus and the Moon transforms the book about Libussa and Bohemia that he had "really" been reading in the frame narrative.

The one protagonist does not simply replace the other, however. Libussa—the sorceress, beloved ruler and dynastic mother of Bohemia—seems less an alter ego of our author-narrator, or of his dream-self, Durocotus, than of Durocotus's mother: both figures are sorceresses, mothers, authorities. Why does this dream-mother die as soon as Durocotus (who narrates the book Kepler reads in his dream) introduces her?—"Her recent death freed me to write" (11). This is not a surprising statement at the level of psychological realism, nor in the context of Kepler's exegetical equation of

---

45. Rudolph II, the increasingly unstable king of Bohemia and Holy Roman Emperor at the time of the first composition of the *Somnium*, was forced to abdicate in favor of his brother, Matthias, archduke of Austria, in 1611. Rudolph's Prague (then capital of the empire) had been a haven for intellectuals and virtuosi, and Rudolph himself was a notable collector. Neither brother was able to fend off the impending conflict of the so-called Thirty Years' War, which began in Prague in 1618, though both had negotiated with the Protestants and granted them religious freedom in their lands. For more on the diplomacy of science, see Moss, "Interplay of Science and Rhetoric."

Durocotus's mother with "Ignorance."[46] But Kepler's equation doesn't actually hold up very well: the mother, supposedly Ignorance, is described right away as tactically wise, learned in herb lore, scornful of "vicious people who malign what dull minds fail to understand" (12), and possessed of a key piece of data withheld from Durocotus: his own father's name. In fact, she knows quite a bit. It is an alternative knowledge (soon to pass away as both content and a relation to knowing, though Kepler may not have anticipated that).[47] In fact, almost the entire book, subtitled "The Lunar Astronomy," is presented as a transcript of oral, daemonic lore to which Durocotus's mother provides him (and therefore us) the access. When the son returns from his long apprenticeship with Tycho Brahe she tells him that now she can die, "since she was leaving behind a son who would inherit her knowledge [!?], the only thing she possessed" (13). A serious conflict is registered here where this figure, who represents a kind of knowledge still highly valuable *to a scientist* (Kepler, Durocotus, his astronomer-readers), must die before the scientist is free to write. The mother's knowledge is the writing's content at the same time as the mother is its censor.

Kepler's appetite for the hybrid, his ability to tolerate ambivalence, are high, higher than a fully rationalized science can express. The mother, Fiolxhilde, is depicted in relation to a knowledge and a form of representation disconnected from those of the New Science—this had fatal consequences for Kepler's real mother, oddly supporting the concept Fiolxhilde embodies, of language as magically productive rather than passively descriptive. Although her son Durocotus has traveled in the body and she has not,[48] Fiol-

---

46. See Kepler's "Note 4": "untutored experience or, to use medical terminology, empirical practice [*empirica exercitatione*] is the mother who gives birth to Science [*Scientam*] as her offspring. For him it is not safe, so long as his mother, Ignorance [*Ignorantiam*], survives among men, to reveal to the public the deeply hidden causes of things" (36). On the relationship in Kepler's time, and in Kepler's case, between science and such a mother-image, see Eileen Reeves' forthcoming article "Old Wives Tales and the New World System."

47. Perhaps more precisely, soon to pass away from the light of legitimacy. See chap. 2, especially, of Keith Thomas's *Man and the Natural World*: "Sir Joseph Banks, the future President of the Royal Society, as a schoolboy paid herb-women to teach him the names of flowers. Physicians and apothecaries had long depended for their supplies upon such persons, what William Turner called 'the old wives that gather herbs'" (73). The chapter goes on to detail the imposition of Latin names and taxonomies on the plants identified by herb-women, fowlers and former soldiers, so that eventually "farmers who still used 'vulgar, provincial names' . . . found themselves unable to communicate with the naturalists" (87).

48. When Durocotus returns to tell his mother about his earthly travels and adventures, she gives voice to one potential consequence of reading/hearing about alternative worlds: "We [here in Iceland] are burdened with cold and darkness and other discomforts, which I feel only now, after I have learned from you about the salubriousness of other lands" (14).

xhilde has her own kind of travel to offer, her own, or at least her "teacher's" kind of representation, presumably alternatives to the travel and scientific education her son has just been narrating. This teacher is a wise and gentle spirit, who "is evoked by twenty-one characters [the letters in the title of Kepler's *Astronomia Copernicana*]. By his help I am not infrequently whisked in an instant to other shores . . . or if I am frightened away from some of them on account of their distance, by inquiring about them I gain as much as if I were there in person. . . . I should like you to become my companion on a visit" (14).

Fiolxhilde's teacher manages a powerful kind of representation: the "visit" to the moon will take the form of listening to him/her (the pronouns are inconsistent), in the ritual setting of a crossroads on the night of the new moon. This preternatural vivaciousness of mimesis brings to mind Michael Taussig's study, *Mimesis and Alterity*, inspired by the mimetic practice of the Cuna Indians of Central America, whose carved wooden curing dolls began to represent Westerners during the period of the Cuna's early contact with Western information technology. *Mimesis and Alterity* would easily absorb such a transportational notion of narrative: Taussig's suggestive book is obsessed with the way the magical practices of what he calls "copy" (and I call representation) and "contact" (manipulation of actual substance of the object to be affected) slide into a single identity (the fingerprint is a good instance; so of course is the voodoo doll containing hair or nail clippings). Although he is speaking mostly about the visual mimesis central to this nonalphabetic culture, he makes it clear that the Cuna do not have exclusive rights to this kind of thinking. They represent only one, strongly unanimous, example of the intensification of mimetic power felt in the encounter with new forms and objects of representation.[49] The spiritual or "pneumatic" magic of Ficino, Pico, and Bruno might be another. Taussig, invoking Walter Benjamin's writing on the medium of film, spreads the transformational net of mimesis more widely when he speaks of "the unstoppable merging of the object of perception with the body of the perceiver and not just with the mind's eye" (25). Commentators have often complained of Kepler's sloppiness in labeling the narrator of the astronomical part of the *Somnium* the "Daemon from ["ex"] Levania," since the daemon makes it plain that his native planet is the earth and that his discourse on the moon is that of a traveler. But perhaps in this case, not only

---

49. Bacon seems to be thinking along Taussig's lines in book 1 of the *Advancement of Learning*: "But the images of men's wits and knowledge remain in books, exempted from the wrong of time and capable of perpetual renovation. Neither are fitly to be called images, because they generate still, and cast their seed into the minds of others, provoking and causing infinite actions and opinions in succeeding ages" (Spedding 3:318).

does the "copy" (the Daemon's lunar description) partake of the substance of the thing copied, but the copier too (the Daemon) participates in the radioactive spread of mimesis: this astronomy is not only *about* the moon but, as the subtitle says, *lunar*, even a little lunatic. There is a strangely permeable membrane between the scientist and his object, here acquiring many of the characteristics of a subject.[50]

About half the text of the *Somnium* is constituted by lavishly detailed accounts of the geography and climatology of "Levania," along with the disposition of the heavenly bodies as perceived from several places on the moon's surface. Technical as these portions may be, their presentation is grounded in the magnetizing principles of desire and lack. The features of the moon and its skies are always represented as things *seen* by "Subvolvans" or "Privolvans" (the respective inhabitants of the two lunar hemispheres) — which lends an eerie emptiness to the landscape so described, as we are not introduced to these inhabitants until the last couple of pages.[51] This is of course an increasingly familiar arrangement of information in earthly voyage literature written in the context of colonial acquisition. In Kepler's narrative, the inhabitants' absence and the landscape's human emptiness irradiate the lengthy discussions of what is visible on their horizons; we are identifying vividly with lunarians, or at least seeing through their eyes, well before we know what they look like. And we are identifying, for the purposes of "lunar astronomy," with both Subvolvans *and* Privolvans — groups which the frame of the *Somnium* suggests would be at perpetual odds on earth (like the emperor and his brother, or Libussa and her barons).

The two hemispheres of earth are discussed as well, in a moment of breathtaking aestheticization. Earth itself is alternative in this book — as "Volva,"

---

50. Taussig says, for instance, of the magical "copy" of Western ethnographer Stephanie Kane (not a user of the "ethnographic present," but a narrator of singular events): "Kane's mode relies not on abstract general locutions such as 'among the Emberá it is believed that . . . ,' but instead concentrates on image-ful particularity in such a way that . . . she creates like magical reproduction itself, a sensuous sense of the real, mimetically at one with what it attempts to represent. . . . Can't we say that *to give an example, to instantiate, to be concrete*, are all examples of the magic of mimesis wherein the replication, the copy, acquires the power of the represented?" (*Mimesis and Alterity*, 16).

51. In this, Kepler's book reads like Hariot's *Virginia*, which also saves its concise remarks on the inhabitants and their customs until the very end, perhaps under the influence of the hexameral tradition of "chronological" commentaries on the seven days of creation in Genesis, where the human comes last. Pratt ("Scratches on the Face of the Country") has offered a more sinister explanation of this tendency in colonialist voyage-writing; it is often, after all, the land and the climate that European governments and settlers want, and a depopulated description permits, at some level, depopulation. Or at least it permits the extreme marginalization of any indigenous competition in the homebound reader's imagined landscape. Here Kepler parts ways with his fellow voyage-writers.

the giant moon that stands in for the Moon on the Moon where the inhabitants are, as the Daemon puts it, "completely deprived" of the Moon. The political geography (or "selenography") of the Moon seems based on Volva as an object of erotic fascination which "they enjoy ["fruuntur"] to make up for our moon, of which they . . . are completely deprived. From the perennial presence of this Volva this region is termed the Subvolvan, just as from the absence of Volva the other region is called the Privolvan, because they are de*priv*ed ["privati"] of the sight of Volva" (21). Since Volva rotates daily on its axis, the Subvolvans tell time by the position of its "spots" (our continents): the Daemon chooses to describe the visible disk of the planet at a moment when familiar parts of both (artificial) hemispheres are visible at once. The Old World ("the eastern side") "looks like the front of a human head cut off at the shoulders and leaning forward to kiss a young girl in a long dress, who stretches her hand back to attract a leaping cat." South America "you might call . . . the outline of a bell hanging from a rope and swinging westward" (24).

These images constitute the first detailed factual description of the planet earth from an external vantage point: of earth as a specular object. This is surely an occasion for vertigo, when to quote Fredric Jameson (out of context) on the genre of romance, "the *worldness* of *world* reveals or manifests itself" (*Political Unconscious*, 112). Whatever the allegorical narrative of these cameos, most salient and most significant is the fact of a world, "the transcendental horizon of our experience" (Jameson, 112), shrunk to a shiny medallion, an amusingly suggestive ornament hanging in the cabinet of some lunar Rudolph (Kepler's employer was famous for his *Wunderkammern*).[52] Of all the imaginative place-changing in this text, the shift of our human world, staggering with the weight of its violent histories, now ineluctably visible and ornamental in the sky of another planet, seems the most revolutionary in its conception and presentation. Another such image, this time photographic, fed revolutionary thought and feeling in 1970: the translation of the sublime into the beautiful (in the aesthetic terms of this chapter) can be experienced as a great disburdening of the tragic sense of life, with its links to monumental and imperial history—although ironically that image became a sign of America's imperial grandeur as well as of the lightness and loveliness of this (un)scepter'd isle, this Earth.

"In general the Subvolvan hemisphere is comparable to our cantons, towns and gardens, the Privolvan to our open country, forests and deserts" (28).

---

52. See Kenseth, "'A World of Wonders.'"

How does this account of the nonexistent lunarian ecosystem fit in with an otherwise textbook-accurate account of things visible and knowable? The question is interesting because in every other feature of Kepler's selenography the thrill is that we're seeing up close or from the Other perspective something (the Moon) plainly visible to us from childhood on. One answer is that the genre of the "voyage" demands it—and as a parodist Kepler is more urgently required to obey than if he were writing an account of actual exploration. The lifeworld of the potential colony, however, was classically rendered in itemizable form, making the unusable aspects of it conceptually detachable. Kepler's brief account of lunar flora and fauna, augmented in the "Selenographical Appendix" where he describes the Levanian method of building what we now call craters, makes size, shape, longevity, diet, habitat, and climate mutually expressive and interdependent. This reads like unremarkable common sense to modern students, but produced in the time of the Jesuit polymath Athanasius Kircher's attempt to recalculate the size and interior design of Noah's Ark so as to accommodate the fauna of the New World, it represents a less familiar and less easily commodifiable view of organic life.[53]

Kepler's lunar ecology blurs two other lines of identity: he does not specify human status for one or another lunar species, yet building activities alluded to in the narrative and dwelt on in the "Appendix" make it clear that some "moon-dwellers" are rational creatures (and thus redeemable, though he makes no mention of that). "The Privolvans have no fixed abode. . . . In the course of one of their days they roam in crowds over their whole sphere, each according to his own nature: some use their legs, which far surpass those of our camels; some resort to wings; and some follow the receding water in boats" (27). And despite the possession of reason, the arts, and enough political organization to build and live in fortified "towns," those "moon-dwellers" capable of conquest and colonization show no interest in it. Every month all the water on the planet is drawn to the Subvolvan hemisphere for two weeks by the combined attraction of the sun and Volva, but the hemispheres have not organized the conflict this would naturally provoke on earth. The fortification of the towns described in the "Appendix" protects them mainly "from the mossy wetness" of this monthly deluge and "from the heat of the sun" (151). It is not just that the moon-dwellers don't think of military conflict or colonial aggression—Kepler doesn't mention these things himself. They are absent from his dream, as Volva is absent from our skies.

---

53. For more on the amazing Kircher see Joscelyn Godwin, *Athanasius Kircher*; Don Cameron Allen, *The Legend of Noah*; and Findlen, *Possessing Nature*.

That Kepler's response to the "worldness" of the moon inspired him to write a book so fully fictional and scientific at once — that we are invited, as Taussig would say (thinking of the camera's "enchanted glass"), to "see down into" the data of the new astronomy — is not the least of the inclusivities sponsored by his sense of the moon. The abrupt ending of the narrative is part of the project of making science signify; as the Levanian Daemon is telling Duracotus about Subvolva's "constant cloud cover and rain," "a wind arose with the rattle of rain, disturbing [Kepler's] sleep and at the same time wiping out the end of the book" (28). The sad way that deeply meaningful dream rain turns out to be nothing but arbitrary actual rain is a familiar letdown to anyone who has ever had a dream — the inevitable lack of closure in dream narrative always refers us to the incompatibility between a world of meaning and a world of brute physics. The end of Kepler's *Somnium* replays that letdown, which "wipes out . . . the end of the book," a stark reminder of what the loss of a signifying science might feel like. The underground possibility this book embodies involves an idea of science as a self-transformative practice of seeing, of *looking*. Kepler balances on a sword-bridge between the conditions of meaning within which Bruno read Copernicus's diagram of the solar system as a hermetic hieroglyph and the approaching quantification and purification of astronomy. It is worth thinking about his *Dream*, also, as a moment in the intertwined histories (here *fused*) of fiction and anthropology. Where Galileo could read only the eros and politics of Renaissance Italy in his sky, Kepler's lunar anthropology imagines something neither closely parallel nor directly opposed to what he knew.

FICTION:
THE MINUET OF FONTENELLE

"Quickly, help me to a definite opinion on the inhabitants of the Moon. Let's preserve them or annihilate them forever and not discuss it anymore — but let's preserve them if possible. I've taken a liking to them. . . ."

"But," the Marquise interrupted, "always by saying 'Why not?' are you going to put people on all the planets for me?"

"Don't doubt it," I replied. "This 'Why not?' has a power which allows it to populate everything."

—Fontenelle, *Entretiens* (1686)

When Fontenelle composed the first edition of *Entretiens sur la pluralité des mondes* (1686), his important work of cosmological popularization, he was mostly a poet and playwright. He was the playwright Corneille's nephew and occasional collaborator, and his first solo published work was a comedy called *La Comète*, of which Nina Rattner Gelbart tells us "newspapers advertised that now going to the theatre could cure the fear of comets" (Fontenelle, *Conversations*, xiii). Fontenelle's sources for the first edition of *Entretiens* were mostly respectable but not primary; after his election five years later (following the book's brilliant success) as the lifetime secretary of the Académie des Sciences, he came into regular contact with astronomers and their research and upgraded the factual content of the many later editions. Thus his modern editor Robert Shackleton can say: "The study of Fontenelle's sources in the *Entretiens* is . . . the study of the evolution of the man"—from "pastoral poet and playwright," that is, into "a man of science" (*Entretiens*, 19–20).[54] The *Entretiens* is a pivot between Fontenelle's identities, the kind of work a writer of fictions was much more capable of writing in 1686 than in 1986.

If science fiction is "the extrapolative art," postmodern, postcolonial readers of Fontenelle are inevitably science fiction writers as well: it is hard not to extrapolate the subsequent interlinked narratives of colonialism, capitalism, anthropology, and fiction from this tremendously dense, witty, and also smug proto-Enlightenment dialogue—the book that put the plurality of worlds on every European coffee table (including some Russian ones).[55] One could almost pull the rest of the present book out of this one little hat. But this chapter has been chasing down a narrative of slightly smaller scope—the story of what Bruno's new space was *for*, what was good to think with it. Fontenelle's book, as the chief disseminator of the ideas espoused in the other works we've glimpsed here, will help us summarize and prophesy. And it has its own historical moment; it registers a different kind of possibility from its predecessor among fictions, Kepler's scientific *Dream*.

Fontenelle's work echoes Bruno throughout, in its image of the cosmological theater (an image less dismissible, and more useful, to the author of plays and operas): "From the Earth, where we are, what we see at the greatest distance is the blue heaven, that great vault, where the stars are fastened like nailheads" (*Conversations*, 12–13); "I have always thought that nature is very

---

54. It is worth noting that Kepler, too, revised his *Somnium* several times, over four decades of development in astronomical thought. The fictional format was probably, for both writers, a usefully flexible medium for works whose freights of fact were unstable. Not only did early stages of the great revisionist cosmology *seem* more obviously fictional than after the Copernican model became a "black box" (in Bruno Latour's sense), but the fiction allowed the writer to diffuse the source of propositions among several characters and to evade the full requirement of truth value.

55. The still-unchallenged classic work on science fiction is Darko Suvin's *Metamorphoses of Science Fiction: The Poetics and History of a Literary Genre*.

much like an opera house" (11).[56] It sweeps Galileo's courtly matrix of metaphor into its own, and draws out explicitly his projections on the Medicean stars and Jupiter: "I'd wish," says the Marquise, "that the inhabitants of the four moons of Jupiter were like its colonies, that they'd receive from it . . . their laws and customs, and that consequently they'd give it some sort of homage" (Van Helden, 56). Kepler is not quoted (or plagiarized), but this work shares his sense of the mutuality of the Earth and Moon (also imagined by Galileo in the stunning image quoted above) and the "worldness" of the Earth, as well as the unrepresentable difference of the sentient Other—when the question is put, the Marquise is unable to picture the starry people so brilliantly hinted at and elusively sketched in the *Somnium*: "'I couldn't describe them to you, but nevertheless I see something.' 'Let me suggest,' I answered, that tonight you give your dreams the task of devising those shapes" (47). But "her dreams weren't at all successful; they kept providing something that resembled what one sees here on earth" (48).

The trajectory we've observed moves fantasy on the astronomical scale in the direction of the sociable. Bruno, at home with sublimity, was happy to postulate the other sentient beings and even to claim that they were probably more excellent and intelligent than we are (a "we" in all these works that for the first time in the intellectual history of Europe includes everyone on Earth as a single ethnos). But he didn't think they should meet each other or us: in the sublime, an Other swallows or destroys. Galileo populates this rather remote heaven metaphorically with celestial bodies themselves pursuing Venetian romances and conquests. Kepler postulates a moon *culture*, in fact two of them—an infralunar sociability which (compensating for an imminent civil war) fails to bear out Bruno's gloomy prognostications on the contact of widely diverse beings. But Fontenelle was writing in the France of the Académie des Sciences and the *Journal des sçavants*, and of an almost institutional salon culture, writing the *Entretiens* in the same year as his notable anthropological work *L'histoire des oracles* and the very brief "Relaçion curieuse de l'île de Bornéo." The history of European colonial development was fifty years farther along than it had been at Kepler's death, one hundred years older and more organized than when Bruno was hobnobbing in London with the "School of Night," though he knew figures deeply implicated in the emergence of both France and England as colonial empires. In the case of his most celebrated progeny, then, Bruno's *infinito* has led us to the "world-

---

56. I quote from the translation of Hargreaves, *Conversations on the Plurality of Worlds*, checked against the French of Shackleton's edition.

ness" of the earth and the imagination of a projected celestial diversity—in other words, to consideration of culture and ethnos.

That it is an *ancien régime* consideration is abundantly evident from the very start: not only is the dialogue held flirtatiously over several moonlit nights at a château, between a charming courtier and a beautiful, lively, young woman, but the central motive, both of discussion and belief, is pleasure. "'I'm ashamed to admit it, but I have a peculiar notion that every star could well be a world. I wouldn't swear that it's true, but I think so because it pleases me to think so. The idea sticks in my mind in a most delightful way. As I see it, this pleasure is an integral part of truth itself'" (10). How relaxed is this neoclassical France; how little Bruno or Galileo (or Kepler's mother) could have made of this pleasure during their many moonless nights of imprisonment in Italy and Germany. How canny and suave that little throwaway, "I wouldn't swear to it." One hears already the charm and arrogant privilege of the imperial subject.

Fontenelle cites Wilkins (anonymously) on the subject of Aristotle's secret knowledge of the inhabited Moon: "'he never wanted to speak of it for fear of displeasing Alexander, who would have been in despair to see a world which he was unable to conquer'" (64). This despair might have threatened the aspirations of France in 1686 as well (and as we have seen, Kepler and Wilkins both thought their nations *could* conquer the moon). The *Entretiens* speaks always in the totalizing mode; it is the subject of an empire who pictures so clearly the ethnic pageant of the turning world: "'I sometimes imagine that I'm suspended in the air, motionless, while the Earth turns under me for twenty-four hours, and that I see passing under my gaze all the different faces: white, black, tawny, and olive complexions. At first there are hats, then turbans; woolly heads, then shaved heads . . . in all, the infinite variety that exists on the surface of the earth'" (20). After a couple of pages of flirtatious ethnological fantasizing, the Marquise calls a sudden halt: "'a serious difficulty has occurred to me. If the Earth turns, we change air every minute, and are always breathing the air of another country'" (21). The narrator reassures her that the Earth is surrounded in a cocoon of air, like a silkworm's, that rotates with it. All that "infinite variety" can stay put, enticingly visible from above but not contaminating our lungs or blood. This is something like an inversion of Mary Louise Pratt's vision of a world incessantly visited, even surveyed, by imperial Europeans, but where there are no "woolly heads" visible in the representation, no heads but the heads of the Europeans who "like to watch"—an Earth all territory, unpopulated. As though Fontenelle, or colonialist Europe, had taken the Marquise up on the second option quoted in the epigraph, above, to "annihilate them forever and not discuss it anymore" (38).

Countering—mirroring?—this ethnic pageant of the earth is a celestial one. It is first outlined by Fontenelle's alter ego as manifesting a logical, Earth-centric spectrum of sheer difference: "'Here, for example, we use the voice; there one only talks by signs; farther away one never talks at all'" (46). In the dialogue of the fourth evening, fantasies about the nature of Venusians and Mercurians are dominated by the planets' positions and presumed climates: "'Venus is closer to the sun than we are, and receives a stronger, hotter light from it.' 'I'm beginning to see,' the Marquise interrupted, 'how these Venusians are made. They resemble our Moors of Grenada, a small, black people, sunburnt, full of verve and fire, always amorous, writing verses, loving music'" (49). The level of flirtation and fantasy is at its highest in this dialogue, as "Fontenelle" had anticipated the night before: "it would be no common pleasure to see many different worlds . . . it would be far better than to go from here to Japan, crawling with great difficulty from one point on the earth to another to see mere men'" (43). As mirror of earthly ethnography (about which the author of the "Relaçion curieuse de l'île de Bornéo" might be expected to have thought),[57] the distant, unreachable heavens bring out clearly the potential long afterwards mined for pure pleasure by French writers of fiction for children, the Babar books and *Le petit prince*.

The power of fiction to "populate everything" is obviously very close here to the power of that ethnographic or cosmographic writing that brought the Marquise her reference point, the Grenadian Moors, in the first place. In that sense, the celestial fantasizing is redundant where not pointedly parodic—there is no difference for the reader between imaginary Venusians and imaginary Muslims. But it is difficult to ignore the sharp contrast or ominous sequence between "let's annihilate them" and the "power to populate everything." Populating the heavens compensates for both the growing representational depopulation of potential colonies (registered in Pratt's *Imperial Eyes*) and the distressing rumors of actual depopulation present and, clearly enough, to come. Going out on a limb, one could look at the birth of European prose fiction, especially "psychological fiction" based on character, like the *Princesse de Clèves* to which Fontenelle compares the *Entretiens*, as a repop-

---

57. The "Relaçion" is actually part of a letter, written from the East Indies. It concerns the people's attitudes towards government, especially by women, and a lineal dispute (pitting "ocular evidence" against authoritative testimony) between two queens and their supporters (Fontenelle, *Oeuvres complètes* 1:521–23).

ulation of a world being simultaneously depopulated by colonial policy, while the genres that directly attended these developments—travel accounts (*récits de voyage*) and cosmography—increasingly failed to cover or compensate. At any rate, the *hunger* for population is intense: "'A tree leaf is a little world inhabited by invisible worms'"; "'even if the Moon were only a mass of rocks I'd sooner have her gnawed by her inhabitants than not put any there at all'" (45).

The Marquise has shown the most extreme and personal signs of sociability (and occasionally animism): "'I love the Moon for staying with us when all the other planets abandoned us'" (16); "'I could imagine with pleasure these telescopes [on Jupiter] aimed at us, as ours are towards them, and the mutual curiosity with which the planets consider one another and ask among themselves 'What world is that? What people live on it?'" (57). She imagines routinely in terms of mutuality and social pleasure (rather than the collector's pleasure more often displayed by "Fontenelle"). Responding to her mentor's fantasy of Moon people fishing for humans from the sky, she says "'Why not? . . . As for me, I'd put myself into their nets of my own volition just to have the pleasure of seeing those who caught me'" (40). This character's gender is put to use in the erotic structure of the *Entretiens* in a number of unsavory ways. If the Marquise is Fontenelle's way of representing women as capable of thinking about science, as well as, more broadly, his way of manifesting the functions of pleasure and desire in learning, she is also his way of dictating the contents and status of feminine knowledge and of splitting certain mental operations and attitudes off from others, gendered (in the privileged, authorial voice of her interlocutor) as male. A gender analysis of this text would make a productive study. We have only room for one fact and a glimpse at one passage, whose implications will be taken up seriously in more extended treatments of texts *by* women.

The fact: Shackleton mentions two works like Fontenelle's that preceded his in 1680—an anti-Copernican and antifeminist *Entretiens de Philemon et de Théandre sur la philosophie des gens de cour* by the Abbé de Gerard, and an *Entretiens sur l'opinion de Copernic touchant la mobilité de la terre* by Jeanne Dumée. The brief notice of the latter that appeared in the *Journal des sçavants* tells us it was Copernican and argued in favor (as does, incessantly, Fontenelle) of Descartes's vortices.[58] This set of dialogues by a woman writer somehow slipped out of public attention and literary history. Shackleton, without irony, suggests Dumée's immediately prior authorship of a book like his in-

---

58. See Shackleton, ed., *Entretiens sur la Pluralité des Mondes*, 8–9. On Jeanne Dumée, see La Lande, *Bibliographie astronomique*.

spired Fontenelle's character of the intelligent, but ignorant, Marquise. But he has turned her from a producer of knowledge, like Dumée, into a student.

The passage (which ends the text): "'Well!' she cried. 'I have the whole universe in my head! I'm a scholar!' 'Yes,' I answered, . . . 'and you've the advantage of being able to believe nothing at all of what I've told you, when ever you choose. I only ask of you, as payment for my trouble, that you never look at the Sun, the sky, or the stars, without thinking of me'" (73). Making one's way past that last extraordinary demand, it is remarkable to consider this consumerist sense of choice about belief (as if it were a jewel, like the shiny medallion of Kepler's Earth) in the context of Bruno's muzzled death by fire.

The arithmetic sublime of Bruno's *infinito* and the global coverage, island by island, of Thevet's cosmography, have in this work been drawn into the vortex of a slender erotic fiction in which wonder, curiosity, sympathy, and speculation belong to the female speaker, and knowledge, numbers, authority and the power of demystification belong to the male. "Pleasure," even whim, have replaced the death-defying Faustian impulses of the earlier writers, and astronomy is domesticated for the consumption (widespread) of women, the incarcerated aliens of the bourgeois home. Fontenelle does not want to talk about the innumerable worlds beyond our home system: "You may put more systems there or not, it's up to you. They're properly the province of the philosophers, those great invisible countries that may be there or not as one wishes, or be whatever one wishes" (73). As we will see, many women writers wished those great invisible countries to be there.

V • A WORLD IN THE MOON
Celestial Fictions of Francis Godwin and Cyrano de Bergerac

At the leastwise it may please God that I doe returne safe home again into my Countrie to give perfect instruction how those admirable devices . . . may be imparted unto publique use. You shall then see men to flie from place to place in the ayre; you shall be able, (without moving or travailing of any creature,) to send Messages in an instant many Miles off, and receive answer againe immediately; you shall be able to declare your minde presently unto your friend, being in some private and remote place of a populous Citie, . . . but that which far surpasseth all the rest, you shall have notice of a new World, of many most rare and incredible secrets of Nature, that all the philosophers of former ages could never so much as dreame off [sic]. But I must be advised, how I be over-liberall in publishing these wonderfull mysteries.

—Francis Godwin, *The Man in the Moon* (1638)

. . . all these phenomena are important. One must make a distinction, however: when dragged into prominence by half poets, the result is
                                                                not poetry,
nor till the poets among us can be
  "literalists of
    the imagination"—above
      insolence and triviality and can present

for inspection, "imaginary gardens with real toads in them,"
                                                                 shall we have
  it.

—Marianne Moore, "Poetry" (1921)

THE SEVENTEENTH- and eighteenth-century literature of space travel came into existence in a world of readers accustomed to the appearance of the heavenly bodies in poetry (narrative as well as lyric), accustomed to accounts of travel to a "new world," accustomed to an astronomy focused on the moon and sun. By the time Francis Godwin's posthumous picaresque moon voyage was published, in 1638, that audience was also accustomed to the idea of experiment (often called "experience") as a method of testing theories. Godwin's book even had predecessors as a fiction about the moon: not only Kepler, whose also posthumous *Somnium* had come out in Latin in 1634 (a year after Godwin's death), but Plutarch and Lucian had written books which anticipated the notion of a habitable, Earth-like moon and, in Lucian's case, narrated an earthly protagonist's voyage to it. But these anticipations, although doubtless known to Godwin (except for Kepler's) and Cyrano de Bergerac, have little bearing on the meaning, in the present context, of Godwin's and Cyrano's books, or of the host of others that followed them in France, England, and even that imperial nonstarter, Germany.[1] This chapter offers readings of these early modern moon voyages in their immediate context, among the more contemporary books they copied and diverged from, amid the new genres they grafted themselves onto or didn't quite.

America changed the moon forever, and both Godwin and Cyrano make that point in many ways. Godwin's lunar emperor is descended from an earthling, and it is to a high hill in America that the lunar children are transported who look as though they might grow up to be troublemakers. Cyrano's first voyage lands him, not on the Moon as intended, but in French Canada, where he risks being suspected of sorcery by the Jesuits who will later count Joseph Lafitau, opponent of libertine thought and a father of cultural anthropology, among their number. The colonial appropriations in vogue in seventeenth-century Europe's relations with the New World are not possible with respect to the Moon (though the transportation of wicked children to America by the lunar legal system has a familiar ring to it), but the voyages take place in that stream of events and encounters. Godwin's protagonist, Domingo Gonsales, is so pleased with his first discovery—the island of St. Helena, still on earth—that he "cannot but wonder, that our King in his wisdome hath not thought fit to plant a Colony, and to fortifie in it" (14). And in the ancient and urbane China where his voyage ends, the Jesuits have begun to live and preach, precursors of a colonial takeover that, in the case of Hong Kong, remained official policy until 1997.

---

1. This large literature has been located and annotated by Marjorie Hope Nicolson, Phillip Gove, and Geoffroy Atkinson (see Chapter 4, n. 4); for a collection of early works see Faith Pizon and T. Allen Comp's *Man in the Moone*. It has most recently been discussed in Karl Guthke's *Last Frontier*. Several important but obscure English works in this tradition have been published in facsimile editions in Garland's "Foundations of the Novel" series, e.g., David Russen's *Iter lunare* (based on Thomas St. Serf's 1659 English translation of Cyrano de Bergerac's *L'autre monde* [1703]); "Captain Samuel Brunt's" "A Journey to the Moon" (the second part of his *Voyage to Cacklogallinia* [1727]); and "Murtagh McDermot's" *Trip to the Moon* (1728).

The idea that these works can be seen usefully as early avatars of modern science fiction suffers from the flaw of all such ideas about forerunners, the idea that a form preexists the historical conditions in which it emerged. But voyages to the moon do have one thing in common which has some transhistorical content: the moon. That is, the moon understood as at least potentially an alternate globe, as "counter-terrestrial" in the phrase of Plutarch (46–120 C.E.—in his *De facie lunae* he calls the side of the moon facing us the "House of counter-terrestrial Persephone" [944D]). Such an idea would obviously be invigorated as well as given density by the discovery of a counter-terrestrial "world" on earth—so much density that, by the time of Kepler's and Godwin's little books, the moon was no longer useful for such purposes as Ariosto's in *Orlando Furioso* (1516), in which Orlando is sent to the moon to indicate his love-lunacy. The poem would not have benefited from the distracting utopian and dystopian connotations that that planet was to acquire under pressure from New World literature.

Godwin's and Cyrano's lunar narratives are often and readily seen in the company of the scientific works from which they borrow their astronomical layouts and with which, in the cases of Plutarch and Kepler, they share their fictionality to some extent.[2] Both are full of passages in which the narrator/protagonist offers first-person testimony to the visual (and tactile) experience of the new astronomical models, and they are full as well of passages in which scientific ideas are expounded or debated by characters in dialogue set-pieces. Godwin's Gonsales more than once articulates a relationship of evidence to theory in which his narrative can cooperate with those of the new scientists: "I will not go so farre as Copernicus, that maketh the Sunne the Center of the Earth, and unmoveable, neither will I define anything one way or the other. Only this I say, allow the Earth his motion (which these eyes of mine can testifie to be his due) and these absurdities [of contrary motion of the spheres, etc.] are quite taken away" (60).

However, a cursory comparison to Kepler's or Plutarch's texts makes clear that we are in a differently valenced world on Godwin's and Cyrano's Moon. Their books are much more fully exploitative of, and concerned with, their fictionality—for Godwin and Cyrano the important thing, the fundamental point, is that we can't know what is up there, can't make voyages there and bring back even the unverifiable testimony of medieval travelers to the Far East. The Moon's value is as a real, visible, and *unknowable* reflecting surface.

---

2. See, e.g., David Knight, "Science Fiction of the Seventeenth Century"; Siegfried Mandel, "From the Mummelsea to the Moon"; or for Milton's cosmology in *Paradise Lost*, John Tanner, "'And Every Star.'"

The narrator's authority is the magical and total authority of a fictional *auctor*. The Moon is for such writers a very different narrative object than it was for writers publishing in America during the "Space Race" of the 1960s and 1970s. Kepler and the English bishop and scientist John Wilkins both believed the Moon was technically reachable — and their books are far less fictional. Wilkins's *Discourse Concerning a New World and Another Planet* (1638) is not fictional at all.[3]

I do not mean here to reinstate an arbitrary and for my purposes problematic division between the fictional and the scientific, or between reading for pleasure and reading for knowledge/power. I mean only to point out that one of the overt fascinations of both the two writers under discussion is the epistemological thrill of fictionality itself, and the challenge of its application to the task, under such intense revision in the seventeenth century, of envisioning the world. On the title page of Godwin's *Man in the Moon* the author is given as "Domingo Gonsales, The Speedy Messenger." Cyrano's protagonist, Dyrcona, is an anagram of his own name, and shares his circle of friends, many of whom are mentioned by name in *L'autre monde* (1657). At the beginning of the sequel, *Des estats et empires du soleil* (1662), Dyrcona is identified as the author of *Des estats et empires de la lune* (another of the titles by which *L'autre monde* is known). And so on. If the moon's visibility and proximity make it an obvious object for the new astronomy, its alternativity and newly-proclaimed "worldness" make it an equally obvious setting for self-conscious fiction, for fiction concerned with the nature of fiction. Both texts under discussion play with the issues of undecidability and disorientation raised by the new and growing knowledge of the Earth and the heavens, and raised as well by fiction's constitutive heedlessness of truth value.

This chapter considers, then, a more than fortuitous three-way intersection (such intersections traditionally sacred to the moon goddess Hecate, or Diana): a New World, a new *world*, and an Other World, or America, the moon, and the Moon. If, as many historians claim, the first two did not occupy the attention of more than a small elite, the third was a bestseller.

---

3. Wilkins was an influential member of the Oxford Philosophical Society and the later Royal Society in London. His book is a clear popularizing account, in a series of thirteen elaborated and illustrated propositions, of the plurality and habitability of worlds and, in particular, the "worldness" of the Moon. The 1640 edition includes a summary of Godwin's speculative fiction that further disseminated that work as a model for the genre. See Nicolson's *Voyages* for the relations between these texts and the ensuing plethora of lunar narrative (in prose, verse, drama, and even opera).

Through this literature a far larger number of people was brought to imagine new discoveries about the second. And people who were not consciously occupied with figuring out a new geography or cosmology were still very often occupied in buying, selling, eating, wearing, toting, and printing the products of the discoveries in the first. Recent critical consensus has challenged the older assumption that the impact of the New World was relatively slight on the daily life of European people in the sixteenth and seventeenth centuries.[4] What may have been slight was the impact of the modern academic idea "the New World," an idea too general and cosmographical to seize a busy or uninstructed mind. But these exemplars of popular or unscientific Moon literature will demonstrate at least one way in which America as an imaginative opportunity was elaborated and shared by people outside the pale of adventure capitalism or government.

Perhaps Thevet's books will make a useful point of comparison in examining the worlds of these lunar fictions. What they share with him, perhaps take from him, is fundamental: the imaginary adventuring narrator and the island as paradigmatic unit of description. In Cyrano's case, the complex relations of plagiarism also make an appearance, although there is a world of difference between the writer who leaves his book unpublished and the writer who obsessively pursues the fame and dignity of originality, while stealing whatever lies at hand. Both writers seem to be paranoid, but Cyrano's paranoia is framed by a punk nihilism that seeks an elite obscurity, while Thevet plays out the anxious quest for public glory of the self-made man.[5] Perhaps the textual phenomenon I am pointing to is the difference between fiction proper and the inadvertent fiction of the imaginative liar. But we will see more shared perception here than not.

THE MAN IN THE MOON:
IMAGINARY GARDENS, REAL TOADS

Speedy Gonsales is the man in the moon of Godwin's title, the starry messenger whose English picaresque adventures were not published until the bishop, who first drafted them in his college days (after hearing Bruno give

---

4. See J. H. Elliott, *The Old World and the New*, or Michael Ryan, "Assimilating New Worlds." Even Anthony Grafton's more recent *New Worlds, Ancient Texts* emphasizes the greater power of the European humanists' canon (vs. the American discoveries and encounters) in shaping a new *episteme* in seventeenth-century Europe.

5. On Cyrano's motives see Joan DeJean, *Libertine Strategies*.

a talk), had been dead for a year.⁶ The title recapitulates the tendency of colonial travel writing (adumbrated in Columbus's first *Letter* about his *insulae novae repertae*) to conceive of the visiting European as the crucial consciousness in the newly discovered place. That the experiences recounted in the little book are Gonsales's is all the glue available for the topics and events appearing between its covers. As time in Thevet's Canada was measured by the position of the sun in his hometown of Angoulême, so the imaginary career of an ambitious Spanish midget serves as the common denominator for the "knowledge" (Gonsales's obsessive term) offered by *The Man in the Moon*.⁷

The comic, episodic genre of the picaresque, born in Spain with *Lazarillo de Tormes* (1554), was "a product of nascent capitalist mentality" (Gutiérrez, *Reception of the Picaresque*), the same mentality in many ways as the French version expressed in Thevet's cosmographies and as that manifested in the adventure capitalism of New World exploration and trade. It is a "mentality" that takes for granted the possibility of a nonaristocratic point of view, and that requires self-centered aspiration. Godwin's choice of a picaresque narrative structure was not so much inspired as inevitable. New worlds are the only worlds in which a poor, bare, unaccommodated *pícaro* can hope to succeed, or "rise," and the new American world was turning out to be a godsend for "youngest sons" (like Gonsales, youngest of seventeen!) and commoners on the rise (Cortéz, for instance, or Thevet, or Walter Ralegh, whose book on Guiana it would seem Godwin had read).⁸ The Moon provides a perfect parodic extension of these supposedly opportunity-rich earthly islands, for

---

6. Dorothy Singer, in her chapter on Bruno's literary and philosophical influence on later European writers (*Giordano Bruno*) almost takes for granted that Godwin's *Man in the Moon* was germinated at Oxford after Godwin, as a student, had heard Bruno give a lecture on the plurality of worlds. Other historians also think the work was first written then, but on different speculative grounds. The published version is cognizant of later events and intellectual fashions. See Nicolson, *Voyages*, and Grant McColley, "The Date of Godwin's *Domingo Gonsales*."

7. Like Bacon in the *New Atlantis*, Godwin makes his utopian explorer(s) Spanish instead of English. In Godwin's case it strengthens a link not only with the ur-explorers of the Renaissance but with the Spanish-born and Spanish-dominated genre of the picaresque.

8. Godwin borrows Ralegh's device, made even more incongruous in the lunar context, of reporting to an audience imagined partly as Queen Elizabeth the natives' (here, lunar) adulation of the Queen. Ralegh: "I shewed them her majesties picture which they so admired and honored, as it had been easie to have brought them Idolatrous thereof. . . . So as in that part of the world her maiesty is very famous and admirable, whom they now call *Ezrabeta Cassipuna Aquerewana*, which is as much as *Elizabeth*, the great princess or greatest commander" (*Empyre of Guiana*, 7). Godwin: "*Pylonas* . . . required of mee but one thing, which was faithfully to promise him, that if ever I had means thereunto, I should salute from him *Elizabeth*, whome he termed the great *Queene* of *England*, calling her the most glorious of all women living, and indeed he would often question with mee of her, and therein delighted so much, as it seemed hee was never satisfied in talking of her" (112–13).

Fig. 14. Domingo Gonsales flies to the Moon with his *gansas*. From Francis Godwin's *Man in the Moon* (London, 1638). By permission of the Houghton Library, Harvard University.

Fig. 15. Airborne by means of evaporating dew. From Cyrano de Bergerac's *L'Autre monde* (image taken from London edition of 1659, *Selenarkia*, trans. Thomas St. Serf). By permission of the Houghton Library, Harvard University.

of course to go there one must literally "rise." (All the better—so both Godwin and Cyrano seem to have thought—that to arrive there one must then literally fall.) The frontispieces of Godwin's and Cyrano's books depict their hapless heroes "on the rise"—borne aloft by a flock of birds in Gonsales's case and by vials of evaporating liquid in Dyrcona's (see figures 14 and 15).

The *pícaro*'s consciousness is as far as possible from the (ideal) scientist's—like Thevet, he is looking for the main chance—the breakthrough, and his narrative is represented as written for the sake of glory and advancement:

"I doe with patience expect; that by inriching my country with the knowledge of hidden mysteries, I may once reap the glory of my fortunate misfortunes" (*Man in the Moon*, 126). He cannot tell us anything factual or factlike about the Moon that isn't introduced for the purposes of one-upmanship ("Amongst many other of [our philosophers'] vain surmises, the time and order of my narration putteth me in mind of one which now my experience found most untrue" [65]) or as an invitation to our identification with him: "Yes, my *Ebulus* will afford you that which I dare say will make you prefer him before . . . all the *Diamonds*, *Saphyres*, *Rubies*, and *Emeralds* that our world can yield" (99). There is no unsituated knowledge in a picaresque novel. On the other hand, the temptation to commodify the life of the other world, as the protoethnography of Hariot does, is undermined by the technical impossibility of exploiting the resources of another planet. So the self-centered protagonist has a more humanized ethnography to offer: he meets lunar people as characters in his personal drama rather than as substances or systems to be catalogued.

What topics and events then do come together under the rubric of Domingo Gonsales's career? First, a background narrative of typically amoral youngest-son shenanigans, made more pointed by the fact that Domingo is a midget ("I must acknowledge my stature to be so little as no man there is living I think less" [6]) and *extremely* defensive about his "honor." His social stature—youngest of seventeen children of a man "that was near kinsman on the mother's side unto Don Pedro Sanchez, that worthy Count of Almenara" (1–2)—does not provide that honor with a firm base. Misadventures culminating in manslaughter send him off—in the same way that, less spectacularly, bourgeois youthful ambition will send off his successors Gulliver and Crusoe—into some adventure capitalism in the East Indies, where sickness and shipwreck put our hero in the way of inventing a method of bird-powered human flight. At this point the work begins to stake out new territory for the picaresque, offering two stabs at utopia (including the earthly paradise common in earlier New World writing), quite a bit of "stenographia" (Godwin's work on codes and communication devices, the *Nuncius Inanimatus*, came out in 1629, nine years before *The Man in the Moon*), fictional confirmations of some aspects of Copernican astronomy, and an ethnography of lunar culture. Finally, and importantly, Domingo Gonsales makes an inadvertent trip to China as he tries to get home from the Moon, and the manuscript of our text is sent by him from China through the good offices of the Jesuit fathers in Peking, where Gonsales remains stranded but hopeful at the end of his narrative.

The book represents for our comparative gaze four distinct locations: the familiar cynical Europe of picaresque realism, the island paradise of St. Helena (a trope also familiar from descriptions in early New World travel writing) where Domingo is left by his ship's crew to recover from an illness, the Moon to which he escapes from the island (complete with its own, unvisitable, island paradise), and the China of the Jesuit *Relations*. Bibliographers of the earlier part of this century believed Godwin's book to depend heavily for its lunar ethnography on available accounts of China, in particular the diaries of Matteo Ricci.[9] Many seventeenth-century intellectuals caught up in the fervent quest for a "real character," a transparently denotative and universal language for representing a reality newly imagined as objective, felt that Chinese might be that language or might at least offer a model. In that sense, then, China had utopic connotations, like but unlike an island such as St. Helena (which Godwin also described from actual published accounts).[10] China was the utopian extreme of urban, bureaucratized civilization, St. Helena of idyllic nature. China was ancient and complete (maintaining no diplomatic relations and sealed against the entry of foreigners), St. Helena to-be-completed ("I cannot but wonder, that our King in his wisdome hath not thought fit to plant a Colonie" [14]). Between these earthly poles—most often territorialized as the Far East and the New World of the western ocean—the Moon is made to intrude a mysterious bulk of constitutively unreachable (but physically actual) felicity.

As a utopian fiction, *The Man in the Moon* has a refreshingly happy tone, more folkloric (or as Marjorie Hope Nicolson puts it, "arcadian" [*Voyages to the Moon*, 80]) than Morean—although it contains a powerfully hierarchical state which is comfortable banishing children to the North American wilderness, Godwin's Moon is by and large a place of plenty, longevity, beauty, and natural virtue. The prince Pylonas and his overlord Irdonozur are generous and friendly, and no one treats the midget space alien as a freak, despite the fact that the natural stature of lunar people is anywhere from ten to twenty-seven feet high (*Man in the Moon*, 89–90) and that both virtue and longevity are understood to be directly proportional to height in lunar society. Even before he arrives on the Moon words like "pleasure," "delight," "admiration," "joy," "happiness," "glorious," and "wonderful" appear on every page, once Domingo has left European civilization (almost) behind him. A sweet passage in an article by Thomas Copeland points out the

---

9. See Nicholas Trigault, ed., *China in the Sixteenth Century: The Journals of Matteo Ricci*. The "bibliographers" in question are mainly Paul Cornelius and Grant McColley.

10. See McColley, "The Date of Godwin's *Domingo Gonsales*," and James Knowlson, "Note on Bishop Godwin's *Man in the Moone*."

change that comes over the rapscallion protagonist when he is set free of Europe: "In this uninhabited retreat . . . Domingo knows his first release from the evils of the world; here is no war or treachery, no sickness or want. . . . Here for the first time in his life, Domingo has the leisure and peace of mind to be creative."[11] The structure of the narrative has a Dickensian quality, taking a small and powerless person out of a world where might makes right and letting him "rise" by his own devices (literally) to a counterworld where scarcity has had no chance to work its bleak alchemy on the social system and the psychology of its members.

But this text, if in some ways utopian, is not really fascinated by sociology and statecraft.[12] The most wonderful and pleasurable moment in the Speedy Messenger's representation of the Moon is his account of the "lunar color" (72), a phenomenon of pure difference and unallegorized sensation—if one which arrives at a moment usually given over in earthly exotic travels to racialist condescension or repulsion:

> their colour and countenance [was] most pleasing, and their habit such, as I know not how to express. For neither did I see any kind of *Cloth*, *Silke*, or other stuffe to resemble the matter of that whereof their Clothes were made; neither . . . can I devise how to describe the colour of them, being in a manner all clothed alike. It was neither blacke, nor white, yellow, nor redde, greene nor blew, nor any colour composed of these. But if you ask me what it was then; I must tell you, it was a colour never seen in our earthly world, and therefore neither to be described unto us by any, nor to be conceived of one that never saw it. For as it were a hard matter to describe unto a man borne blind the difference betweene blew and Greene, so can I not bethinke my selfe any meane how to decipher unto you this *Lunar* colour, having no affinitie with any other that ever I beheld with mine eyes. Onely this can I say of it, that it was the most glorious and delightfull, that can possibly be; neither in truth was there any one thing, that more delighted me, during my abode in that new world, then the beholding of that most pleasing and resplendent colour. (70–72)

This color (here attributed to the lunar people, or to their clothes, or both?) appears again in some precious and magical stones given to Domingo by the Emperor of the Moon, Irdonozur, and is once again sublimely desirable: "To say nothing of the colour, (the Lunar wherof I made mention before, which

---

11. Thomas Copeland, "Francis Godwin's *The Man in the Moon*: A Picaresque Satire," 158.
12. For discussion of the utopia as "the *sociopolitical subgenre of science fiction*" see Darko Suvin, *Metamorphosis of Science Fiction* (61). Elsewhere in that work Suvin makes a distinction important to the context of this chapter: utopia proper "is *located in this world* . . . an Other World immanent in the world of human endeavor, dominion and hypothetic possibility . . . [Utopia] is a nonexistent country on the map of *this* globe, a 'this-worldly other world'" (42). See also Raymond Williams, "Utopia and Science Fiction."

notwithstanding is so incredibly beautifull, as a man should travell 1000 Leagues to behold it)" (99).

Compare this reaction to those Thomas Browne discusses and maligns in his chapters on Blackness in *Pseudodoxia Epidemica*. This first human meeting is one of Godwin's most successfully counterterrestrial moments; although we may all perhaps have fantasized a beautiful color as yet unseen, most available narratives of earthly encounters with a new color are sites of aesthetic repudiation. Where there is physical attraction, the narrator must first explain (as Ralegh does in *Guiana* [1596]) that the beautiful native looks just like a European.[13] It is interesting that the color here is abstracted from its usual location—as signifier of (hierarchically defined) difference it always belongs to the human body alone, while the "*Lunar* colour" is simply a general possibility. It can belong to human clothes or skin or to precious stones and it is the *color*, not the human or mineral resources, which is supposed to draw men "1000 Leagues" to see it. (An early conception of tourism?)

Godwin's Moon is thus established as a place of difference not necessarily tied to hierarchy, though usually a source of delight. It is exotic without being exploited, better without being properly comparable. The cognitive emotion drawn on by this text is clearly wonder, rather than the aggressive judgmentalness of satire (a literary mode in which the moon had obvious uses). Writers have thought less often of the earth in relation to the moon than of the moon in its wonderful absence of "affinities." The (single) language of the moon is also described as having "no affinitie with any other that ever I heard" (93). One can only imagine how Gonsales learns it ("within two moneths space" [95]!); perhaps he is motivated by the delight of a language of song. For this is the language of poetry's Arcadia: "you have few wordes but they signifie divers and severall things, and they are distinguished onely by their tunes that are as it were sung in the utterance of them" (93–94). The language of the Moon (which like earthly music can do without words altogether) is one of excessive meaning or expressivity, and it must be performed; it must be beautiful to be understood at all.

This would seem as far a cry from the usual conceit of the "real character" or universal language as can be imagined, although Godwin was seriously interested in that idea, as were many who quoted Domingo in their works. The real character was to fulfill the hopes expressed by Sprat in the famous passage from his *History of the Royal Society* (1667), "so many things, almost in an equal number of words" (113). It would permit scientists to speak across na-

---

13. "I haue seene a Lady in England so like hir, as but for the difference of colour I would haue sworne might haue beene the same" (*Guiana*, 55).

tional barriers with a previously impossible degree of precision and pure denotative power, freed from the "mists and uncertainties" (Sprat, *History of the Royal Society*, 112) occasioned by natural language's historicity and its tendency to trope.[14] But the affective powers of music and the exaggerated performativity of a language which must be sung do not accord well with the focused beam of the longed-for real character and its technical uses.

There is in fact an actual earthly cousin (affine) to the lunar language: Mandarin Chinese as understood by intellectuals of the seventeenth century. The serious European interest in this language as a model for the real or universal character was based in a perception of ideographs as originally literal depictions of the things to which they referred; in other words it was written Chinese that excited them philosophically (some to the point of declaring it the original alphabet—displacing Hebrew!).[15] But Cornelius and others who have emphasized the important Chinese analogy to Godwin's lunar language have not considered the difference between that ideographic *character*, in utopically general use throughout much of Southeast Asia, and the system of musical tones that distinguishes sememes in (often mutually incomprehensible) *spoken* dialects of Chinese.[16] In any case, it was the musical, affective, oral, and performative aspect of Chinese that Godwin gave to his Arcadian Moon, not the written character of a large imperial bureaucracy. John Wilkins was later to pick up on Domingo's suggestion that "this . . . great Mystery . . . [is] worthier the searching after than at first sight you would imagine" (95), but only to produce a set of tonal equivalents to the letters of the European alphabet (minus 'q')—a cipher rather than a language. And I think he missed Godwin's point.

There are other points of comparison between Godwin's Moon and the China of his protagonist's last adventure, and the inclusion of China in the text's group of adventure locations suggests that these comparisons are functional (as does the explicit analogy Domingo makes between their languages). Besides being almost as far away as the Moon, and almost as lacking in "affinities" to European culture (or even nature), the powerful of China are powerful in the imperial style—gracious, secretive, gently constraining of the *pícaro*'s freedom, culturally superior. The mandarin into

---

14. The definitive work on these matters in their connection to the extraordinary voyage literature remains Paul Cornelius's *Languages in Seventeenth- and Early Eighteenth-Century Imaginary Voyages*. See my Chapter 3, n. 16, for general works on the "real character."

15. See Cornelius, chap. 4, "A 'Real Character' for Europe Based on the Chinese Model," which includes lengthy extracts from such hard-to-find texts as Athanasius Kircher's *Oedipus Aegyptiacus* (1652–54) and George Dalgarno's *Ars signorum* (1661).

16. See, e.g., Knowlson, "Note on Bishop Godwin's *Man in the Moone*."

whose hands Domingo is delivered treats him as well as Pylonas had, and takes the same "delight" in regular conversation with him. As on the Moon, "I lodged well, fared well, was attended well, and could not fault anything but my restraint." The differences then are significant, such as they are. (Most of the description of China, "the disposition of the people and the policy of the country," is postponed for "my second part" [125] which is, as so often the case in earthly voyage writing, never written.)

Maybe I should say "*the* difference"—which, while having ingredient details, is really a single thing, a matter of atmosphere. China is obviously "real"; as such, it can be deferred for another book, and in fact Godwin lets you know *which* book if you want to investigate. It is Father Pantoja who sees that his manuscript is sent to Macao for posting to Spain (126), and Father Pantoja's account of China is available to anyone who can get her hands on Samuel Purchas's ethnological voyage collection, *Purchas His Pilgrimes* (1625).[17] China (like St. Helena and Spain) has real people in it, people we've heard of in other accounts of travel, exploration, or shipwreck. They add verisimilitude, say some, to a wildly imaginary tale, but the point of that is not self-evident—why not just read Father Pantoja, or *Hakluyt's Voyages*, if the real (the actual) is what you want? Why would a writer of fiction lure readers of fiction with a promise of—the unfictional?

In Godwin's book it is clear to see that the bracketing of the Moon by earthly places and historical figures has the effect of highlighting the difference between the real and the imaginary. China is different from the Moon in attributes, such as the arbitrary class system that requires both a "vulgar" tongue and the mandarin, or the absence of magical knowledge (his mandarin host hopes to acquire some from Domingo); but its chief difference for our purposes is in the manner of presentation. It has far less power than the Moon to capture the attention (thus our narrator can defer description till "my second part"), and it is rendered only so far as it intrudes on the immediate experience of a protagonist who is eager to get home. It arrives in pieces, unexplained and unsystematized—in minor events and remarks. China has the verisimilitude of a travel diary, or a travel novel. It is far away in a straight line, but there is no gulf between where I sit reading and the world or reality in which the mandarin takes Domingo with him to Peking when he has business there. Domingo is as good as home when we leave him in this hugely distant spot, and that is the ironic point of its distance. In such a syntax China no longer holds the place of the exotic. As in Kepler's moon's-

---

17. For Father Pantoja (Pantoia), see Samuel Purchas, *Purchas His Pilgrimes* 12:328, 331–410, and 485.

eye view of the planet Earth, all of this world is compressed into a large-scale version of the concept "home" (a parallel with the half-conscious telos of empire, as Lafitau's great ethnological work will confirm).

There is a depressed quality, too, to the account of China, brief though it is. It is not just random and unsignifying; the narrator seems to find China also faintly creepy in its failures to match the Moon. The Chinese peasants who find him are more ignorant and vehement (even xenophobic) than the lunarians who first encounter Domingo. Though they do the same things, one gets the sense that it is knee-jerk fear of authority that motivates the Chinese, whereas the lunarians are simply doing what they should (their virtues being somehow inherent rather than learned or imposed). The passage on how well lodged and well attended Domingo is can't compare in exuberance to the quality of paradisal abundance attributed to his lunarian quarters, and he mentions the constraint on his movement here explicitly, while leaving it merely implicit in the lunar narrative (in fact, obscuring it with accounts of pleasurable journeys). The effect is similar to that of the final scene of the film version of *The Wizard of Oz* when Dorothy, waking from her marvelous dream, recognizes all its materials in the relatives and farmhands who surround her bed and the local events that preceded the storm. The weight of the real, its resistance to individual desire, its *independence* of our projections comes like a blow, for all Judy Garland's performance of joyful relief.

To understand what these imperfect utopias of the actual are doing on either side of Godwin's imaginary Moon it is necessary to examine the other bookend, the Isle of St. Helena, earth's "only paradise" (14). This island is placed at a different site in the syntax of cynicism, aspiration, bliss, and frustration that structures the book. It follows on the hustle and depravity of European picaresque reality and prepares Domingo for an even more cornucopian adventure in the heavens. It is a promise, soon fulfilled, that there are more things in heaven and earth than have been dreamed of in his philosophy. Left there to recuperate from an illness by the ship on which he is returning from the East Indies, Domingo finds it fantastically abundant, healthful, and beautiful, equipped with a "pretty Chappell," "fair walks made by hand," and best of all a "Blackmoore" slave with a name very like his own, Diego (16). He is almost unable to express, however, what I take to be the crucial charm of the place: it is "about 3. leagues in compasse, having no firme land or continent within 300. leagues, nay not so much as an *Island* within 100. leagues of the same, so that it may seeme a miracle of Nature, that out of so huge and tempestuous an Ocean, such a little peece of ground should arise and discover itselfe" (16).

The island St. Helena (a real place in the South Atlantic, midway between Angola and Brazil) seems to represent Domingo, in the fashion of Thevet's (ironically imaginary) Isle of Thevet.[18] It is tiny, lost in a huge turbulence, far away from everything and everyone, but nonetheless able to arise and discover itself. Domingo is not alone on the island that signifies him;[19] it is part of his geographical projection that he have someone (or something) to command, someone more precisely to rise above. When he is first prepared to test the bird-powered flying device he has invented, Diego wants to be the experimental subject. But while using a person of low status for such a purpose might seem natural enough to nineteenth- or twentieth-century researchers, Domingo in this symbolic spot cannot cede Diego the honor of rising above him in the air: "for I hold it farre more honour to have been the first flying man, than to be another *Neptune* that first adventured to sayle upon the Sea" (26). Diego's blackness here means what Domingo's littleness does not: that he can be subjected to the desires of another. In fact, Domingo uses his littleness as a rationalization of his prior right to take the first flight: "I onely told him . . . that all my *Gansas* [birds like swans] were not of sufficient strength to carry him, being a man, though of no great stature, yet twice my weight at least" (27). This same littleness will change its significance for Domingo when he reaches the Moon, where it signifies what blackness so often did in Renaissance Europe.

The major events of the island experience belong to the category of applied science: the invention of flight and Domingo's experiments with long distance communication. If Diego functions in the invention of flight only to provide Domingo with someone to rise above, he functions in relation to Domingo's many ingenious communication devices (light signals, carrier partridges, *gansas* trained to fly towards a white cloth) to give him someone to command. As with the lunar language, Godwin's interest in ciphers and long distance signaling were quite serious. His *Nuncius Inanimatus* provided the spur for Wilkins's later classic *Essay on the Real Character*, and he inserts here, as in Domingo's discussion of the lunar language, a clue to his own seriousness: "But this Art containeth more mysteries than are to be set downe in few words: Hereafter I will perhaps afford a discourse for it of purpose, assuring my selfe that it may prove exceedingly profitable unto mankind" (22). The wonder of these methods, in which "according to a certaine rule and agreement between us, I certified him at pleasure what I list" (21), lies in their unimpeded, instantaneous effectiveness: all frustrations of speech-

---

18. St. Helena was an "important staging [post] on the Cape and Indian trading routes," a paradise "to sailors exhausted and weakened by long voyages"; for its part in the development of environmental awareness in colonialist Europe, see Richard Grove, *Green Imperialism*, 42–7.

19. Godwin seems to have read a recent report of a man who had been left alone to recuperate on St. Helena and gone mad before the next ship came to pick him up. See McColley, "Date of Godwin's *Domingo Gonsales*."

less, desiring infancy are forgotten in this smooth sequence of desire, utterance, understanding, and fulfillment. All that's required is an intelligent slave.

This is an honest portrayal of the utopia implicit in Europe's system of yearnings and deprivations. And to the degree that that system has survived, it retains some of its effectiveness. Elizabeth Bishop's famous poem "Crusoe in England," written as recently as the 1970s, still offers most of the crucial aspects of this "onely paradice," if more leavened with ambivalence and irony—solitude, sublime vastness, a single companion who is socially subordinate (and foreign), inventions and language games, rescue and subsequent loss of the companion. In a word, the childhood in which we learn to walk (fly) and communicate our wishes. But it is a revamped, ideal childhood, in which that subordinate companion does not replace us in anyone's regard, and from which we are so successfully rescued that we can dispense with even the compensatingly subordinate companion. Historically speaking, the fantasy took forms very like the ones offered by this fiction—the men of the emerging political class of Britain, at any rate, would one day find it normal to live in isolation with servants (and such other subordinates as wives) to dominate and express wishes to, on estates complete with "pretty Chappells" and "fair walks made by hand." See any male protagonist of Jane Austen's, for instance (or even *Emma*).

However, this fantasy is not what the book offers. It is afforded Domingo only to be retracted, replaced at first by pirates and shipwreck, then by a lunar world of felicities with little "affinitie" to those he has been educated to desire. Godwin's Moon does not for the most part have the relation of critique (as Cyrano's will) to the distant Earth. It is alternative, rather than parodic. It is better in many ways, and while he is there, Domingo himself is better, but the affinities are too few for the Moon's superiority to function systematically as critique or even rebuke.

Ironically, the main affinity between our disparate worlds is represented as a literal, biological one: the lunar population is partially descended from an earthly conqueror (the first Irdonozur) (76), and some Earthlings are changeling Lunars: "their ordinary vent for them [children judged partly 'by the stature' as 'likely to be . . . wicked' (104)] is a certaine high hill in the North of *America*. . . . Sometimes they mistake their aime, and fall upon Christendome, *Asia* or *Affrique*, marry that is but seldome" (105). This provides a twist on the most unpleasant and familiar of lunar society's features: the belief that a certain group of people physically marked as different (in this case small) are inferior. The deportation of short and potentially wicked children before they become malefactors is said to obviate the need for a legal death penalty in lunar society (104), and Lunars replace their population loss by exchanging them for American children whom they apparently raise as a class of servants. Unlike colonists, the deported lunar children are supposed to go native. They are taken care of at first away from the North Americans "till that the ayre of the Earth may alter their colour to be like unto ours" (105). Then they become culturally American Earthlings, and their counterparts, the changelings from Earth, "base" Lunars.

The physical anthropology of this situation is complex and dense with implication. The Moon is inhabited by a race of mixed autochthonic and exotic strains; miscegenation continues on Earth between lunar and earthly individuals. On the Moon, although the mixture is the result of a conquest by an earthly king whose dynasty still rules with honor, the earthly strains in the lineage (short stature, brief life span, inability to tolerate bright light) are despised and rejected as signifying, even embodying, moral and intellectual inferiority. Yet lunar people have no cause in their idyllic social or physical environment that would explain either this need to reject, or such inferiority as supposedly provokes it. If vice is an inherited trait, so must virtue be; if all moral character is biologically determined, then the culture cannot have any utopic lessons for us on Earth. It is simply different, as foxes are different from finches.

There will be many subsequent fictions and fantasies in which alien cultural systems are understood as hard-wired, in one or another form of that conception, while the implicit term of comparison, "earthly" (European) society, is imagined as still the product of choices.[20] And there is of course a model in older fictions and cosmologies: the model of the angels, arranged in their heavenly hierarchy of immutable ranks, too intelligent and too rational to imagine a rebellion (especially once they had seen how it went with Lucifer). This is a model Bishop Godwin must have known well, and the angels as archetypal messengers must have appealed in many ways to the author behind the "Speedy Messenger" (not to mention the author of the *Nuncius Inanimatus*). Masters of flight and communication, at home in the upper heavens, the angels are huge, beautiful versions of the tiny *pícaro* Domingo as well as of the gigantic and long-lived Lunars with their automatic virtue. The lunar ethnography is in many ways an ethnography of angel culture, but Godwin points more than once to Domingo's own participation in the angelic species — in his identification as "messenger" on the title page, in his flight through the heavens, in his implied descent from lunar parentage (like their

---

20. Despite the social conservatism of much "cyberpunk" science fiction (see articles by Kathleen Biddick and Tyler Curtin, on William Gibson's *Neuromancer*), that is one tendency it successfully challenges — in these works, it is the Earthling protagonists who are the hard-wired puppets of huge totalitarian information systems (i.e., civilizations).

other earthly changelings, he is small and of "imperfect disposition"). It will not be long before Milton represents in extraordinary detail (complete with a carefully imagined cosmic flight to Earth from Heaven) the rebellion of angels who wish, like Domingo, to rise above their station.

Claude Lévi-Strauss's famous cultural categories of "cool" (customary or traditional) and "hot" (in dynamic relations with others, developed) seem usable here, however specious in their original application, in a significant moment for the very history of Europeans' perceptions of themselves and others to which his categories belong. One might see both *Paradise Lost* and *The Man in the Moon* as narratives in which his "cool cultures" are seen heating up, and the cultural stasis of "tradition" (represented by "manners and customs") becomes the viciously mobile "history" characteristic (in Lévi-Strauss's spatialized view) of "hot cultures." Do these celestial social groups, displaced images of earthly human ones, function as self-fulfilling prophecies of the various fates implied in colonial annexation and development? Certainly Domingo loses paradise—twice, since he puts both St. Helena and the Moon in that category.[21] But (the longest way round is the shortest way home) unlike Satan he rises out of it as well as falling out of it: from St. Helena up to the Moon, then down to China.

How does such a sequence go with the transition from the real to the imaginary, and with the affective shifts I began by sketching? The desire—the need, even—for "glory" takes the *pícaro* Domingo where he goes; and though it can be *earned* there, glory cannot be enjoyed on a desert island or on a planet where he is outside the social system. It can only come from Castile. (At least for this version of the adventurer—some later protagonists go native and win their glory in a wild alternative world they have subordinated or abandoned their own values to join.) The glorious model, especially for Iberian climbers, is Cortés. Columbus found only a paradise (islands like St. Helena); leaving Cuba behind him, Cortés found an imperial bureaucracy, and gold. So far, Domingo shares the career pattern and the motives of "stout Cortez." In the moral reading of *The Man in the Moon*, this motive and career are shown to require the imaginary, the Moon, and the effort of imaginary experience is too great to maintain—the fiction falls back to Earth (*sans* "gansas") as Dorothy does to Kansas. In this reading, America, land of opportunity (in its imperial Aztec avatar), is the signified of the imaginary Moon, and Europe, like Kansas, is humble but Home. Domingo wants to go back—"We're going home, Toto!" He can't really imagine anyplace else as

---

21. Copeland complains of Domingo that he never recognizes what's good for him, is always leaving good places in pursuit of spurious glory. This is true, but a strange complaint—how else could the novel continue? Or any novel? Protagonists, like baby birds, leave their nests.

answering his endogenous desires, whether they are for wife and children or for glory.

But this reading does not get at the most important way in which the Moon offers itself to the fiction makers of the seventeenth century (any more than the end of the film version of *The Wizard of Oz* has satisfied twentieth-century children). Godwin's moon voyage was a popular book, widely read, reprinted, alluded to, and imitated for a long period in several countries—it even became an opera and contributed to a play by Aphra Behn. Neither opera nor Behn are much given to pious lessons or the reality principle. What stands out about the Moon as alterior New World is precisely that there is nothing to be done with it. St. Helena can (and will) be "fortified" and a colony planted (it remains a "British Dominion"). China too has suffered from centuries not only of Jesuit missionaries but of European economic exploitation and domination; Hong Kong was a British colony until 1997. Although Domingo promises flying machines and telephones when he returns, what we get instead (because he never returns) is "that which far surpasseth all the rest . . . notice of a new World."

What would one do with that? What would it provide the reading imagination that the traditional objects of vicarious voyage-lust—Western paradise islands and the ancient, learned East—do not? Well, if the major plot device of this fantasy is any clue—escape. Over and over what Domingo experiences, despite the pleasure afforded by some of his ports of call, is escape—from his parents, from the military, from Europe (and his wife and kids), from the ship where he has fallen ill, from the isle of St. Helena, from a shipwreck, from the Earth, from the Moon, from the Chinese peasants who find him after he lands back on Earth. The last event of the narrative is the flight of the text itself, in Father Pantoja's hands, to Macao and thence to us; that we are reading it is evidence of the text's successful escape from a land in which it would have been illegible. America was a land of many actual escapes: Consider Ralegh, for instance, literally let out of a thirteen-year imprisonment in the Tower to seek the gold of Guiana one more time. Or the Pilgrims, mythic escapee Founders. Even Cortés was fleeing the law when he set out for Mexico and took it. As a fiction raising the ante of readerly submission, *The Man in the Moon* can take that function of America and hyperbolize it: Domingo escapes humanity, gravity, yea, the great globe itself and all which it inherit.

The pleasure afforded by escape is intimately bound up in whatever the escape rejects. So the question of what the protagonist escapes *from*, and what the represented spectrum is of such places or situations, becomes crucial. Although medieval romance is often taught as "medieval escapist fiction," that is an inaccurate description. For one thing the romances do not represent the successful evasion of reality—the "reality" they portray is successfully *joined* by their protagonists (like their first readers, aristocrats). The plots are about fitting in. The heroes *come back*. Only the French Grail romances leave their heroes in some other or "counter-world," and those texts are bucking the genre's norm to make a religious point.

In contrast, nonsatirical moon voyages *are* "escapist," that is, they proselytize for escape from the reality of the European social system and of its normative cosmography (clearly linked phenomena). This may lead to utopian fiction, or to political satire, but those forms provide shallow satisfactions to those who need to escape the very matter and spirit of the known world. Both Godwin's and Cyrano's narratives end without the protagonist's return, though Domingo is back on Earth (and Dyrcona/Cyrano comes home from the Moon before leaving on his one-way voyage to the sun).[22] Like dreams, they mimic features of the known world, but also like dreams, they are more concerned to explore desire and to test the dreamer's improvisatory skill than to depict a recognizable actuality. The protagonists of both narratives need escape badly—Domingo is a midget without social power and "Dyrcona" (Cyrano) is a libertine atheist who does not dare to publish his novel.

If St. Helena, "the onely paradice" this earth yields, is an island image of the self as self-sufficient (all Thevet, or Marguerite de Roberval, could want), what is that larger island in the sky for which Domingo leaves it? Its depiction as a place with a history and government, and with manners and customs, makes it a collectivity, in (politically organized) space as well as (genealogical) time. Compared to St. Helena, it is like Bruno's universe of infinite worlds in comparison with the Aristotelian/Christian one world. It is a world of plurality, of others, but it is designed specifically around Domingo's fears and desires. This world for which he has left his island/childhood is one in which, once again, he remains small where others are big. Will the others accept him? Will they be worthy of their power? Can he have honor there? The Moon is very different. Is it different enough?

Not quite, perhaps—the world of his visit starts out well, with that indescribable "Lunar colour," but a novel full of indescribables would sooner or later pall, and this one inches gently towards the familiar. What is most familiar to someone like Domingo, legally sanctioned oppression of the physically abnormal, finally cuts the lunar glow. No matter that it is followed by a description of benign proto-Enlightenment values apparently programmed

---

22. Bacon's *New Atlantis* also ends without conclusion, a feature of Renaissance romance in general (another good example would be Spenser's *Faerie Queene* [1596]). On *New Atlantis* as it wavers between the codes of *Amadis of Gaul* and More's *Utopia*, see Denise Albanese, "The *New Atlantis* and the Uses of Utopia."

into the Lunars' biology. That is in some ways more of the same (these values and oppressions coexisted easily enough in Britain, or "Castile"). The novel we read as an outward-leading series of concentric escapes suggests in its structure that there is always more otherness just over the horizon, but also that whatever we *live* in begins to look like our own dreams. If Thevet's narcissism gave us the imaginary protagonist who condenses the world into a single story, vicariously shareable, Godwin's gives us islands and planets that tell us *who his protagonist is*. Domingo rejects, repeatedly, what is already known or accomplished, fleeing the familiar as a form of death. He might seem a good adventure capitalist — never satisfied long, restless, projecting his self-knowledge outward onto places he can visit and manipulate. But he does not want to make use of the worlds he discovers so much as to escape them. In that respect, he is a writer, a modern writer. Usable worlds are digested worlds — writers, with Chauncy Gardiner, just like to watch.

CYRANO DE BERGERAC'S OTHER WORLD — MIMESIS AND ALTERITY

Here is a book, which I will leave with you. . . . It is entitled: *The States and Empires of the Sun*, together with the *History of the Spark*. I am giving you this one besides, which I regard much more highly. It is *The Great Works of the Philosophers* composed by one of the best brains on the sun. In it he proves that all things are true.

— Cyrano de Bergerac, *L'autre monde* (1657)

Where all things are true, the concept of information begins to come apart. Kepler and Godwin, to differing degrees, produced a fiction that communicated information, while also interpreting and situating it. Although Kepler's fiction is more pedagogical than Godwin's, both their works respect the access provided by scientific thinking to a form of truth they privilege, if not above all others (Godwin was a bishop, after all), at least above that of poetry. That is why Godwin's novel, though up to something quite other than Kepler's, still takes its opportunities to "prove" aspects of Copernican theory through the testimony of his imaginary space traveler. Cyrano does not take these opportunities. He is for fiction, at least in *L'autre monde*, as a concept acidly destructive of scientific (perhaps more specifically Cartesian) certainty. However much he may have been inspired to write his books, *L'autre monde* (or *Les estats et empires de la lune*) and *Les estats et empires du soleil*, by such acknowledged heroes and role models of his as Galileo, the books are inimical, as responses to the rearranged vastness of the universe, to those of the astronomers.

And indeed Godwin's picaresque novel seems to have been more directly inspirational than the books of Kepler or Galileo. *The Man in the Moon* was translated into French in 1648; the episodic, earthly and lunar travel novel *L'autre monde*, begun soon after, can be seen as a reading of it as well as a transformation (or correction).²³ It retains quite a bit of matter from Godwin's model: the first-person narration (very rare in France except in the "comic novels" of the libertines), the narrator's "rise" in his home-made vehicle, his initial lateral trip to someplace outside Europe (New France), the presence on the Moon of an earthly paradise (but see also Isidore of Seville, Hrabanus Maurus, et al.), the allusions and links to America, the consciousness of scientific or technical knowledge (Cyrano's *science*) as the fleece to be stolen here, the incarcerating hospitality of the lunar big shots, the musical language. Above all, there is the presence in Cyrano's book of Domingo himself, presented here as the queen's pet, understood by the Lunars to be a monkey.

Cyrano's Moon is far less paradisal than Godwin's, indeed it is less of a stable imagined or fantasized place in general. His narrator's character is as hard to pin down as is the setting—a libertine goofball in early conversations with Elijah and Enoch, following his ejection from the lunar Paradise and capture by "beastly" natives he becomes a prudishly pious Catholic in later debates with the demon of Socrates (as Madeleine Alcover puts it, "Il se pose au début comme un Prométhée, il finit en apôtre" [lix]).²⁴ The text is an endless set of writing possibilities, of opportunities to shock; it is structured by paradox rather than plot. But there is a consistent satiric edge. For instance, the paradisal single language of Godwin's Moon here becomes two (as in Godwin's depressingly actual China), one for the powerful and one for the not. The powerful speak in music (without words) and the common people "by the shaking of the limbs" ("it seems less like a man talking than a body trembling" [37]). Cyrano's narrator is first captured and displayed as a monster by a "charlatan" (31), then brought to court and put in a cage with

---

23. *L'autre monde* and its sequel have difficult publication histories. Both were first published posthumously, *L'autre monde* "sans privilege" in 1650 and authorized, by a friend, in an expurgated edition (1657). (There are two surviving manuscripts, neither of them Cyrano's, and both produce other textual traditions.) *Les estats du soleil*, for which no manuscript of any kind survives, appeared in 1662. *L'autre monde* entered English literature in Thomas St. Serf's and A. Lovell's translations of the expurgated version, 1659 and 1687 respectively. The work goes by many titles: the first edition's *Histoire comique par Monsiuer de Cyrano Bergerac, contenant Les Estats et Empires de la Lune*, playing off the title of Pierre d'Avity's famous cosmography, and since then *L'autre monde* and *Voyage dans la lune* (likewise *Voyage au soleil*). Madeleine Alcover's definitive edition, from which I will, where necessary, quote the French, is called *L'autre monde ou les estats et empires de la lune*. English is quoted from Geoffrey Strachan, *Other Worlds*.

24. See the introduction to her edition of *L'autre monde*.

Domingo Gonsales as "the great ones [had] reached the conclusion that I was doubtless the female of the Queen's 'little animal'" (43).

Lunar society, described at length by the intermittently incarcerated narrator, is familiarly European in its pastimes, its trivialities, its class divisions and characteristic forms of cruelty. People are as large and as long-lived as on Godwin's Moon, but without the hard-wired virtue those qualities were supposed to manifest. But there are some pleasures to be had, and some superior features. For a while at least, the narrator's visit to a house party of lunar libertines is both a sensational and intellectual feast, and the lunar cult of philosophers is admirable, particularly their manner of observing a philosopher's death. The one the philosopher loves the best stabs him while kissing him and inhaling his final breath. Then the philosopher's blood is drunk by his friends and "four or five hours afterwards a girl of sixteen or seventeen is brought to each of them and during the three or four days which they spend in tasting the pleasures of love they are only nourished from the flesh of the dead man, which they are made to eat quite raw. Thus if anything may be born of a hundred embraces, they are assured that it is their friend who lives again" (91).[25]

Like so many episodes or passages of cultural description in *L'autre monde*, this one (a striking anticipation of Sade) has valences with the question What is a man? — the question of anthropology. The scene of cannibalistic feasting is already the master scene of ethnography, and in straight travel writing or accounts of exploration, whatever its particular nuances of context, it always means one thing: they (the Others) are not really Human.[26] They are functionally animals, even if genealogically human, and so (the inevitable if usually implicit corollary) we (the Europeans) can take their property and rights away from them. Its closeness to the symbolism of the eucharist, however, makes it satirically usable for the different purpose of

---

25. "On introduit à chacun au bout de quatre ou cinq heures une fille de seize ou dix-sept ans, et pendant trois ou quatre jours qu'ils sont à gouster les delices de l'amour, ils ne sont nourris que de la chair du mort qu'on leur faict manger toutte crue, affin que si de ces embrassemens il peut naistre quelque chose, ils soient commes asseurés que c'est leur amy qui revit" (Alcover, 184).

26. Essays and books on the cannibalism question are too many to list. Hans Staden's account of his experience with the Tupinamba, for which see Chapter 9, had an extended life in print (and was in fact reprinted, in Richard Burton's translation but without the infamous notes, in *Explorer* magazine in the 1990s!). Many ethnohistorical articles about this and other sixteenth- and seventeenth-century reports of cannibalism have come out since Arens stirred the flames again with his 1979 *Man-Eating Myth*. The most recent work in cultural history is Lestringant's *Cannibals* (1997); see also Hulme, *Colonial Encounters*. Notes to my Chapter 9 provide more bibliography.

mocking European cultural arrogance; it is used this way famously in *Mandeville's Travels* and Montaigne's essay "Of Cannibals."

In Cyrano's astonishingly sexy version of this scene there is much more going on than "the world upside down." The combination of murder, the kiss of the lover and beloved, anthropophagy, and procreative orgy has an energy of detail and hyperbole (*enargia*) that exceeds the needs of that topos by quite a bit. The scene is about how to reincarnate a specific person; it is about procreation, then, but with a specificity that smacks more of magic than of biology. The man made in this scene (written before Leeuwenhoek first saw sperm cells with the microscope) is made of blood, flesh, breath, and affect, digestively transmuted into procreative fluid in the bodies of his male friends and placed, literally, in the matrices of healthy young breeder-females. Of course the time-honored way to provide offspring is more direct — the philosopher could have impregnated the girls himself before his artificial death. The desire this scene plays to is a homosocial desire for male parthenogenesis, for merging substances with the beloved man or men — a desire for losing individual boundary, even if it includes death, or emasculating contact with a woman. The women are pornographic objects, neither persons nor contributors to the personhood of their offspring. It is, despite its thorough misogyny (pervasive in the book), a strangely maternal image of the transmission of knowledge and culture.[27] In place of Godwin's notion of lunar culture as biological program, it offers a notion almost the reverse, of cultural constructions (selves, Donna Haraway would say "cyborgs") powerful enough to infuse matter, to instruct it — as now our genes are imagined by many to instruct our passions and behavior.[28]

A man (I use the word advisedly, with reference to a pair of books at least as misogynist as Swift's) is not only someone made by men, he is also someone who lives collectively with other men, in a culture bound by language. Medieval and Renaissance depictions of the "Wild Man," solitary and speechless, living beyond the edge of town or in the forest, permitted investigation of the centrality of culture to defining the human.[29] The question is always framed in *L'autre monde* as concerning the species difference between

---

27. See the other image of "Nature" in the illuminated letter from Bacon's *Novum Organum* of (Mother) Nature with a child in her lap (Chapter 3, fig. 10). Here and elsewhere in Bacon's book Nature is a mother nurturing or teaching her human children.

28. See Haraway's "Cyborg Manifesto."

29. See Richard Bernheimer, *Wild Men in the Middle Ages* and H. W. Janson, *Apes and Ape Lore*. Cyrano also makes use of the notion of monkeys as mimic-men, beasts whose intelligence and physical similarities to us force on us the question of culture and reason in defining the human species. This is the issue confronted by eighteenth- and nineteenth-century investigations of feral children such as Victor, the "Wild Boy" of Aveyron.

men and beasts. It begins as a question on both sides: in contrast to the pleasant encounter with difference represented in Domingo's first experience of "the Lunar colour," Cyrano's narrator is surrounded by "seven or eight hundred [very large animals with] . . . bodies and faces like ours" (29). "This occurrence reminded me of the tales I had once heard from my nurse about sirens, fauns, and satyrs. From time to time they raised hootings so furious . . . that I almost thought I must have turned into a monster" (29–30). The narrator senses that his task is to show the Lunars (who abduct him) that he has the human capacity for living in a cultural formation by learning to speak to them (Swift's model for Gulliver's adventures as a Yahoo).

He does learn the language, better than "my male," Domingo, with whom he is caged in hopes that they will procreate: "the neatness of my epigrams and the esteem in which my wit was held has already become the sole topic of conversation in company" (54). However, the priests are committed to keeping the earthlings categorized as animals, or even monsters, because of the theological impropriety of their claim that the Moon's moon, our Earth, is a world. (Here the narrator's experience repeats Galileo's with the Church.) The narrator's witty command of the language gets his status (and Domingo's) upgraded to that of wild men (53), but after a public interrogation to decide the controversial question of his humanity, Cyrano/Dyrcona is declared an ostrich (because of his rigidly Aristotelian answers to the judges' philosophical questions). The process is repeated twice more; the second time the narrator is again returned to his cage on account of the unreasonable Christian and Aristotelian dogmatism of his views, but the third time he is rescued by a sophistical oration from his friend the demon;[30] the demon takes him home to his house for a dinner party.

Lunar society is shown as unable to decide the question of humanity in itself — the debates are about the plausibility or irreligion of certain viewpoints, and their vigor is maintained by *realpolitik*. By the end of the controversy Cyrano's narrator is in the same position he inhabits on earth — marginalized and despised for his dangerous intellectual difference and lack of affiliation.

This satirical account of the motives and results to be expected from such investigations — for instance that of the famous sixteenth-century Sepúlveda–Las Casas debate at Valladolid over the humanity of the Caribbean peoples — would suggest a Said-like scorn of academic anthropology on Cyrano's part. He sees clearly that any discourse capable of certifying some-

---

30. The demon is noted by Joan DeJean ("Method and Madness") as "the character who, after the narrator's expulsion from paradise, becomes the narrator's mouthpiece" (227).

one's humanity can take it away as well.³¹ And yet he is part of the century's fascination with imagining cultural formation and displacement. The opportunity the Moon holds out to a writer of being culturally *designed*, from the ground up, is taken here in a way analogous to Cyrano's attempt, within it, to design a method of extremely task-specific procreation. That such activities contribute necessarily to a critique of existing or familiar culture (and biology) does not mean they are merely reactive. In Cyrano's imaginary voyage the reader sees anthropology, before it is institutionally born, conceived as potentially utopian, reconstructive. As long as we can invent rather than merely describe other cultures—other ways of constructing people—we can get a purchase on our own, and perhaps reinvent it. At the least, we can imagine something outside of it. The initial lateral trip to New France may be an allegorical suggestion about how to use the opportunities for imagination-nourishing relativism offered by actual and visitable new worlds. Cyrano's narrator almost literally bounces off New France to his Other World in the Moon, where he is once again discovered and captured by a band of wild-looking natives, but where, this time, his capture results in the public airing of questions fundamental to the conduct of colonial expansion.

But though the lunar world encountered by Cyrano is not the escapist dream of Godwin, it does contain an unearthly paradise, and in fact that is the location of the narrator's initial fall from the sky. In a tendentious twist on the narrative pattern of Godwin's voyage—confinement and escape—Cyrano's pattern tends to be confinement and *expulsion*. Having been accidentally expelled from New France, by the explosion of the firework rockets soldiers of the garrison had attached to his flying device, the narrator falls into paradise with a plop, smashing an apple with his face. He will soon be expelled from this place too, after blasphemous encounters with one of his five earthly predecessors, Elijah, and the theft of an apple from the Tree of Knowledge. But first Cyrano offers a long description of paradise, stolen from one of his own published letters.³²

The description is stylistically unlike anything else in either of his pair of imaginary voyages, and as DeJean points out in her acute essay "Method and Madness in the *Voyage dans la lune*," it is especially uncharacteristic in its monologic quality. According to DeJean's reading, the shifting quality of lunar experience and of the narrator's philosophical positions in this text is part

---

31. For the debate see Chapter 1, n. 17.
32. See DeJean, *Libertine Strategies*, 185–87, who puts the auto-theft into the larger context of plagiarism and quotation in the libertine aesthetic.

of the book's dialogic or even polyphonic character (long associated with Menippean satire, a genre initiated by a moon voyage): "[Cyrano's] goal is not so much to refute certain viewpoints, but rather to integrate a plurality of discourse into a text with no clearly dominating context.... All paradoxes are intentional here.... The sum total of positions expressed resembles a manual of philosophical trends of the mid-seventeenth century" (226).[33] But paradise is not naturally a setting where difference or polyphony makes sense. (For one thing, the experience of describing or inventing a paradise is one of profoundly individual gratification, no matter how much the socially conditioned motifs may repeat from individual to individual—it is in that way analogous to the writing of pornographic scenes.) In this paradise even the birdsongs are all one voice, reinforced by an echo: "The fluttering assembly of these divine musicians is so ubiquitous that it seems as if every leaf in the woods had adopted the shape and tongue of a nightingale, and even the echo takes so much pleasure in their tunes that, to hear her repeating them, you would think she desired to learn them by heart" (16).

But I think DeJean overstates the "monologic" presence of this passage—which becomes the setting for a dialogue (with Elijah) of such serious disharmony that Elijah ends up banishing the narrator from paradise. The physical description of the *locus amoenus* introduces us, at any rate, to two things that will continue to characterize the moon—realized or literalized paradoxes and the potential for enormous sensory gratification. The narrator comes first, allegorically enough, "to a crossroads where five avenues met," and the rest of the description is one of often almost synesthetic sensationalism: "I must confess to you that at the sight of so many objects of beauty, I felt myself tickled by those pleasant pains which the embryo is said to feel at the infusion of its soul" (17). The experience has been so intense that, in a kind of mimetic ardor, his "old hair" falls out and he grows new hair, regaining his ruddy complexion and the "bodily moisture" of his youth (17). As his age becomes youth and an alien planet his mother's womb, so the meadows can be taken for oceans, and the ground seems suspended from the roots of the trees, while their crowns are "bowed down under the weight of the celestial spheres" (16). This is appropriate to a world in the Moon, at least for a culture where "world" and "moon" are a mutually exclusive pair (as the lunar priests will agree). It is appropriate, too, to the consequence of the narrator's debate with Elijah: tossed out on his ear for atheism, he becomes in subsequent dialogues the Aristotelian Christian whose final experience on

---

33. For discussion of those trends and Cyrano in relation to them see Erica Harth, *Cyrano de Bergerac and the Polemics of Modernity*. On Menippean satire, see Northrop Frye, *Anatomy of Criticism*; on Lucian's Imennipus and his lunar connections see James Romm, "Lucian and Plutarch."

the Moon is to argue the immortality of the soul against a young libertine atheist at the house of his friend the demon.

These combined features make up the "lunatic" character and his native epistemology, in terms far more complex than those that delimit Godwin's aspiring midget. What makes Cyrano's nameless narrator a continuous being, here in a world of such fundamental discontinuity and self-contradiction, is his awareness of sensation, or rather the passages in which readers are brought to imagine his sensations. There is no subject in the sense conveyed by a persistence of desire (as in Thevet or in Godwin's Domingo), only in that conveyed by the site of repeated sensations.[34] What other kind of subject could keep from going mad in a world where "all things are true"? It is important that the narration is in the first person, especially notable for the period: important not because the first person narrator shares the right of credibility with actual travel writers, but because "person" here is almost definable as subject of sensation, and sensation, however imaginary, is best reported by its putative sensorium. We are not accustomed now, and certainly the seventeenth century was not, to the idea that sensation is accessible for reporting purposes to any but its subject.[35]

If in Cyrano's text sensation denotes personhood, it turns out that, at least according to several philosophizers the narrator meets on the Moon, there are many more kinds of person than one could find in Domingo's little world of masters and servants. A "learned doctor" at the dinner party to which the narrator is taken from his anti-Galilean trial describes to him this way an infinity not just of worlds but of sensoria:

> It remains for me to prove that there are infinite worlds within an infinite world. Picture the universe, therefore, as a vast organism ["grand animal"—cf. Bruno]. Within this vast organism the stars, which are worlds, are like a further set of vast organisms, each serving inversely as the worlds of lesser populations such as ourselves, our horses, etc. We, in our turn, are also worlds from the point of view of certain organisms incomparably smaller than ourselves, like certain worms, lice, and mites. They are the earths of others, yet more imperceptible ["la terre d'autres imperceptibles"]. So, just as each single one of us seems to this tiny people to be a great world, perhaps our flesh, our blood, and our minds are nothing but a tissue of little animals, nourishing themselves, lending us their movements, allowing themselves to be driven blindly by our will.... For do you find it hard to believe that a louse should take your body for a world, or that, when

---

34. See Chapter 3 for my discussion of Christopher Wren's cyborg-scientist and Chapter 6 for that of the sensational detail of Hooke's micrographic observations.

35. The reading of (nonsubjective) reports of pain will be an important topic in Chapters 8 and 9, where its peculiar effectiveness in invoking identification will be discussed in reference to both fiction and ethnography.

one of them travels from one of your ears to the other, his friends should say that he has voyaged to the ends of the earth ["au deux bouts du monde"]? . . . Why, doubtless this tiny people take your hair for the forests of their country, your pores full of sweat for springs, your pimples for lakes and ponds. (75–76)

The fantasies of "animated nature" go on and on, cornucopia-wise, and another philosopher discoursing of atoms extends the realm of the animate even deeper towards the material. Although the hierarchy of size is not discounted (the assumption of it lies behind the comic effect of most of these fantasias), it is still handled in a manner quite different from the oppressed and oppressive paranoia that informs Godwin's Domingo and the naturally virtuous giants of his Moon. If all things here are true, it is among those truths that all things are alive and subject to experience. The polyphony of the text, its constantly shifting point of view, help to manifest the picture of a world in which nothing has been demoted to the level of the inanimate. This vision will be seen again in the next chapter, in Margaret Cavendish's *New Blazing-World*.

Such visions are rarely the mental property of members of the power elite. We have seen that Godwin represents even his comparatively staid impulse toward otherworldly experience as the adventure of a commoner, a youngest son, a midget. Cyrano's narrator is metaphorically and associatively (as is Cavendish's literally) gendered female. As one of his editors, Maurice Laugaa, has pointed out (in an article called "Lune ou L'Autre"), Cyrano's narrator is first introduced to us as pregnant (with ideas about the moon) and when his not very like-minded friends accuse him of having "a quarter of the moon in his head" it is an upside-down honor he shares with his female predecessor, Noah's daughter Achab, whose friends call her out the same way as she jumps out of the Ark into the waters of the Moon (21). When he finds himself at last in the heart of lunar culture, at "the Court," he is presumed to be a female monkey, given to the Queen, and mated with Domingo in hopes that he will become pregnant in fact and not just in fancy.

In a misogynist and homoerotic "voyage" whose cultural description does not include women except intermittently as (usually unpleasant and undesirable) sexual objects, this gendering is especially notable. Cyrano has here transformed the bodily basis of Godwin's narrator's identity into a social ground—a pointedly disembodied notion of gender, since the narrator retains his physical maleness. What compels or at least suggests this choice of gender, and what can it tell us about the imaginative activity of going to the Moon?

Even the familiar satirical use of the Moon to defamiliarize European custom and corruption is an act of rupture with the ground bass of consensus thrummed in the ears of literate or urban monarchical subjects all day. Both actual and imaginary exotic travelers owe a debt of explanation or reparation to the cultural whole they cracked open by departing.[36] It is an explanation that can never be discursively sufficient because he or she can never come all the way back. As DeJean so memorably says of the "libertine lunatic": "the secret that sets him apart from his race is not a power but a voyage. The lunatic has been somewhere almost no one else has been" (*Libertine Strategies*, 120). This voyage may very well come to seem like a power — both Domingo in China and Cyrano's narrator in New France (as well as later, back from the Moon in old France) are suspected of sorcery. The prophet Tiresias is a capacious figure for that power in classical literature — travel advisor to Odysseus in *Odyssey* (bk. 23), he has also traveled to the other world of the dead and to the world of female experience. Cyrano sends his narrator in Tiresias's footsteps in sending him to the Moon (home for such as Plutarch, and Kepler, of the spirits of the dead) and sending him as, in many ways, female — sending him, that is, across the bounds of the actual as it is constructed by any known culture.

Once America had become a part of the European economy it was less useful as a setting for such culture-cracking voyages as Cyrano's or even Godwin's (though such a setting would not be needed outside of the history that changed America's meaning). And as can be seen in almost any moon voyage, the classic "lunatic" protagonist proleptically shows the mark of Cain or Tiresias in an identity socially marked as marginal or holy. In Fontenelle's curious Marquise and Cyrano's combination of "female monkey" and sensorium one can see approaches to a solution that the novel's increasing penchant for verisimilitude will enforce as simply "natural" — the female protagonist. The insufficient discursive explanation for the disruptive voyage is supplemented with easy relevance by the fact of the disenfranchised gender. The less of a mirror the moon is cast as, the more it is cast as an alternative, the more the voyage proposes the possibility of revolution. All the moon voyages we have looked at, as well as Bacon's earlier voyage out to Atlantis, were posthumously published. Before Bacon's *New Atlantis* was written, the author of *Utopia* was imprisoned for his political views, then hung, drawn, and quartered.

---

36. On travel accounts as ritual moments in the returned traveler's "reaggregation," see Michael Harbsmeier, "Spontaneous Ethnographies."

VI • OUTSIDE IN
Hooke, Cavendish, and the Invisible Worlds

If that most instructive of our senses, seeing, were in any man a thousand, or a hundred times more acute than it is by the best microscope, things several millions of times less than the smallest object of his sight now would then be visible to his naked eyes, and so he would come nearer to the discovery of the texture and motion of the minute parts of corporeal things . . . but then he would be in a quite different world from other people.

—John Locke, *An Essay Concerning Human Understanding* (1690)

Just like unto a Nest of Boxes round,
Degrees of size within each Boxe are found.
So in this World, may many Worlds more be,
Thinner, and lesse, and lesse still by degree.
Although they are not subject to our Sense,
A World may be no bigger than two-pence.

—Margaret Cavendish, *Poems and Fancies* (1653)

WORLDS CAN BE CONSTRUCTED in other dimensions as well as on other planets. In England, immediately after the restoration of the monarchy instigated some world (re)construction there, two antithetical writers and thinkers produced nearly simultaneous accounts of invisible worlds in spaces not yet adequately described or easily imagined: the submicroscopic and (in the psychological sense especially) the interior.

Both of these newly located dimensions provoked developments in the arts of prose that must be of keen interest to the historian of significance. Construction is most active and inventive where there is little vocabulary, bibliography, or taxonomy on which to build. Although Robert Hooke was not the first natural philosopher to describe what he saw through the microscope, his superbly illustrated *Micrographia* (1665) was the first published work adequate to its important novelty. Margaret Cavendish was not the first prose writer to conceive of the human "interior" as setting or medium for narratable events (as I have noted in such travel writing as Columbus's

Letter or Léry's *History of a Voyage to the Land of Brazil*), and high lyric poetry had assumed an interior stage at least since the songs of the troubadours had been refined by the *stile nuova* of the *trecento* Italian poets. But with daring and almost literal éclat Cavendish's *Description of a New World Called the Blazing-World* (1666) made of the interior an articulated world, not merely the assumed emotional sensorium of the subject of experience.

Reading a pair of mutually opposed texts by these two philosopher-writers will allow a look at newly developing capacities of prose representation, and what John Rogers has called the period's "discursive commingling" of politics, science, and imaginative representation.[1] Beyond that, because the two reporters from invisible worlds are in fact politically different mainly around the issue of feminism and the capacity for feminine identifications, they can together provide another view of the gendering, not only of early modern science, but of its cousin, the novel. More specifically, they provide views of the dialectical gendering (and emerging division) of the material and immaterial worlds under construction in a literary culture newly open to the value of close description.

What the texts have particularly in common with each other (as well as with Bruno's and Kepler's, and the many pamphlets on the comets of 1618 and 1664/65) is a fascination with the sublime. It is a sublime registered in the sensational, not the grand contemplation of Bruno's *Infinite Universe and Innumerable Worlds*. In Hooke's case it is a sublime deflated in the moment of its production, in a bathetic catharsis. But its presence is a reminder of the literary and readerly stakes of world production.

The *Blazing-World* was published appended to Cavendish's *Observations upon Experimental Philosophy*. In both works she writes in opposition to Hooke and the methods and principles of the Royal Society, but it is in her novel that she demonstrates most clearly and consequentially the values she sees as undercut by the materialism of the new philosophy. Hooke investigates interiors, but his optical instrument deconstructs the notion of "interior" — all it finds is further surfaces, as I will show. Cavendish is interested in the immaterial interior of the person, equally elusive (and in a sense illusive) but representable nonetheless. Her defensive sense that her interest and method are alternatives to Hooke's, rather than simply coexisting around a pun, may come in part from a sense that the immaterial and unverifiable are losing status as objects of knowledge under the pressure of the cyber-certainty of in-

---

1. John Rogers, *Matter of Revolution*, ix. This book contains a clear and cogent discussion, in chap. 6, of the animistic atomism of Margaret Cavendish and its political implications. My readings differ from his in part because I emphasize her later work, after she had rejected the atomism of the materialists.

strument-based (prosthetic) perception—a perception to which women had little or no access.

TRUTH EVEN UNTO
ITS INNERMOST PARTS:
MICROGRAPHIA AND THE DETAILS OF
THE CORPOREAL EYE

[To] use Kenneth Burke's dramatistic terminology, we might say that in romance the category of Scene tends to capture and to appropriate the attributes of Agency and Act, making the "hero" over into something like a registering apparatus for transformed states of being, sudden alterations of temperature, mysterious heightenings, local intensities, sudden drops in quality, and alarming effluvia....

—Fredric Jameson, *The Political Unconscious* (1981)

"Graphic detail" and "explicit language": these suggestive phrases of the modern censor encapsulate the rhetorical aims of western and especially northern European science in the seventeenth century, as well as those of the historically parallel emergence of prose pornography. "So many *things*, almost in an equal number of *words*," advises Bishop Sprat in his *History*, echoing the prostitute Antonia's antimetaphorical advice to Nanna, in Aretino's widely circulated and often-translated *Ragionamenti* (1536): "Speak plainly and say 'Fuck,' 'prick,' 'cunt' and 'ass' if you want anyone but the scholars at the University of Rome to understand you" (Aretine, *Dialogues*, 43).[2]

"Detail," arguably (if vaguely) the characterizing feature of narrative realisms, is in English a seventeenth-century word, brought over with "gros" from the French vocabulary of commerce and meaning "retail," as opposed to wholesale, an item on an itemized invoice or catalogue. The illustrated (graphic) catalogue of courtesans had been the basic form of Renaissance pornography, along with the various editions and copies of Romano's *modi*, engravings based on the Roman *Spintriae*, "a series of chits used as a form of payment in brothels and illustrated with similar couplings" (Paula Findlen, "Humanism, Politics, and Pornography," 78). "Graphic" is also a seventeenth-century word, at least in the meanings invoking vivid description, and the use of pen and pencil.

It is not hard to imagine why English would be enriching its lexicons in the areas of commerce and graphics at this time. My interest in detail and the graphic has to do with their involvement in aesthetic production, including

---

2. The *Ragionamenti* came out in two parts, in 1536 and 1556.

scientific production with aesthetic valences. The much observed and eponymously observing seventeenth-century reader Samuel Pepys (1633–1703), around whose needs and habits so much cultural history has been organized, was excited enough by both Hooke's *Micrographia* and the pornographic *École des filles* (1655) to narrate their purchases and perusals in his diary. He stayed up late, right away, to read both. The difference—an important one—is that he burned *L'École des filles* before he went to bed.[3] As Sprat pointed out, in descriptions of experiments, men "may enjoy [the pleasures of their senses] without guilt or remors" (Sprat, *History*, 344).

What Findlen calls the late Renaissance "eroticization of the senses" ("Humanism, Politics, and Pornography," 76) might be more precisely called the eroticization of sight, although it's true that music also had an ecstatic effect, and Pepys is almost embarrassed one evening by the "too much raffined" detail of Hooke's analysis of musical sound: "[he] told me that having come to a certain number of vibrations proper to make any tone, he is able to tell how many strokes a fly makes with her wings . . . by the note that it answers to in musique during their flying" (8 August 1666). Pepys's demurral here is an important reminder that all graphic detail, not just vivid representations of sexual organs and acts, had acquired an erotic charge: of a piece by the Dutch painter Verelst, "the finest thing that ever I think I saw in my life," Pepys singles out "the drops of dew hanging on the leaves, so as I was forced again and again to put my finger to it" (11 April 1669).[4]

The effect of verisimilitude may be, as Roman Jakobson points out, an effect based on a style's relation to inherited paradigms, and therefore both historically determined *and* perennial ("On Realism in Art," 20–24). But it is undeniable that the painting and illustration of the seventeenth century in northern Europe maintain that effect on viewers late in the twentieth, and that the energies of the period's scientific institutions were directed towards description—description stripped, so to speak, of ornamental "mists and uncertainties." Scientists and painters both seem a bit like those "registering ap-

---

3. Pepys records working himself up to buying and then reading *L'École des filles* in a series of three entries in 1668, culminating on 9 February ("*Lords Day*"): "I to my chamber, where I did read through *L'Escholle des Filles*; a lewd book, but what doth me no wrong to read for my information sake (but it did hazer my prick para stand all the while, and una vez to discharger); and after I had done it, I burned it, that it might not be among my books to my shame" (*Diary* 9:59). The viewing, purchase, and reading of *Micrographia* are also recorded in three separate entries, ending on 23 January 1665: "Before I went to bed, I sat up till 2 a-clock in my chamber, reading of Mr. Hookes Microscopicall Observations, the most ingenious book that ever I read in my life" (*Diary* 6:18).

4. The last three quotations from Pepys can be found in 3:126, 7:239, and 9:515, respectively. On Renaissance study of the senses, see Alistair Crombie.

paratuses," Jameson's romance heroes (112; see epigraph above) — in fact, Christopher Wren designed a kind of robot apparatus that would register, when the scientist was asleep, precisely those features of "Scene" enumerated by Jameson.[5]

This period then is one that saw the emergence of some of the most convincing of all representational tricks: if erotic representation physically arouses, if effects described in an experiment report can be recognizably reproduced, then verisimilitude (also a seventeenth-century word) can produce not merely the imitation but the physiological illusion of reality. It can, for better or worse, compel belief. Verbal techniques that deploy this power in the service of "true history" will be called novelistic: "Truth hath no greater Enemy," said Flecknoe in his *Relation of Ten Years Travels* (1656), "than verisimilitude or likelihood" (29–30).[6] It will be a new power, newly deployed through a new *techne*, and it will carry with it the burdens of its other uses, or atmospheric qualities transferred from the other worlds in which it has compelled different, but not altogether different, varieties of belief.

In the discussion that follows I attend to the atmospheric qualities of verisimilitude's other worlds, especially the world called Nature, which is in the process of being constructed through the discourses of natural philosophy by writers mostly outside the world of the court. The normative textual varieties of world construction at this time I take to be the familiar literary and subliterary form of romance (as Jameson says, "precisely that form in which the *worldness* of *world* reveals or manifests itself," 112) and the historically new form of the commercial or colonial travel account — the form that most reminds us of ethnography. Around the productive if sometimes deadly embrace of these two kinds of writing crowd other kinds of writing, of looking and imagining, of feeling. It is hermeneutically useful to look at them all as "kinds of world-making" (to paraphrase Nelson Goodman), and many of them were so explicitly (George Abbot's *Brief Description of the Whole World* [1599], John Wilkins's *A Discourse Concerning a New World* [1638], and so forth). The early modern impulse to bring a world into imagined being seems general, even compulsively so — in the arenas of government, commerce, and academia, as well as in the island-wild and planet-mad imaginations of travelers, astronomers, and fiction writers. It was an impulse underwritten by the

---

5. Plot describes this apparatus in his *Natural History of Oxfordshire*. See Chapter 3, 104. For Jameson on romance heroes, see *Political Unconscious*, chap. 2.

6. By "likelihood" Flecknoe may mean as much probability (and thus credibility) as "likeness." "Verisimilitude" is a carpet under which whole worldviews can be swept.

material possessiveness of state-supported monopoly commerce, a collective attitude expressed with startling concision when a western European trading outfit is named the "East India Company." Such possessiveness, profuse generator of listing genres like the invoice, the catalogue, and the questionnaire, seems bound to foster detail in more discursive forms, including description in natural philosophy and pornography, and in general to respond with pleasure to "the texture and motion of the minute parts of corporeal things," as Locke puts it (see the chapter epigraph above). A world, from this point of view, is a huge accretion of objects or fragments, held together by their common owner, or by a subjectivity modeled on ownership.[7]

Detail then expresses investment, and particularly erotic investment — the reason the phrase "graphic detail" raises parental hackles is that we instantly assume the human body as the mimetic object or source of such detail. "Gratuitous violence" is an especially abhorred subset, not because we abhor violence, but because "gratuitous" or "graphic" details are understood to inspire erotic feelings, and sadism is a tabooed source of sexual excitement. (This was so for Robert Hooke as well, who cannot detail his experimental dissection of a live dog, and will not repeat the experiment when asked: "A Dog was dissected, and by means of a pair of bellows, and a certain Pipe thrust into the Wind-pipe of the Creature, the heart continued beating for a very long while after all the Thorax and the Belly had been opened" [Sprat, 232]).[8] Pornography, then, might seem to be the central forum for the development of verisimilar techniques, or at any rate the genre for which other descriptive genres can be read as substitutions: "What raptures can the most voluptuous men fancy to which these experiments are not equal?" Reading Lynn Hunt's magisterial account of early pornography and its literary and social valences (her introduction to *The Invention of Pornography*), one wonders what the developmental history of the novel might have been if the pornographic had not been so heavily censored early on, but this is an imaginary line of thought.[9] The novel as we know it might best be thought of as what the sensational looks like when the pornographic is excluded; the same

---

7. In his article "Rhetoric and Graphics in *Micrographia*," John Harwood treats Hooke's method of introducing coherence into the accretion of *Micrographia*'s observations, in part via a specifically located *persona* which functioned as well to represent the collective subjectivity of the Royal Society's *virtuosi*. My reading suggests an additional resource, the plot. Harwood also has relevant things to say about Hooke's use of the rhetorical trope of *enargeia*, vividness, a trope whose resources were, as we have observed, increasing in his time and place.

8. Context for this event can be found in Margaret 'Espinasse, *Robert Hooke*, 52.

9. See Hunt, "Introduction: Obscenity and the Origins of Modernity," an introductory essay enriched by the research and speculations of the other contributors as well as by recourse to some major previous works in the area, especially David Foxon, *Libertine Literature in England*; Robert Darnton, *Literary Underground of the Old Regime*; and Angela Carter, *Sadeian Woman*. See

could be said of seventeenth-century empirical science and mechanics. Pornography was known, at least in France, as "philosophical books."

"Fetish" is a term we might consider as the hyperbolic extension of "detail": it is also a seventeenth-century word in English and, like detail, a word borrowed from the vocabulary of commerce (in this case, the Portuguese gold trade in Guinea, where Akan traders weighed the material by means of ornamental brass "amulets"—items, detail—projectively mistaken as objects of worship by the greedy Portuguese!).[10] Like so many features of early capitalist and colonialist expansion, the term and concept have been internalized, imported into the European language of individual and private eros, carrying along their inventory of racist misunderstanding and projection. The Akan "fetishist" then and the later neurotic "fetishist" of *fin de siècle* Vienna were both assumed to be making a materialist *mistake*, worshiping as the thing itself what is "only" an arbitrarily and even whimsically designated signifier. Although both the Akan trader and the Viennese foot fancier were up to things beyond the purview of this European glance, the concept of fetish has its applications: detail in representation *is* describable this way, at least in its sensational and sensationalist early salience, underwritten by such rhetorical fantasies as Sprat's ("so many things in almost an equal number of words") and repudiated a few decades later by the ethical Dr. Johnson (who does not care to "number the streaks of the tulip"). In the case of prose it is read as evidence, a trace of the narrator's avid encounter with the object, and consequently for readers the fetishized detail is observed voyeuristically. This may explain why Hooke describes his dissection of the dog in the passive voice, at a time when the first-person and the active verb were normal in scientific writing.

The relation of prose detail and commerce is summed up nicely, especially as regards the masterwork of Robert Hooke, inventor of the balance-spring watch, by a Rolex ad quoted in Susan Stewart's wonderful book *On Longing*: "Detail illuminates John Cheever's writing. Just as detail inspires every Rolex craftsman. Created like no other timepiece in the world. With an unrelenting, meticulous attention to excellence in a world fraught with compromise. The Rolex Oyster Perpetual Day-Date Superlative Chronometer" (*On Longing*, 28). In the *New Yorker* ad of the 1970s an anonymous copywriter, a Grub Street hack and probably an aspiring Cheever herself, brilliantly fuses the strains of lust, commodity value, literary realism, and mechanistic philosophy that first came into productive contact in Hooke's time, adding

---

also, for some bibliographic context, T. A. Birrell's "Reading as Pastime," on "light literature"—as he names the category often called "curious" in seventeenth-century catalogues.

10. See William Pietz, "The Problem of the Fetish, IIIa."

her time's timepiece to a tradition in fetishistic realism that includes Gulliver's giant pocket watch ("the God he worships," conclude the Lilliputian ethnographers) and the terrifying cacophony of the condemned man's watch in Ambrose Bierce's "Occurrence at Owl Creek Bridge."[11]

Detail has a relation to epistemology as well as to sensational reading, especially in the case of scientific writing—I don't mean to say that the *Micrographia* is *only* a respectable alternative to *L'École des filles*, and indeed by the end of the century there was considerably more pornography than micrography being written. As Hooke lamented in a lecture to the Royal Society, microscopes "are now reduced [!] almost to a single votary, which is Mr. Leeuwenhoek" ("Discourse Concerning Telescopes and Microscopes," 738).[12] The fragment, the datum, had a reforming function in an overly theorized epistemological realm like that inherited from Renaissance natural philosophy. Though of course no datum is or can be free of theoretical context, the point was to keep the *salience* of theory down, to channel the energies of natural philosophy away from the philosophy part, "whence," Sprat says, "it has flown away too high" (119). To motivate an attention to the *material*, a business fraught with narcissistic peril for the gentlemen of a metaphysical and mysogynist culture—not to mention a society in the throes of armed class struggle—was a task that required the promise of some pleasure.[13] A century later, Kant was still placing satisfactions taken in small or ornamental particulars at the very bottom of his aesthetic hierarchy—the fetish is not even beautiful, it is "trifling" (*läppisch*), and so, therefore, are the Africans whose imaginary megaculture he represents by means of that metonymic detail.[14] Hooke anticipates this attitude in the preface to his *Micrographia*, asking that his book be no more compared to "the *Productions* of many other *Natural Philosophers* . . . then my *little Objects* are to be compared to the greater and more beautiful *Works of Nature*" (gii^v).

I have taken Hooke at his word, and compare his *Micrographia*, not to the

---

11. "Striking through the thought of his dear ones was a sound which he could neither ignore nor understand, a sharp, distinct, metallic percussion like the stroke of a blacksmith's hammer upon the anvil; it had the same ringing quality. . . . Its recurrence was regular, but as slow as the tolling of a death knell. . . . The intervals of silence grew progressively longer, the delays became maddening. With their greater infrequency the sounds increased in strength and sharpness. They hurt his ear like the thrust of a knife; he feared he would shriek" (Bierce, 306–7). Swift's Lilliputian description of Gulliver's pocket watch is in *Gulliver's Travels*, pt. 1, chap. 2.

12. He goes on to say they've become "a diversion and pastime," i.e., a toy for women—that class of person whose scopophiliac needs must always be "diverted" (738). (Swift never had a microscope; when he did eventually buy one, it was for Stella to play with.)

13. See Daston and Park, *Wonders*, 305–16, esp. 315.

14. *Observations on the Feeling of the Beautiful and Sublime*, 111.

productions of other natural philosophers, but to other graphias of his moment. Reading his beautiful book as a kind of ethnography of the miniature, a motionless voyage into another world, a permissible catalogue of the technically ob-scene, what can we discover about the function of graphic detail? What kind of world is it equipped to make?

The curator of experiments for (and the only paid professional scientist in) the Royal Society, trained during the Interregnum in Oxford's brief scientific Golden Age (when brilliant dissenters staffed and ran so many colleges), Hooke shares Milton's sense of living in a fallen body in a fallen world.[15] The attitude of the *Micrographia*'s preface is utopian and revisionist, despite the Restoration: "as at first, mankind fell by tasting of the forbidden Tree of Knowledge, so we, their Posterity, may be in part restored by the *same* way, not only by beholding and contemplating, but by tasting too those fruits of Natural knowledge, that were never yet forbidden" (another instance of what seems at times a conscious exclusion of the pornographic from his sensational text; *Micrographia*, bii^{r-v}). The twist on this "tasting" is Hooke's sense, painfully thematized throughout the work, of the embodied subject's insufficient sensitivity. "The first thing to be undertaken in this weighty work [the reformation of natural philosophy], is a watchfulness over the failings and an enlargement of the dominion, of the Senses" (aii^r). The register/hero of Hooke's romance, then, is imagined cybernetically: "as *Glasses* have highly promoted our *seeing*, so 'tis not improbable, but that there may be found many *Mechanical Inventions* to improve our other Senses, of *hearing, smelling, tasting, touching*" (bi^v).

The preface promotes a strange entanglement of prosthetic distancing with sensationalist intensity: On the one hand, "I do not only propose this kind of Experimental Philosophy as a matter of high rapture and delight of the mind, but even as a material and sensible pleasure" (dii^r). On the other hand, "the roughness and smoothness of a Body is made much more sensible by the help of a Microscope, then by the most tender and delicate Hand" (cii^v). In fact, as the passage makes clear, it is the sense of touch whose dominion has been most successfully (and prosthetically) enlarged by 1665, by the development (sometimes Hooke's own work) of such instruments as the

---

15. For the strongest view of Puritan influence on (or "commingling" with) scientific development in the seventeenth century, see Charles Webster, *The Great Instauration*. Dissenting voices among historians include James Jacob and Margaret Jacob's "Anglican Origins of Modern Science" and Lotte Mulligan's "Puritans and English Science." John Rogers's literary account of the "commingling" absorbs both views of influence but remains Websterian in its fundamental allegiance to connections between the two "revolutions" in England.

thermometer, the barometer, the microscope, and the burning glass. The book itself is included here as a prosthetic device for the brain's function of memory—a reminder of the disembodied, spectral quality reading must generally have had in the early days of print culture. It was not only voice that print replaced, but the bodily experiences of memory and fantasy, both of which have come to seem far less bodily over time.

The whole of the preface is structured to produce the registering apparatus-hero (by means of elaborate descriptions of various actual and potential means for "enlarging the dominion of the senses") and to motivate him (or it); the plot in which our cyberscientist will perform is—surprise!—a plot of conquest, set in a New World. Hooke explains his dependence on instruments as "a way . . . herein taken . . . to promote the use of Mechanical helps for the Senses, both in the surveying of the already visible World, and for the discovery of many others hitherto unknown, and to make us, with the great Conqueror [Alexander], to be affected that we have not yet overcome one World when there are so many others to be discovered, every considerable improvement of Telescopes or Microscopes producing new Worlds and Terra-Incognita's to our view" (dii^v). This is not just an easily available metaphor, though it appears in every English micrographic work of the period.[16] The title *Micrographia* is modeled—parodically, in Bakhtin's lofty sense of that term—on the sixteenth-century *Cosmographia*s of such authors as Waldseemüller, Münster, and Thevet because in fact it is one, or the microscopic beginnings of one. The (etymologically) *ob-scene* world of cells and microbes was newer than new: antipodean peoples had been guessed and fantasized but protozoa arrived unheralded. Here was a whole physical *dimension* to absorb, through the categories of place and creaturely inhabitant that structured cosmography and the nascent enterprise of mercantile ethnography. Everywhere the cyberscientist turned his enhanced sensing equipment, more places and creatures sprang innumerably to his attention. As Jameson says of the novel: "events take place within the infinite space of sheer Cartesian extension, of the quantification of the market system: a space which like that of film extends indefinitely beyond any particular momentary 'still'

---

16. Nehemiah Grew, for instance, in the dedicatory letter to Charles II of his heavily micrographic *Anatomy of Plants* (1682), calls plants one of the "Terrae Incognitae in Philosophy," and himself "the first, who have given a Map of the Country." Other English works of the period include Henry Powers, *Experimental Philosophy* (1664), illustrated with only a few crude drawings. Hooke's serious competitors were continental—Malpighi in Italy and Swammerdam and Leeuwenhoek in Holland. These continental philosophers made important anatomical discoveries with the microscope, although it seems to have driven Swammerdam mad. For more on the early texts of microscopy see especially Hellmut Lehmann-Haupt, "The Microscope and the Book," and Catherine Wilson, "Visual Surface and Visual Symbol" and *The Invisible World*. The best illustrated recent discussion is Barbara Maria Stafford's sumptuous *Body Criticism*.

or setting or larger vista or panorama and is incapable of symbolic unification" (111).

I find Jameson's formulations eerily, because so literally, applicable to the works I am considering vis-à-vis Hooke, but his distinction between novel and romance here seems mainly heuristic; at least it is possible to consider *Micrographia* as both. Clearly the narrator of the preface is overcome by the infinite extension of his new "world." But as in most of the fictions that will follow Hooke's opus, and novels generally, this infinite space *is* in fact capable of "symbolic unification." It is not that detail in representation registers a world prior to or in excess of its aesthetic or sentimental framing—I am suggesting that detail merely invites a more physiological participation in the always partly fantastic process of "getting to know." The extreme form of such a carnal knowledge, as the seventeenth-century translators of the Bible knew, is sexual; the extreme of verisimilitude is pornographic. As Fran Lebowitz said in a recent *Paris Review* interview, "Writers have problems writing sex scenes, because writing one really well is pornography" (183).

Naturally there is a female love-interest named Nature whose footsteps are especially "to be trac'd, not only in her ordinary course, but when she seems to be put to her shifts, to make many doublings and turnings, and to use some kind of art in indeavouring to avoid our discovery" (aii^r). The "secret workings" and "inward motions" of this coy mistress, equally interesting to voyeuristic authors and readers of seventeenth-century pornography, with its "School of Girls" and "Academy of Ladies" and its various private dialogues between women of pleasure, would eventually become the focus of the domestic novel, banished both from the premises of the laboratory and from the role of pornographic protagonist (a role increasingly taken by male figures). At this point, though, she functions, outside pornography, mainly in *informational* prose, as a figure for America, the Moon, and Nature in general. Her presence in the body, so to speak, of Hooke's text is mainly indicated by the preponderance there, amidst the cornucopian physical detail, of the language of desire and repulsion, in which language we can trace the structure of a subterranean erotic plot, moving the registering hero from aversion and artifice to a sublime and despairing climax, and finally to the genial domesticity of the kitchen.

The plot of *Micrographia* is enriched by a double task: not only to effect a resolution of erotic conflict but to restructure the reader's values. The small cannot remain trivial in this expensive book. We must be converted to another scale besides that social scale in which the chief objects of microscopy figure, as filth and vermin, at the devalued bottom. This conversion has a political meaning: "detail increases stature," as the *Dictionnaire de botanique chrétienne* put it in a description of the periwinkle.[17] Discussion of the small, the humble and the common in such *dense* detail, accompanied by the stunning art of the engravings and the sheer etymological fact of magnifying, requires a complex decorum in a book dedicated to King Charles II, in the wake of the restoration of the monarchy. Hooke's figural solution, which ends the dedicatory letter, links the imperial and the erotic in the notion of surveillance:

> Amidst all those *greater* Designs [the improvement of manufactures and agriculture, the increase of commerce, the advantage of navigation], I here presume to bring in that which is more *proportionable* to the *smallness* of my Abilities, and to offer some of the *least* of all *visible things*, to that *Mighty King*, that has *establish't an Empire* over the best of all *Invisible things* of this World, the *Minds* of Men.

If we take it seriously, the feudal figure offered here, of the mirroring servant giving gifts to the beneficent Lord, the King's ominous power over the "Invisible . . . Minds of Men" has its parallel in the scientist's power to penetrate the visual surfaces of the "least of visible things." And so the Terra Incognita of the microscopic, that "dominated world" (in Bachelard's phrase for the miniature), is also an Empire.[18] The analogue to Hooke's *microscopic* subjects (the Louse, the Mite, the snowflake) is the *human* subject, the political person's interiority, with its lattices of sensation and emotion—the subject of sensational texts and of the rhetoric of the sublime. Interiors *must* be valuable, since they are the spaces of the King's "Empire." *Micrographia*'s greatest contribution to the vocabulary of biology is the word "cell," a metaphor in which a human-scaled architectural form of absolute interiority is applied to the fundamental and microscopically miniature unit of organic life.

But there is more here than a figural paradigm; as I said there is a *plot*, the structure of which serves both erotic and political allegories. The plot of desire begins in narcissistic repulsion, in a characteristically self-reflexive opening chapter describing various microscopic encounters with "graphia" itself (including micro-graphia, tiny writing). It is especially stern on the subject of full stops (see figure 16):

> I observed many both printed ones and written; and among multitudes I found few of them more round or regular than this which I have delineated in the third figure . . . , but many more abundantly disfigur'd . . . the most curious and smoothly engraven strokes and points looking but as so many furrows and holes,

---

17. The phrase from the *Dictionnaire* (a volume in the 1851 *Nouvelle Encyclopédie théologique*) is quoted in Chapter 7 ("Miniature") of Gaston Bachelard's *Poetics of Space*, 155.

18. "I feel more at home in miniature worlds, which, for me, are dominated worlds" (*Poetics of Space*, 161).

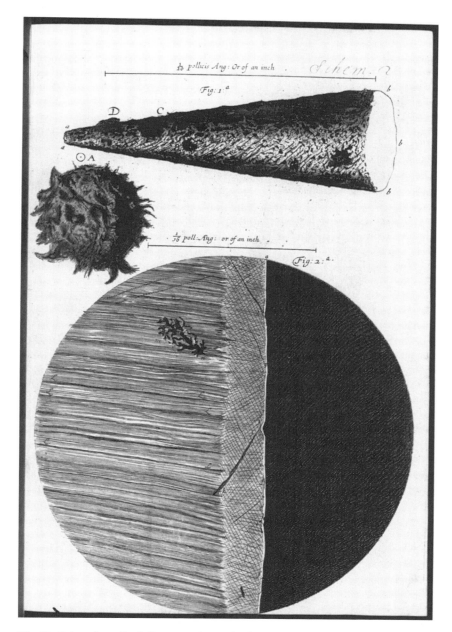

Fig. 16. Point of needle, full stop, and razor's edge. Schema 2 of Robert Hooke's *Micrographia* (London, 1665). By permission of the Houghton Library, Harvard University.

and their printed impressions, but like smutty daubings on a matt or uneven floor with a blunt extinguish't brand or stick's end.... It appeared through the *Microscope gray*, like a great splatch of *London* dirt. (3)

Many repressions seem on the verge of returning in this passage, with its horror of dirt and disfigurement, of "furrows and holes," of the body of writing figured ultimately in what seems an allusion to the raped and mutilated Lavinia of Shakespeare's *Titus Andronicus* (1594), writing her catastrophe on the ground with a "stick's end." Moving from text to textile, Hooke offers next his observations of fine linen, taffeta, and watered silk; here the "liquefaction of [Julia's] clothes" is closely analyzed and diagramed as proceeding from "the variety of the Reflections of light, which is caus'd by the various shape of the Particles, or little protuberant parts of the thread that compose the surface" (8).[19] Like Swift, Hooke dislikes the pores and furrows discoverable in smooth surfaces by "ocular observation," and worries that this graven quality may pervade even the natural world: "future observers may discover even [the fluid bodies of nature] also rugged ... and may find reason to think there is scarce a surface *in rerum natura* perfectly smooth" (5). With the benefit of hindsight it is easy enough to leap ahead from the body horror of these passages to that of Swift's micrographic rendition of the texture (or topography) of Brobdingnagian female skin: "Their skins appeared so coarse and uneven, so variously coloured when I saw them near, with a Mole here and there as broad as a Trencher, and hairs hanging from it thicker than Pack-threads" (95).[20]

But if the opening chapters of *Micrographia* present a strangely depressed and squeamish entry beneath the visible surface of the world, our hero's anxiety diminishes as he turns his attention to more and more elemental substances, diving deeper, beneath Nature's paint and laces, into the inward but visible cells of Truth. The salt crystals in urine, ice crystals in snow, or quartz crystals in flint are beautiful to him, as finally even charcoal is (less appealing in its earlier appearance as an instrument, a "blunt extinguish'd brand"

---

19. "Whenas in silks my Julia goes, / Then, then (methinks), how sweetly flows / That liquefaction of her clothes"; Robert Herrick, Cavalier poet, wrote these lines (from "Upon Julia's Clothes") in 1648.

20. For a brilliant investigation of finely depicted texture in relation to bodily pleasure and horror in prose fiction, see Ranu Bora, "Outing Texture," which opens with a resonant passage from Henry James (*The Ambassador*): "Chad was brown and thick and strong; and, of old, Chad had been rough. Was all the difference therefore that he was actually smooth? For that he *was* smooth was as marked as in the taste of a sauce or in the rub of a hand. The effect ... had given him a form and a surface ... as if ... put in a firm mould and turned successfully out ... marked enough to be touched by a finger" (94).

that makes graphemes); he likes its "infinite company of exceedingly small, and *very regular pores, so thick and orderly set*" (107, emphasis mine). Increasingly our author represents the objects of his and now our observation as objects providing occasions of beauty, desire and satisfaction. A method of making "small Globules . . . of Lead" is "in it self exceeding pretty" (47); the iridescence of mother-of-pearl "have I also sometimes with pleasure observed even in Muscles and Tendons" (53). Blue mold, he finds, "has a very pretty shap'd Vegetative body" (125) and as for moss, "it may compare for the beauty of it with any Plant that grows" (131). Perhaps most communicative of pleasure and wonder are the passages in which one can detect the hero in the action of playing with the tiny things of this world: "rotten Wood, rotten Fish, Sea water, Gloworms, etc. have nothing of tangible heat in them, and yet . . . they shine some of them so Vividly, that one may make a shift to read by them" (55).

The gradual progress from the repulsively irregular cultural artifact to the desirably lovely work of Nature is echoed in the progressive order of the book's "observations," as the chapters are labeled (an order reversing the hexameral tendency we saw in Hariot's and Kepler's accounts of America and the Moon). Along with the register-hero's erotic trajectory toward the natural, another of Hooke's methods of restructuring his readers' values is his destabilization of scale. The observations advance in a more or less traditional, if inexplicit, chain of being, from the most dead things — in this case, human marks and artifacts — to the most living — the Louse, anthropomorphized into a bit of socially mobile low-life that "fears not to trample on the best, and affects nothing so much as a Crown" (211). If the book ended here, with a Louse in the syntactical position of the King on its foreshortened Chain of Being, it would suggest a clear allegory of the world-upside-down, so recently "righted" in England with the Restoration. This is surely not Hooke's open intention, in a book dedicated to Charles II, but the materials of microscopy tend toward that allegory of their own accord in a culture where the opposition small/great has such prominent social meanings. The preface ends with an interpretive prophylactic, already briefly quoted. Calling his book a "Mite, cast into the vast treasury of *A Philosophical History*," Hooke asks "that these my *Labours* will be no more comparable to the *Productions* of many other *Natural Philosophers*, who are now everywhere busie about *greater* things; then my *little Objects* are to be compar'd to the greater and more beautiful *Works of Nature*, A Flea, a Mite, a Gnat, to an Horse, an Elephant, or a Lyon" (gii^v). The careful analogies and parallelisms and the closing zeugma of this sentence, however, tend to operate against its expressed demand, making one end of the scale equal to, because parallel with, the other, and implicitly reversing the polarity of "greater" and "little" as we translate the modesty topos into its lightly concealed and customary source in ambition.

But the book does not—could not—end after the sensational and climactic display of the Louse. It wobbles into another depression, making its transition to two concluding telescopic observations through a series of increasingly small and nameless objects terminating in a long chapter on air. Air, of which the particles are *invisibly* small, provides a physiological limit to Hooke's investigations—and appropriately the chapter on air is especially concerned with horizon. The self-enclosing "worldness" of this book's world is manifest in Hooke's gesture of turning that impenetrably transparent object into a giant lens, a lens subject to all the impurities and limitations that haunted the "most curious wrought Glasses" discussed in his first chapter: "since the Invention . . . of Telescopes, it has been observ'd by several, that the Sun and Moon neer the Horizon, are disfigur'd . . . and are bounded with an edge every way . . . / ragged and indented like a Saw" (217–18). This raggedy disfigurement of the supreme celestial bodies Hooke thinks to be the fault of what he terms "the *inflection,* or *multiplicate refraction* of those Rays of light within the body of the Atmosphere," and he can see it easily enough without a telescope, in the twinkling of candles or, at sunset, "the tremulation of the Trees and Bushes as well as of the edges of the Sun" (219).

Micrographia, like pornographia, tests as well the limits of Sprat's and Aretino's linguistic fetishism. The test Hooke provides is an occasion of the sublime, in a passage that explodes the denotative constraints of thick description partly through its very excess of thickness, illustrated by an engraving that similarly and simultaneously bursts the informational norms of scientific engraving. He introduces the internal logical limit to the pleasures of the microscopical eye in the preface, in reference to an object which is itself an eye, but a thoroughly object-ified eye, a dead one that can't look back:

> It is exceedingly difficult in some objects, to distinguish between a prominency and a depression, between a shadow and a black stain, or a reflection and a whiteness in the colour. . . . The Eyes of a Fly in one kind of light appear almost like a Lattice, drilled through with abundant small holes. . . . In the Sunshine they look like a Surface cover'd with golden Nails; in another posture, like a surface cover'd with Pyramids; in another with Cones. (fii^v)

This eye is dramatized, visually and verbally, in plate 24 and chapter 39: "Of the Eyes and Head of a Grey drone-Fly." The engraving is a sensational object, with a number of effects (see figure 17). It positions us—after we unfold the outsized page on which it is printed—in a relation of initially

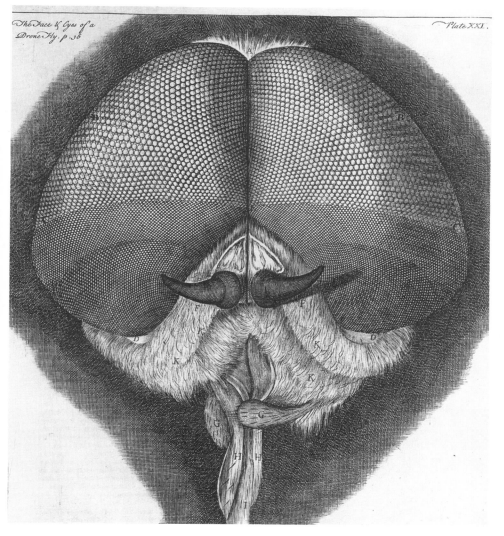

Fig. 17. Eye of a gray drone-fly. Schema 24 of Robert Hooke's *Micrographia* (taken from *Micrographia restaurata*, 1745 abridged reissue of original plates [there labeled plate 21]). By permission of the Houghton Library, Harvard University.

horrifying intimacy with the tiny creature whose decapitated head is here exposed to our gaze, and in yet another flip-flop of scale reverses the real size difference between our kinds; an eye that size must belong to a fly larger, at least with extended wings, than ourselves. Then in a nanosecond we realize that this huge and complex organ of sense is in fact dead, and therefore blind, and another wave of sensation takes over at the prospect of such extravagant capacity rendered impotent—hyperbolically visible, hyperbolically blind at once.

The extended climax of the verbal description—six folio pages of minute detail—emphasizes the mechanism of vision in such an eye, but manages only to convey a sense of surface beneath surface, each of them reflecting back to the unexamined eye of the *scientist* an image of himself and his own domestic "interior" (physically an exterior, socially an interior). After first introducing the "multitude of small *Hemispheres* . . . rang'd over the whole surface of the eye in very lovely rows," he goes on to offer through those hemispheres the very raw material of "realism":

> Every one of these *Hemispheres*, as they seemed to be pretty near the true shape of a *Hemisphere*, so was the surface exceeding smooth and regular, reflecting as exact, regular and perfect an Image of any Object from the surface of them, as a small Ball of Quick-silver of that bigness would do. . . . In so much that in each of these *Hemispheres*, I have been able to discover a Landscape of those things that lay before my window, one thing of which was a large Tree, whose trunk and top I could plainly discover, as I could also the parts of my window, and my hand and fingers, if I held it between the Window and the Object; a small draught of nineteen of which [hemispheres], as they appeared in the bigger Magnifying-glass to reflect the Image of the two windows of my Chamber, are delineated in the third *Figure* of the 23. *Scheme* [see figure 18]. (175–76)

My window, my chamber, my hand; a landscape of things. Like Robinson Crusoe's, this is an owned world; even other creatures are owned. If Hooke had not decapitated it, the drone-fly would be able to see it all, for "he [hath] an eye every way." Swift will rectify the matter early in the next century, when his narrator Gulliver takes up the position of the fly, in Glumdalclitch's chamber. For now, we are given to see it not as the fly experiences it but as it is visible to the owner in the dead mirrors of the fly's "corporeal eye." Unlike Hermione or Galatea, the object of this erotic gaze does not come to life and satisfy a passionate desire. More like Defoe's Friday, or the Hottentot Venus of nineteenth-century anatomists, it is turned into a possession and dies.

The kind of world Hooke's details seem best equipped to make is a private, domestic, vernacular world. It would seem to be the interior of the bourgeois

Fig. 18. Detail from drone-fly's eye. Figure 3 of schema 23 of Robert Hooke's *Micrographia* [1665]. By permission of the Houghton Library, Harvard University.

home that the microscopists have discovered as, in Nehemiah Grew's words, "we come ashore into a new World, whereof we see no end. It may be, some will say, into a new Utopia" (dedicatory letter, in Grew, *Anatomy of Plants*) — but there's no place like home. Hooke closes the *Micrographia* with a gesture toward the great beyond but in fact, though his observations veer suddenly away from the domestic environment in which he has encountered his paradoxically scintillating vermin, mold, and urine, the new inverted scale of values seems to be holding steady. He winds up with a few little stars, carelessly graven on plate 38 as asterisks, and finding "a pretty large corner of the Plate . . . void," he fills it with "one small specimen of the Moon" (see figure 19).

As it happens, it's the Moon's Mt. Olympus he chooses to describe, which turns out not to be a mountain at all but a valley (as with the hemispheres on a fly's eye, it's hard to tell the prominencies from the depressions). The valley is pear-shaped, and "most of [the] encompassing hills may be covered with so thin a vegetable Coat, as we may observe the hills with us to be, such as the short Sheep pasture which covers the Hills of Salisbury Plains"

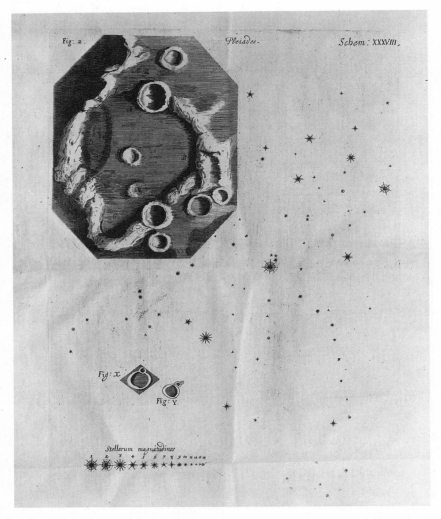

Fig. 19. A patch of the Moon and "a few stars." Final schema, number 38, of Robert Hooke's *Micrographia* [1665]. By permission of the Houghton Library, Harvard University.

(243). The thing that interests and baffles him most is the origin of the craters. We leave him making models of them, in "experimental ways": "The first was with a very soft and welltemper'd mixture of Tobacco-pipe clay and Water, into which, if I let fall any heavy Body, as a Bullet, it would throw up the mixture round the place, which for a while would make a representation, not unlike these of the Moon" (243). This is not the courtier-astronomer Kepler had in mind when he apostrophized the telescope earlier in the century: "O telescope, instrument of much knowledge, more precious than any sceptre! Is not he who holds thee in his hand made king and lord of the works of God?" ("Preface," *Dioptrice*). It's an industrious and inquiring English bourgeois playing with mud in his kitchen, a guy who like Bachelard seems, for the moment, "more *at home* in miniature worlds." His little models of the Moon's craters please him quite a bit, but Cynthia, for all her radiance, leaves him cold.

The mundane moon is an astounding innovation, dragged in oxymoronically to serve what seems to me the most detumescent piece of narrative closure in seventeenth-century English literature. It is among other things a *clever* move, feeding the end directly back into the visually analogous and equally oxymoronic full stop (also cratered) with which the text begins; it thus preserves the sense of a coherently circular horizon, the "*worldness*" of the micrographic world. As a gesture in the plot of the "world upside down," the domestication (and investigation) of the heavenly body is obviously demystifying, in every way. As a destination for the romance plot it is wittily parodic—the questing registrar reaches the Moon, but it's just like home. Mt. Olympus is not a mountain, it's a valley, in fact it's a sheep pasture. The whole business is visible from the backyard, reproducible in the kitchen. This is not a consummation, it's a marriage, aimed at (mechanical) reproduction ("finding a pretty large portion of the Plate . . . void"). All that "high delight" and "material pleasure" were left behind, after the sublime and bathetic encounter with the microscopic eye of a gray drone-fly.

This plot that must reject or lose its object (the Other) anticipates the familiar depressed structure of many early novels—*La Princesse de Clèves* (1678), *Oroonoko* (1688), *Robinson Crusoe* (1719), *Gulliver's Travels* (1726); its scandalous precision of physical observation reads like a riposte to *Don Quixote* (the two parts of which had first been translated in 1612 and 1620) and a forerunner to *Pamela*'s minute invoices of the sensible world (1741). But I may be the only person who ever read it for the plot, and even I am really just pretending. *Micrographia* is *more* highly plotted, more artificially *shaped* than other works in its genre and thus functions more easily to channel features of style or perception into more characteristically shapely and plotted genres like the novel. But its basic appeal is shared with other micrographic works as well as with analogous pornographic and ethnographic worldmaking; it is the appeal, often experienced as erotic, of the present tense. The musical working out of rules or patterns slides synesthetically into the visual and spatial, away from sequence and temporality: Buddhist monks, chaos theorists, poets, and autistic children also know the calm and pleasure that rise from fixation on a dynamic pattern. Graphic detail of any sort holds your attention, which wants very much to *be* held. Focal objects like genital sex or violent trauma satisfy additional libidinal imperatives derived from specific social conditions—booksellers were willing to take a *loss* on pornography during the heavily censored Restoration in England, or the post-Fronde paranoia in France. But the bottom line is simply arrest (not *stream*) of consciousness, and this need not be—indeed "was never yet forbidden." This is the phenomenon of consciousness experienced as "wonder" and so often associated with silence, paralysis, the suspension of rationality.

As students are always reminding us, the "descriptive bits" impede the narrative. The sudden efflorescence, in the century after Elizabeth/Cynthia/Astraea's unprecedentedly mystified reign, of genres, both visual and verbal, designed to arrest and fix attention, generally at one or another level on domestic life (the pornographic *modi*, the ethnographic "manners" and "fashions," the micrographic lice, mold, and sperm)—this efflorescence testifies to a need Marx and Benjamin would tell us found an imperfect satisfaction also in "things." In *Micrographia*'s restructuring of values we can see a turn from the sublime, the transcendent, the celestial to the concrete, the physiological, the mundane—a depressed turn but a decided one. Hooke's plot adumbrates the history of scientific attention; his details introduce us not only to the Virtuoso but to the Consumer—whose things include novels and whose novels are full of itemized things.

MARGARET CAVENDISH'S *BLAZING-WORLD*

> What, said the Empress, can any mortal be a creator? Yes, answered the spirits; for every human creature can create an immaterial world fully inhabited by immaterial creatures, and populous of immaterial subjects, such as we are, and all this within the compass of the head or scull; nay, not only so, but he may create a world of what fashion and government he will, and give the creatures thereof such motions, figures, forms, colours, perceptions, etc. as he pleases, and make whirlpools, lights, pressures and reactions, etc. as he thinks best; nay, he may make a world full of veins, muscles, and nerves, and all these to move by one jolt or stroke; also he may alter that world as often as he pleases, or change it from a natural world, to an artificial; he may make a world of ideas, a world of atoms, a world of lights, or whatsoever his fancy leads him to. And since it is in your power to create such a world, what need you to venture life, reputation and tranquility, to conquer a gross material world?
>
> —Margaret Cavendish, *Description of a New World, Called the Blazing-World* (1666)

I don't know about you, but I suffer from a terrible case of self-envy.

—k. d. lang (Wang Center, Boston, February 1996)

Other World

The Duchess of Newcastle, that thrice-noble Princess, prefaces her consciously inimitable scientific (anti)romance, *The Description of a New World*

Called the Blazing-World (1666), by distancing it from the genre of the lunar voyage, extremely popular in the wake of Godwin's *Man in the Moon* and recent translations of Lucian's *True History* and Cyrano de Bergerac's *Estats et empires de la lune*.[21] "It is a description of a *new world*, not such as Lucian's or the *French*-man's world in the moon; but a world of my own creating" (124).[22] The text was published (by her husband) as an addition to her *Observations on Experimental Philosophy*, a philosophical work that opposed the particular philosophy of its title; it seems to have been occasioned by the 1665 publication, for the Royal Society, of Hooke's more covert romance, the *Micrographia*, to which she reacts with eloquence, humor, and formal ingenuity.

Cavendish dissociates herself from Lucian and Cyrano, not or not only out of her well-known desire for singularity, but because her choice, to invent a world rather than to project on one already richly present in European literature and science, was in fact generically significant. The Moon (especially in a mental world as attuned to analogy and as prone to allegory as that of seventeenth-century England) comes equipped with the burden of actuality as well as with the ineluctable symbolic features of oppositeness and mirroring (and perhaps surveillance). As detailed in earlier chapters, it had been functioning for decades as a hyperbole of America, and in the most successful and influential of the moon voyages the lunar inhabitants were said to be descended from earthlings—in Godwin's case seemingly North American ones. The opportunities for satiric or utopian relevance to affairs of the actual were irrefusable.

Mimesis seems to have been only intermittently interesting as a motive to Cavendish. It is alterity that energizes the production of the Blazing World, a world more *alter* than *idem*, to use the terms of Bishop Joseph Hall's title, *Mundus alter et idem* (1607), the only other fully imaginary world of the seventeenth century (though his was in fact far more parodic, thus mimetic and *idem*, than the *Blazing-World*).[23] Lisa Sarasohn in an important early article

21. Lucian's *True History* was first Englished in 1634 and Cyrano's book in 1659. Godwin's *Man in the Moon* came out in 1638 and 1657; Wilkins's *Discourse Concerning a New World* also appeared in 1638, with second and third editions by 1640. For other lunar publications in the period preceding Cavendish's novel, or elaborations of the plurality of worlds, see Nicolson's bibliography in *Voyages to the Moon*. Cavendish's debt to them is easily apparent in, for instance, her magic stone (163—cf. Godwin's "ebelus") or the scab-armor (156), borrowed from Kepler, of her long-lived other-worlders. On the romance/antiromance structure of *Blazing-World* see Lee Cullen Khanna, "The Subject of Utopia," and Marina Leslie, "Gender, Genre, and the Utopian Body."

22. Quotation from *Blazing-World* is from Kate Lilley's 1994 edition; quotation from other works by Cavendish is from original editions.

23. On Hall in the context of utopian writing in the period, see Leslie, *Utopian Ventures*. Charles Cotton's "anonymous" *Erotopolis: The Present State of Betty-land* is a later (1684) and more

invoked the *topos* of "the world upside down," often invoked for Cyrano too, but even that is too *idem* to explain the nuclear force of the *Blazing-World*.[24] I don't mean to imply that this text is silent about colonial appropriation in America and imperialist warmongering at home (for example, the Dutch War is seen here as the concerted attack on the sovereignty of "EFSI," the original home country of the protagonist).[25] But the Blazing-World is the product of a narrative consciousness that is too disenfranchised to maintain for long the common ground with a worldly audience that satire assumes, even in the progressive wish-fulfillment mode of utopia.

Unlike the utopia, the writerly and readerly hunger this narrative aims to satisfy is the hunger for *something else*—not "better," necessarily, in the terms of a virtue generated after all by a patriarchal, monarchist, colonialist status quo—but other.[26] The escapist possibilities of the moon are limited by its long preexistence in the imagination of those who constitute and are nurtured by the status quo. The Blazing-World is Otherness *by and for* the Other: privately printed, vengefully compensatory, written as if to that ghost friend Cavendish's 1664 *Sociable Letters* poignantly outline as her (imaginary) correspondent.[27] A different kind of otherness here from the bracing and threatening America of travel narrative.

Travel narratives about voyages to "new worlds" helped Europe and various European polities to define and idealize themselves as cultural wholes

---

pornographic world-upside-down that would profit from comparison, especially in the context of representing bodies. Cotton's antifeminist request for a recantation of Agrippa's treatise "On the Excellence of the Feminine Sex" refers directly to Hooke's representations: "If [Agrippa's] Admirers object the incomparable Fabricature of that particular part where human Offspring is concern'd, 'tis no more then if you should admire that most curious piece of Nature's workmanship, the head of a Fly, which is all the while but the head of a Fly" (56).

24. See Sarasohn, "A Science Turned Upside Down," and, for Cyrano, Jean Lafond, "Le Monde à l'envers."

25. It occurred to me one evening in 1996, while giving a talk on Cavendish, that the long-baffling EFSI stood for "England France Scotland Ireland"—the realms claimed by the British throne. Present Cavendishians were swayed.

26. My emphasis seems to fly in the face of much recent feminist work on the *Blazing-World*, for example Lilley, "Blazing Worlds"; Khanna, "The Subject of Utopia"; and Leslie, "Gender, Genre, and the Utopian Body." These essays widen and subtilize our understanding of genre as transformed by female authorship, and I do not mean to contradict them. The word "utopian" nevertheless still commonly connotes the notion of an alternative world which in some pedagogic way models the actual, and I do not see that structure as particularly generative in this work.

27. "Women's minds or souls are like shops of small-wares, wherein some have pretty toys, but nothing of great value. I . . . should deserve to be chidden [for this opinion], if all women were like you; but you are but one, and I speak of women, not of one woman . . . but I wish with all my heart, our whole sex were like you" (*Sociable Letters*, 14–15).

by contrast, exclusion, repudiation or domination. As I have pointed out, voyages to new *imaginary* worlds in space had a version of the same task, at the species level: anthropology aims, however crookedly, at a complete depiction of the human, only conceivable as anything but a biological entity if one can conceive other sentient species and sketch the boundaries that make them separate from our own (thus it requires the "primates"). This work of hypothesizing other sentient species entails at the same time an assessment of the human as a single collectivity. Despite the pluralizing titles of Cyrano's two voyages, his Moon, like Godwin's, had only a single culture and his Earth, though divided between Europe and America, was all French—when his initial flight lands him in Quebec, the narrator thinks at first that he is near Paris. The narrators of both lunar voyages are of course put in the position of visiting aliens—neither normal nor at home—and so their anthropological meditations on human culture are *manqués*, fictionally framed as *lunar* ethnographies—studies of what is not human or at any rate homely.

That confusing play of mirrors effects at one extreme Christiaan Huygens's striking *Cosmotheoros* (1698), in which the important amateur astronomer (first discoverer of Saturn's rings and a correspondent of Cavendish's) proves logically that the inhabitants of each planet must be nearly identical physically, culturally, and morally to Enlightenment Europeans: "If . . . the Principle we laid down before be true, that the other Planets are not inferior in dignity to ours, what follows but that they have Creatures not to stare and wonder at the works of Nature only, but who employ their reason in the examination and knowledge of them, and have made as great advances therein as we have?" (61).

Cavendish inhabited a point of view, that of the unenfranchised dependent and social misfit, that led her to a very different fantasy. As she could not produce satire, wanting any real solidarity with her audience, so she could not manage anthropology, or at least could not *want* it. An other world that merely functioned to produce the Earth and Man as back-formations could not satisfy, it seems, the needs of a resident alien for a truly alien textual alternative to the actual world of her incarcerations. The *Blazing-World* is the most "scientific" of the period's imaginary voyages and utopias because Margaret Cavendish has to construct it from the atom up.[28] Not only will embodiment be a different matter here, so will the dust of which bodies are

---

28. In fact, Cavendish had left the fold of atomists by the time she was composing the *Blazing-World*, because of her increasingly firm belief in the life and sentience of all matter. She uses the word "atom" a fair amount anyway (as in the epigraph above to this chapter section), but without the usually accompanying assumption of material senselessness. Along with her own

made. For reasons to be discussed later, *Observations on Experimental Philosophy* promotes a thoroughly animate and even sentient matter. Cavendish has to make a world in the same fundamental sense that an infant in the "mirror stage" does; she has to make a boundary not between Europe and America but between interior and exterior, "own" and all other. Both must be created together, as they are functions of each other, and in this, Cavendish may be more truly a forerunner of the modern novelist than either Godwin or Cyrano (or even their more canonical contemporary, Madame de Lafayette, whose characters and milieus are given, historically actual in living memory). Cavendish's is the situation, as well, not of Defoe, Swift, and Richardson, but certainly of their mobile and marginal narrator-protagonists, Crusoe, Gulliver, and Pamela.

Inner World

So the *Description of a New World Called the Blazing-World* represents three crucially interrelated phenomena: (1) the wished-for Blazing-World itself, including the way "into" it, (2) the protagonist as exterior spectacle—the Blazing Empress—and (3) the interiority of the Empress, in a psychomachia constituted by her intimate triangular relationship with the author (her "Scribe") and her Scribe's husband the Duke, back home in Nottinghamshire. We'll take a look at striking passages from all three representations and then turn to a consideration of a question rarely asked: what is this narrative doing appended to an antimicroscopic critique of experimental philosophy? It is surprising that the question does not get taken up, for here, between one set of covers at a moment of white-hot transition in the world of letters, are the texts of what would soon enough become, what are in them engaged in *becoming*, the infamous "Two Cultures."[29]

The paired modes of interior and exterior generate structure and setting from the start. After her abductors die in a shipwreck at the Pole, the young Lady protagonist survives a winter alone on the open ice "by the light of her beauty" (126). Next, she is escorted to the adjoining Blazing-World through a nearly impenetrable labyrinth "*into* the island, where the Emperor . . . kept

---

Observations on Experimental Philosophy, see Robert Kargon, *Atomism in England from Hariot to Newton*, esp. 73–76.

29. The recent work of Anna Battigelli will correct this negligence and help to place Cavendish back in the lineage of natural philosophers, to which she was passionately ambitious to be joined. On Cavendish and Hooke, see "Between the Glass and the Hand"; on Cavendish and Hobbes, "Political Thought/Political Action"; in general, see *Margaret Cavendish and the Exiles of the Mind*.

his residence" (emphasis mine), which island is itself a labyrinth leading to a palace divided in turn into "outward" and "inner part[s]" between which "a stranger would lose himself without a guide" (131).[30] In the midst of this maze lies the Emperor's apartment, and in the midst of that his bedchamber, a vision of the exterior Blazing-World (so-called on account of the extreme brilliance of its stars): "the walls whereof were of jet, and the floor of black marble; the roof was of mother of pearl, where the moon and blazing stars were represented by white diamonds, and his bed was made of diamonds and carbuncles" (132).

The most interior cell of this world, then, precisely replicates the blazing night sky for which the larger world is named. We will see this substitution of outside for "inside" again at the level of the individual person; we are certainly "inside" the Empress when she is visited by the soul of the Duchess of Newcastle. We have been told repeatedly, here and in the *Observations*, that a spirit must have a material vehicle, so the Duchess's soul may actually be indwelling in the Empress's body during these visits, but if not, the immaterial visitor must still be apprehended internally, that is, as a mental representation. For "the perception of Sight, when awake, is made on the outside of the Eye, but in sleep on the inside; and as for some sorts of Thoughts or Conceptions . . . they are to my apprehension made in the inner part of the head. . . . For there are Perceptions of interior parts, as well as of exterior" (*Observations*, 238–39).

So what we find "inside" the Empress is her own author, and a series of stimulating encounters between the two powerful women from different worlds, in which they discuss the creation of textual worlds such as—and apparently including—the one we're reading.[31] Later, the souls of both Ladies take a trip to England, checking out the Court and the London theater scene before paying a visit to the Cavendishes' chief residence at Wellbeck. Here they enter the Duke's body and the three have a delightful time together inside his material envelope: "had there been but some such souls more, the Duke would have been like the Grand Signior in his seraglio, only it would

---

30. This antiromance heroine may begin as usual in an abduction, but loses her pursuers here successfully, unlike Hooke's Nature, when "she seems to be put to her shifts, to make many doublings and turnings, and to use some kind of art in indeavouring to avoid our discovery" (*Micrographia*, aii^r). On the other side of the labyrinth she becomes the creator of strategic illusions (abductions, captations in psychoanalytic terms) herself—an Empress, not an abductee or rescued princess.

31. "Inside" with the meaning here used is yet another seventeenth-century word: the *Oxford English Dictionary* gives as its earliest instance Phillip Massenger et al., *The Old Law* (1599): "Come, come, here's none but friends here, we may speak / Our insides freely."

have been a platonic seraglio" (194). In fact they get so merry that the Duke's and Empress's souls "became enamored of each other," which makes the Duchess jealous. It is not easy, even here, to evade the specular, but she reminds herself "that no adultery could be committed amongst Platonic lovers" (194–95).

Such interior bodilessness makes for some peculiar exchanges between the two ladies at moments of high emotion or leavetaking, when they are moved to offer "immaterial kisses" and "shed immaterial tears" (202). But however hard these gestures may be to imagine they are postulated as a reality microscopes have already been denounced for being blind to: "they could as yet by no means contrive such glasses, by the help of which they could spy out a vacuum, with all its dimensions, nor immaterial substances, non-beings, and mixed beings, or such as are between something and nothing" (145).[32] Cavendish's irritability with the incipient positivism of natural philosophy in the first "Age of Consumption" could be read, as well, as a call for a wider ontological spectrum in anthropology (imagine the difference it would have made for the objects of ethnology, had there been categories of person for such disembodied beings as *obeahs* and *legbas*, early on classed as phantasmatic "unclean spirits").[33]

This *inner* inner world, "within the compass of the head or scull" (185), offers a challenge not only to the microscope but to modernist or postmodern descriptions of ego dynamics. How do we account for the amount of disembodiment in a work so rife, at the same time, with moments of hyperbolic specularity around the person of the Empress? And what could it mean that the body all three characters inhabit together is neither the protagonist's nor the author's but that of the author's *husband*? Most surprising of all from the point of view of both psychic and aesthetic economies, why is the ego-ideal character of the Empress duplicated by the extremely similar character of the Duchess who has fantasized her? (So similar that critics writing about the text often write "Duchess" when they mean the Empress.[34]) For the Em-

---

32. Cavendish's idea of herself was clearly that of a "mixed being" (others of her heroines are more pointedly hermaphroditic than the Empress—for instance, the cross-dressing Travellia of "Assaulted and Pursued Chastity"). In her introduction to *New Blazing-World* Lilley points out as well the isomorphic hermaphroditism of the form of *Blazing-World* (already cross-dressed as appendix to the nonfictional *Observations*). "As Cavendish argues in *Nature's Pictures*, 'descriptions are to imitate and fancy to create; for fancy is not an imitation of nature, but a natural creation. . . .' A non-imitative or fantastic description is therefore a hermaphroditic foundation for a text" (xxiii–xxiv).

33. Brackette Williams, cited earlier, has begun that work of imagination with her study of the "Dutchman Ghosts" of Guyana.

34. E.g., Sandra Sherman, "Trembling Texts."

press to work as a character with whom her author satisfyingly *identifies*, as she dazzles and rules and starts religious sects and conquers worlds, one might have assumed she would have to fill all the protagonistic space, and to stay carefully isolated in a world that replaces rather than adjoins the world of Margaret Cavendish. The author's appearance in a text is the most disruptive gesture possible against the radiant coherence of narrative verisimilitude; it withholds that optical illusion Walter Benjamin ironically claims we have a right to expect from a work of art, the "aspect of reality which is free from all equipment" ("The Work of Art," 236).[35]

This appearance of the "authoress" is a strong argument against typical readings of Cavendish's fiction as wish-fulfillment fantasies that can be quoted as easily and directly as her autobiography in support of biographical claims. The location or constitution of the biographical subject in her fiction seems a vexed question, for starters, and the *Verfremdungseffekt* of the foregrounding of authorship necessarily complicates our sense of the narrative as a simple case of the imaginary. Whatever detailed reading we might be able to produce of the protagonist's "interior" from this most important of all instances of finding the exterior at the heart of things (the heavens in the Emperor's bedroom, the social crowd inside the single body of the Duke, the Author inside the body and story of the protagonist), the first thing to notice is the sheer, brazen complexity of it, as if Cavendish were *taunting* the optical impotence of experimental science in the face of such impacted *invisibilia*.

In fact we do have a brilliantly responsive reading of that complex interior, in Catherine Gallagher's essay "Embracing the Absolute." For her the interior world of which Cavendish desires to be absolute monarch (actual and authorial in her reading) is a picture of infinite regression: the self must appear as ruler within a microcosm which is also a representation of the self, and of which she is thus only a part, requiring the generation of yet another microcosmos "in order to meet the demands of absolutism," which characteristically "imagines the self as a microcosm" (32). Indeed, says Gallagher, in a one-liner equal to k. d. lang's, in the epigraph above, "we might call Cavendish the seventeenth century's Ms. en abyme" (32)!

---

35. Kaja Silverman (on whose treatment of visibility and subjectivity see below, 210–11): "Crucial to the encouragement and maintenance of a 'heteropathic' identification [see n. 36 below] is the designation of the scene of representation as radically discontinuous with the world of the spectator" (*Threshold of the Visible World*, 86). Also, "'Distance . . . would seem to necessitate a foregrounding of the frame separating an image from the world of objects, and the marking of it as a representation. . . . These discursive strategies . . . mark the otherness or alterity of the image with respect not only to normative representation, but also to the viewer, and . . . thwart the drive toward possession. Through them, the viewer apprehends the image very precisely in the guise of the 'not me'" (99).

I hate to tamper with such a good one, but I think there is another way to see this complex generativity, which in fact is *not* represented as infinite. There is a way to understand the Chinese box of interior privacy as represented in the narrative *along with* the Empress's sublime specularity, blazingly displayed at moments of both erotic and military conquest. Kaja Silverman's recent book, *The Threshold of the Visible World*, revisits Lacan's mirror stage via the work of two lesser-known contemporaries of Freud, Paul Schilder and Henri Wallon. She is looking for a way to "reeducate the look" (5) and finds sustenance in accounts of ego development which do not assume an "incorporative" identification with the mirror image.[36] I would like to suggest that the double presence of Empress and Duchess is significantly *double*, rather than the start of an infinite series, and that it represents a sense of what Silverman refers to as "identity at a distance." According to Silverman, Wallon suggests "that the infant initially responds to the reflection of its body as a separate thing, like its mother or father. At the same time . . . that reflection provides an image in relation to which it somehow orients itself. The reflection offers what . . . I will call 'identity at a distance.' Such an identity is . . . inimical to the very concept implied by that word, which literally means 'the condition or quality of being the same' (*OED*, 881). Identity-at-a-distance entails precisely the opposite state of affairs—the condition or quality of being 'other'" (*Threshold*, 15).

The consequences of such a form of self-recognition include an ability to "identify" noncannibalistically (instanced here by among other things the Duchess's rejection of jealousy when her double's and her husband's souls mingle so pleasantly) and with a "deidealizing image" (by which Silverman does not mean "Fat Ladies" or any other queer or punk body forms but simply women and nonwhite men). I imagine they would also involve a less fundamental reliance for identity on the external body-image. Silverman recommends the intervention of "aesthetic texts" as we try to reeducate our "looks" (I would say our attention generally as well as our optical experience

---

36. Silverman outlines several ways of looking at incorporative identification in the section of chap. 1 called "The Self-Same Body" (22–27). She first cites Freud: identification behaves "like a derivative of the . . . *oral* phase . . . , in which the object that we long for and prize is assimilated by eating and is in that way annihilated as such" (Silverman, 22–23, quoted from Freud, *Standard Edition* 18:105). Then she cites German philosopher Max Scheler retelling a story from Schopenhauer as "an allegory about heteropathy and ideopathy" (Silverman, 23): "A white squirrel, having met the gaze of a snake, hanging on a tree and showing every sign of a mighty appetite for its prey, is so terrified by this that it gradually moves towards instead of away from the snake, and finally throws itself into the open jaws" (Silverman, 24, quoted from Scheler, *Nature of Sympathy*, 22).

per se): "We cannot decide that we will henceforth idealize differently; that activity is primarily unconscious, and for the most part textually steered. We consequently need aesthetic works which will make it possible for us to idealize, and, so, to identify with bodies we would otherwise repudiate" (2).

Repudiation has famously been the reaction of even Cavendish's feminist readers until quite recently (who can forget Woolf's "giant cucumber"?).[37] I think the jubilation my students and I feel, *au contraire*, in reading Cavendish is connected to this other structure of identity and identification—signaled by the copresence of the self (the Empress) and her intangible mirror image (the "Duchess") in a text that represents the microcosm of the self. Silverman's language about identification is full of figures reminiscent of Cavendish's Other World: "illuminated," "lit up," "irradiated." Of course her book is about looking and visual images. But there is an emphasis on brightness that shares a tonal register with the shining and sparkling and splendor of Margaret Cavendish's "Blazing" microcosm of the self: perhaps the jubilant shining is the quality of idealization released from the author's distanced mirror image—as from nuclear fission—and shared by us when we can free ourselves or are luckily born into a freedom from the impulse to repudiate, which Woolf and others have felt so strongly.[38]

Cavendish's blazing is not manifested in any traditional erotic *blazon* of ro-

---

37. Virginia Woolf: "What a vision of loneliness and riot the thought of Margaret Cavendish brings to mind! as if some giant cucumber had spread itself over all the roses and carnations in the garden and choked them to death" (*A Room of One's Own*, 65 [1929]). Sandra Gilbert and Susan Gubar: "Finally the contradictions between her attitude toward her gender and her sense of her own vocation seem really to have made her in some sense 'mad'" (*Madwoman in the Attic*, 63 [1979]). Sylvia Bowerbank: "The tedious chaos of the 'plot' is an obvious feature of this work which has been attacked elsewhere" ("The Spider's Delight," 402 [1984]). Sara Mendelson: "She was not a true champion of her sex, but an egoist who happened to be of the female gender" (*Mental World of Stuart Women*, 55 [1987]). And so on. One happy result of feminist gains in the academy in the past fifteen years has been a gradual decrease in the felt need to distance oneself from work not easily available to traditional aesthetic appreciation. Bowerbank and Mendelson themselves, who helped give us a Cavendish to learn to love, have now (1999) finished a textbook/anthology for use in facilitating such appreciation in the undergraduate classroom, *Paper Bodies: A Margaret Cavendish Reader*.

38. Talking about the somewhat different "jubilation" many feel at moments of (necessarily temporary) identification with an "idealizing image," Silverman concludes: "The aspiration to wholeness and unity not only has tragic personal consequences, but also calamitous social effects, since it represents one of the most important psychic manifestations of 'difference'" (26–27). Certainly on the large scale of colonial "incorporation" and "possession," not to mention the imperial "universalizing" of Enlightenment philosophy and anthropology, the social effects are "calamitous." It is worth noting that novels urgently detailing the "personal consequences" of this psychic pattern (*La Princesse de Clèves* leaps to mind) are emerging along with the growth of colonial "incorporation." The novels of such early feminists as Cavendish and Behn (both Roy-

mance or of lyric poetry, despite the etymological link.[39] The wonderful passages in which the Empress is described as a show rather than a complex subjectivity are so bright they are blinding—it is the clothing and ornamentation and sometimes the accompanying torches of "fire-stone" that blaze, and the body is once again hidden (as it is in the drawings of princesses I and my friends produced by the thousand in our childhoods):

> On her head she wore a cap of pearl, and a half-moon of diamonds just before it; on the top of her crown came spreading over a broad carbuncle, cut in the form of the sun; her coat was of pearl, mixed with blue diamonds, and fringed with red ones; her buskins and sandals were made of green diamonds: in her left hand she held a buckler, to signify the defense of her dominions; which buckler was made of that sort of diamond as has several colours; and being cut and made in the form of an arch, showed like a rainbow; in her right hand she carried a spear made of white diamond, cut like the tail of a blazing star, which signified that she was ready to assault those that proved her enemies. (133)

Some of these attention-getting features, I am sorry to say, were not part of my princess doodles. My princesses, for instance, were not dressed and armed like the warrior heroes of tragedies. I did, however, share the author's passion for celestial objects, here represented in metaphorical clothing after the manner of a heraldic *blazon* or escutcheon that might also be worn by such a warrior/hero. It is a provocative get-up for someone so elusive to the gaze as to spend most of her represented time doing the looking herself, often escaping the look back altogether in her travels and encounters as a disembodied spirit. She is dressed, figuratively, as the object or rather the field of objects that most held the gaze of the natural philosophers whose ranks Cavendish would have liked to join: the heavens of Galileo's telescope, Hariot's, Kepler's, Wilkins's, Huygens's, Hooke's.

Despite her obvious if ambivalent desire to *be* a heavenly body, however, the Empress's remarks on astronomy and telescopes, during her initial interrogations of the Blazing-World's virtuosi, are severely negative; she even orders at one point that the astronomers' telescopes be broken. She has been frustrated by the astronomers' (the bear-men's) inability to agree about whether they saw one "blazing-star" rising and setting rapidly or three separate ones at the "Pole" where the Blazing-World and her home world adjoin.

---

alists) are fascinated by positive images of internal multiplicity, or negative ones of physical disintegration and dismemberment.

39. See any of several articles by Nancy Vickers on the implications of this poetic topos, the piecemeal visual description of the (usually female) body, especially "The Body Re-Membered." See also Patricia Parker, "Rhetorics of Property."

This is only one of many passages where the possibility of quarrel or faction provokes distress and disapproval (the Restoration had only recently recalled Cavendish and her husband from the exile of the Civil War). But it is a strangely detailed and specific little matter, of the three stars or one; it reminds me not only of the three worlds in which the novel's action takes place but of the three souls in the Duke's one body that will appear later on. In that connection, then, it is an inversion of the imperial bedroom as night sky: an image of the invisible interior encountered in the observation of the most outward of all visible exteriors, the heaven of stars. The bear-men's confusion over the identity of the phenomenon is also a reminder that an image is an event and not a thing, an event witnessed in time, by a particular person, "in the inner part of the head."

What the Microscope Missed

The exasperation the Empress expresses toward telescopes prompts the astronomers to offer "several other artificial optic-glasses, which they were sure would give her Majesty a great deal more satisfaction. Amongst the rest they brought forth several microscopes, by the means of which they could enlarge the shapes of little bodies, and make a louse appear as big as an elephant" (142). As might be expected, these "distorted" images are even less pleasing to the Empress, and lead to the paragraph of optical disillusionment quoted earlier; the microscope cannot observe or reveal "interior corporeal, figurative motions" (150 and elsewhere). Cavendish here takes on directly the most astonishing of Hooke's examples of micrographia, the eye of the gray drone-fly with its 140,000 hemispheres, describes it carefully, and offers a different interpretation from Hooke's: perhaps they are "glassy pearls, and . . . not eyes" (143—"*Those are pearls / Which were his eyes*"). The ornamental image resonates with that of the pearls in her blazing outfits: Cavendish is identifying with the eye *beneath* the microscope, and not the eye behind the gaze.[40]

---

40. Perhaps the most familiar instance of her face-off with Hooke is the passage in which she is shown a "flea, and a louse" through the microscope of the bear-men: "The Empress after the view of those strangely shaped creatures, pitied much those that are molested with them, especially poor beggars, which although they have nothing to live on themselves, are yet necessitated to maintain and feed, of their own flesh and blood, a company of such terrible creatures called lice" (144). Even the humble artisan/genius Leeuwenhoek (discoverer of spermatozoa) had a more empathetic relation to microscopic observation than Hooke: "I saw in this water, or on the duckweed, many wonderful animalcules, some of them getting their food from it, and others (as I imagined) using it as a skulking place, to avoid being devoured by little fishes" ("Letter 25," in Leeuwenhoek, 283).

Micrographia is a coffee-table book as well as a scientific text—a piece of Royal Society propaganda for experimental science—and, as such, designed to provoke what Patricia Parker analyzes as the dilated "gaze of wonder," as well as to inform.[41] It is a gaze of wonder very different from the look solicited by the blazing getup of the Empress. I have discussed it in terms of the *detail* common to micrographic representation, inventory, pornography, and novelistic realism. The work's estranging elevation, via detail, of ignoble vermin and mold to the level of the sublime was probably offensive to the *arriviste* classism of the duchess, who married up. But there are serious philosophical and aesthetic objections possible for a writer like Cavendish: "the writing of miniaturization," says Susan Stewart, "does not want to call attention to itself or its author; rather, it continually refers to the physical world. . . . If [features such as 'correctness of design' and 'accuracy of representation'] are especially appropriate to the 'lesser theatre of life,' it is because they allow the reader to disengage himself or herself from the field of representation as a transcendent subject" (*On Longing*, 45).[42] Cavendish did not wish to disengage herself—and she *did* want to call attention to herself!

As we have seen, the climax of *Micrographia* is the representation of the dead eye of the gray drone-fly, so magnified that Hooke can see the window of his lab reflected in each of the 140,000 little "hemispheres" of the composite eye. Talk about blazing! Like the strange reversal at the heart of several important descriptions in the *Blazing-World*, where a phenomenon of fundamental exteriority replaces some unimaginable but looked-for revelation of insideness, the most palpably visualized image in Hooke's book is an eye—but the brilliantly reflective *outer surface* of an eye, the eye as pure blind object. Cavendish turns that surface satirically or at least reductively into jewelry (pearls, like the strung-together "worlds" of her text, joined each to each at the poles, or the globes of her poem "A World in an Eare-ring," or the globular pearls of the necklace she wears in both her portrait and her effigy).[43]

---

41. See Parker's "Rhetorics of Property" for the last word (as well as one of the first) on the rhetorical system of imperializing and sexualizing "new-found lands" through blazon, catalogue, inventory, and tropes of possession. Parker is especially illuminating in her expansion of the scope of Vicker's analysis, cited above.

42. Cf. Hooke's "Sincere hand and a faithful Eye," in the preface to *Micrographia* (a2v).

43. Reductive is perhaps the wrong world for the astonishing poem "A World in an Eare-Ring" (*Poems and Fancies*, 45–46), which plays some of the best games with sublime scale-shifting in seventeenth-century lyric (notably fond of that device), and begins: "An *Eare-ring round* may well a *Zodiacke* bee, / Where in a *Sun* goeth round, and we not see. / And *Planets seven* about that *Sun* may move, / And Hee stand still, as *some wise men* may prove. / And *fixed Stars*, like *twinkling Diamonds*, plac'd / About this *Eare-ring*, which a *World* is vast." The poem goes on to invoke a world inhabited in detail.

The gesture points to the mistake for which she repeatedly berates the microscopists—their assumption that the instrument offers them in some sense access to the interior of things. She has done the same to the surface of her Empress, smothering her with fabulous jewels to protect her from the prying, inventorial eye that is looking for the referent of the *blazon*, an interior bodily surface beneath the pearls and diamonds of her royal garments. The interior is not to be reached that way—observational detail is always detail of the surface.

The point for point, image for image response to Hooke in the *Blazing-World* is the response of a competitor. The prominence of the image of the dead eye in both texts shows they were both concerned with the issue of *looking back*. Cavendish's writerly protection of her characters from being closely inspected additionally suggests that she was sensitive to the possibilities of surveillance and reification lurking in the instrumental practice of the new experimental philosophy. "Of this I am confident, that this same Art, with all its Instruments, is not able to discover the interior natural motions of any part or creature of Nature" (7). Even Foucault has some dour words about the function of the microscope in the seventeenth century: "to attempt to improve one's power of observation by looking through a lens, one must renounce the attempt to achieve knowledge by means of the other senses or from hearsay" (*Order of Things*, 133). For him, the century's natural history is a matter of *reduced* information, a quest for data formed from the reduced participation of the senses and of other people. In the wake of such recent discoveries as those of U.S. government–run radiation experiments on socially devalued human subjects in the 1950s, it is hard to agree with Cavendish's biographer that "none of her observations shows that she realized at all the full significance of Hooke's method" (Douglas Grant, *Margaret the First*, 205–6).[44]

Thomas Sprat the following year, in his *History of the Royal Society*, was optimistic about the relations between literary and scientific observation and writing: "Another benefit of *Experiments* . . . is, that their discoveries will be very serviceable to the *Wits*, and *Writers* of this, and all future Ages"—wit being "founded on such images which are generally known, and are able to bring a strong, and a sensible impression on the *mind*" (414). "The Comparisons which these [Experiments] may afford will be intelligible to all, becaus

---

44. See, e.g., U.S. Congress, House Committee on Government Operations, "Cold War Era Human Subject Experimentation," and Massachusetts Task Force on Human Subject Research, "A Report on the Use of Radioactive Materials in Human Subject Research That Involved Residents of State-Operated Facilities . . . from 1943 through 1973."

they proceed from things which enter into all mens senses. These will make the most vigorous impression on mens *Fancies*, because they do even touch their *Eyes*, and are nearest to their *Nature*" (416). His hope might have been well-founded, but nonetheless it was shared by few Wits and Writers. Certainly Cavendish's Fancies had not been impressed, despite her intimacy with so many great scientists.

Cavendish's preface to the *Blazing-World* alludes to a less seamlessly exploitative relation than Sprat suggests: "I added this piece of fancy to my philosophical observations, *and joined them as two worlds at the ends of their poles*, . . . to divert my studious thoughts . . . and to delight the reader with variety" (124). The "Two Cultures" here look already divided. The frustration and parody of the "philosophical" part of the narrative remind us to consider that there is here a *necessary* divide, an antagonism, as opposed to a simple unrelatedness. The old speculative science that relied on the "natural eye" ("the best optic is a perfect natural eye, and a regular sensitive perception") was less hostile to, certainly less different from, the workings of the poetic faculty: many great works of theoretical science in antiquity and the Middle Ages were composed in the form of long allegorical poems. Behind Sprat's "images which are generally known" and "things which enter into all mens senses" is the far more complex process not only of perception but of assembling a subjectivity that perceives. What Cavendish has to tell us about that in the occlusion of even her most brilliant images, and the cohabitation of her central characters in a single body, confirms her remarks in *Observations* about the different knowledges of the different "parts of Nature," which include the different parts of a single body: "each part [of Nature] retains its own life and knowledg. Indeed it is with these parts as it is with particular creatures; for as one man is not another man, nor has another man's knowledg, so it is likewise with the mentioned parts of matter" (preface). The fixing and transparency of both perceiver and perceived in the practice of microscopy do not seem real to her—how could one person see all there was to see about one object in only one posture and kind of light? ("Nay, Artists do confess themselves, that Flies . . . will appear of several figures . . . according to the several reflections, refractions, mediums and positions of several lights" [9].) The confidence of the microscopist, even the very successful one, is a category mistake, for if "Man" himself is only a "small part . . . of Nature," one person is a speck. "And since Nature is but one body, it is intirely wise and knowing, ordering her self-moving parts with all facility and ease, without any disturbance, living in pleasure and delight, with infinite varieties and curiosities, such as no single Part or Creature of hers can ever attain to" (4).

These insights, of perspectivism and relativism, many of them commonplace in our moment, and not without supporters in hers, have in the interim subjected Cavendish in her capacity as a natural philosopher to little but scorn and arrogant ridicule. Much of that ridicule has been directed at her as a "Lady," a bluestocking, or an upstart whose real interest was fancy clothes and gaudy jewelry (admittedly they were a major interest). I think the intensity of it is also explained in the fundamental challenge to microscopic reification represented most famously by her charming and consistently vilified novel.

Here is the key passage: "if the Picture of a young beautiful Lady should be drawn according to the representation of the Microscope, or according to the various refraction and reflection of light through such like glasses, it would be so far from being like her, as it would not be like a human face, but rather a Monster, than a picture of Nature" (*Observations*, 9–10). Swift's micrographic views of Brobdingnagian women will soon bear Cavendish out. Here is Silverman: "It is not possible . . . to be completely 'inside' any other kind of image [than the ideal], even momentarily. . . . At the very least [the subject] refuses to invest narcissistically in the image (unless it can be somehow oppositionally 'redeemed'), and attempts in all kinds of ways to maintain his or her distance from it" (*Threshold*, 20).

The fiction Cavendish considers less serious than the philosophical work to which it is attached bears out, as a kind of nonpornographic *blazon*, the primary, antimicroscopical insight of her natural philosophy: that all matter is alive and in motion, "self-moving," independent and self-aware, and therefore in its *essence* invisible—too subtle and too quick to be caught behind glass in a visual image that is "true." This is crucial to the young beautiful Lady who so identified herself with Nature (Hooke's object) that she fuses the two in the preface to the *Observations* (and elsewhere): "I do not applaud myself so much, as to think, that my work can be without errors, for Nature is not a Deity, but her parts are often irregular."

The continuous fragmentation and dissolution of the protagonist's character (or its border) might best be understood as a scintillation, a glittering, a "multiplicate refraction"—a blazing where we had expected a *blazon*.[45] And if the character comes to pieces that is not such a problem, since all of Nature's parts and pieces are alive and perceptive, even to the point of having knowledge of God. This blaze gets fiery, nuclear, in the world conquest of part 2. The Empress can be, like Oppenheimer (or Cortés), a destroyer of worlds. But tutored by the Duchess and the spirits, she is also a Creator: "She may make a world of ideas, a world of atoms, a world of lights."

---

45. Cf. Hooke, who places it outside, in the great lens of the Air, evidenced at sunset by "the tremulation of the Trees and Bushes as well as of the edges of the Sun" (*Micrographia*, 219)

Either way, the difference between the fate of the disenfranchised Duchess's sublime and the detumescence of Hooke's seems aptly allegorized by a passage from the *Observations*: "When I say that the *Exterior Object is the Agent, and the Sentient Body the Patient* . . . I retain only those words, because they are used in Schools; But as for their Actions, I am quite of a contrary Opinion, to wit, That the sentient body is the principal Agent, and the external Body the Patient; for the motions of the sentient in the act of perception, do figure out or imitate the motions of the object, so that the object is but as a Copy that is . . . imitated by the sentient, which is the chiefly Agent in all transforming and perceptive actions" (preface).

Both writers have a taste for wonder, new worlds, and visual brilliance. But Cavendish's sublime, the "perceptive actions" explicated in her *Observations* and dramatized in her *Blazing-World*, was a register opposed to the self-effacing scientist and beautiful dead material of the subject-object syntax Hooke was helping to construct. When Cavendish joins her "two worlds . . . at the poles" in a single volume, the volume expresses a version of that same gulf that divides her philosophical views from Hooke's. But it also offers a vision of scientific practice—located, motivated, visible, active, transformative—that might have made, if everything else were different too, for a different history of a different science. The rest of this book will examine the troubles and inventions attendant on the subject-object construction of scientific observation and fictional realism in the representation of cultural worlds. We'll begin by taking up the Duchess's other obsession, the adornment of the body.

Fig. 20. Frontispiece to John Bulwer's *Anthropometamorphosis: man transform'd...* (London, 1653). The edition of the following year was subtitled *A View of the People of the Whole World*. By permission of the Houghton Library, Harvard University.

Part III
THE ARTS OF ANTHROPOLOGY

THE NEXT THREE CHAPTERS depict several kinds of passage to imaginary worlds that (like Thevet's America) lead an independent existence separate from, though not unaffected by, their representations. Despite the variety of genres and media they display (dance, cosmetics, fashion plate, map, encyclopedia, pamphlet, parody, novel, ethnography, and a work hard not to classify as anthropology), the chapters tell, in their roughly chronological sequence, a single story. It begins in a jumbled, shocking, and marvelous diversity, hemmed round with high voltages of dangerous significance (the monstrous, the formless, the deformed, the unnatural, the unsexed or double-sexed, the listed and pictured but entirely uncomprehended). It concentrates on the place of the body in the texts that make up the early modern prehistory of anthropology, especially the marked body, increasingly the body in pain. It ends in an Enlightenment serenity of system, bolstered by imperial growth, inspired by the imperial ideal of universalism, but still marbleized with trouble by the shocks of passage.

Chapter 7 offers an overview of some understudied contributions to the early modern accumulation of ethnographic data and its management. The fashions, fashion plates, dance, and (anti-)fashion literature of the mid-sixteenth to mid-seventeenth century provide an aestheticized and body-centered panoply of responses to cultural novelty and its increasingly global circulation; the chapter focuses at greatest length on physician John Bulwer's 1650–54 ethnographic encyclopedia of body-fashions, the *Anthropometamorphosis*.

In Aphra Behn's ethnographic fiction, set in the Surinam of her own stay there in the 1660s, we see more than the anxieties about difference and merging evident in Bulwer's tendentious book. Behn's work illustrates the psychological terror of colonial life, presenting an account concentrated on subjective experience and culminating in the dismemberment of the character—the African prince-slave, Oroonoko—who has been required to exist in the largest number of worlds at once. She registers in this imaginary account of actual experience the difficulty of real other-world travel, with its impossible requirements of empathy and relativism and the conditions, for many, of involuntary passage.

Behn's little fiction shares with Lafitau's anthropological tomes a grounding in personal experience of the colonial New World and a marked concern with physical pain. As her work is recognizably a modern fiction, his *Moeurs des sauvages amériquains* (1724) is recognizably a work of both ethnography and anthropological theory. It shares with Newton's *Principia* (1687), Dampier's *New Voyage around the World* (1697), Swift's *Gulliver's Travels* (1726), and the first edition of Linnaeus's *Systema Naturae* (1735) a world that's whole and single, though very variously construed by the writers of this decisive half-century in European history. It is a law-bound, taxonomized, imperialized world, as an *imago mundi* (though not yet in experiential "fact"). Every place can now be known, theoretically, by its coordinates on the map's grid, every plant and animal by its philogeny, and every native American folkway or religious practice by its deviation from those original *moeurs* of Greece and Israel. The travel writer and the make-believe travel writer listed above, Dampier and Gulliver, claim all four corners of this single world as their witnessed territory, and the anthropologist, Lafitau, proposes these four corners as mirrors of the center, all participating in the same religious logic and many similar practices.

This is the end, then, of a story dependent to this point on more pervasive literary and cultural turmoil and disciplinary insecurity. The "cognitive emotion" was no longer useful or respectable for those who wanted to light up the whole darkness, the whole world. Europe's mercantile empires depended on exoticism and its valent desires, but depended as well on international law and global maritime infrastructure. The Enlightenment's relegation of otherness and wonder to an aesthetic and thus alienated and low-prestige existence suggests a conviction that other worlds are "nothing but" fantasies, dreams, escapes, *bijoux*. It is *this* world that is "the best of all possible worlds," because it is the only possible world.

The virtuoso Christiaan Huygens demonstrates this with unconsciously hilarious clarity in his Copernican treatment of the solar system, *Cosmotheoros* (1698, translated as *The Celestial World Discover'd*). Imagining the "Planetary Inhabitants" of our system's other worlds, the book winds up assured by its own careful logic that "they must have Mechanical Arts and Astronomy, without which Navigation can no more subsist, than they can with Geometry" (83) and that thus they have music, and if they have music there's no reason "why we should look upon [it] to be worse than ours," so they must have half-notes and quarter-notes, the third and fifth intervals and scales enclosed by octaves (86–91). "Nay very credible Authors report, that there's a sort of Bird in *America*, that can plainly sing in order six musical Notes: whence it follows that the Laws of Musick are unchangeably fix'd by Nature" (86).

Of course the phrase quoted above from Voltaire's *Candide*—"the best of all possible worlds"—immediately invokes the protest of artists (and others) against the New World Order and its euphemization of horrors. And in the last chapter we will observe not only the deep fractures in the Jesuit priest Joseph Lafitau's systemization of cultural variety, but the French imposter "George Psalmanazar's" immortal and notorious refusal to get on the boat. Little is easier than to find cracks in Enlightenment facades, and we know, as its producers did not, how it all turned out. But it is a notable and a consequential phenomenon, the splitting of cognitive consciousness into dichotomized, differently valued and mutually disconnected modes—as if only a fragmented mind could comprehend so impossibly unified a world.

7 • ANTHROPOMETAMORPHOSIS
Manners, Customs, Fashions, and Monsters

Then my nephew is such a coxcomb he has studied these twenty years about the nature of lice, spiders, and insects and has been as long compiling a book of geography for the world in the moon.

— Thomas Shadwell, *The Virtuoso* (1676)

THAT "EXPERIMENTAL PHILOSOPHY" was fashionable in the seventeenth century is evident — in the expensiveness of the coffee-table books put out by such serious scientists as Robert Hooke, in the necessity felt by the Duchess of Newcastle of having "Philosophical Opinions," in Pepys's purchases and social adventures, in theatrical parodies by such Restoration dramatists as Thomas Shadwell and Susan Centlivre. The first several characters delineated in the chapter on "Fashion" in La Bruyère's *Caractères* (1688) are naturalists and collectors like Shadwell's coxcomb in the epigraph above. As long as material collection defined both science and shopping, the boundary was permeable between the activities at the individual level; at the collective level, many of the new sciences were absolutely dependent on the state's colonial expansion, none more so than what would come to be codified as anthropology.

But there is a closer connection: anthropology (or in the sixteenth and seventeenth centuries, the collection and contemplation of "manners and customs") is not only an object for the fashion-conscious, it is a mode of being conscious of fashion (*la mode*).[1] Fashion stimulated — even constituted — the growth of this still unborn science in ways that are of particular interest to two main concerns of this book. First, as a form of consumption, it carried out the economic requirements of the commercial culture so much expanded and encouraged by the discovery of new sources of materials and new routes

---

1. See Margaret Hodgen's still unsurpassed history of preinstitutional anthropology, *Early Anthropology in the Sixteenth and Seventeenth Centuries*, esp. chaps. 4 and 5, "*Fardle of Façions*" and "Collections of Customs."

of access to them. The colonialism of Europe's emerging nation-states was enacted at home in the promenading, masquerading, and conspicuous display of bourgeois and noble consumers alike.² Fashion, a system of temporal intensification, demands an alternation of novelty and obsolescence: the so-called Age of Discovery was tailor-made to answer this demand.³ And as I have observed (with the help of Michel de Certeau) in the specific case of Thomas Hariot, the ethnographic discourse (which for our purposes includes the mute discourse of commodities in a grammar of fashion) is one that converts the linear historicity of a foreign culture into a timeless present serving the historicity and narrative identity of the European writer/consumer.

Second then, fashion is a form of time as well as a theater of display, or, in the words of Gilles Lipovetsky, "a form of social change, independent of any particular object; it is first and foremost a social mechanism characterized by a particularly brief time span and by more or less fanciful shifts that enable it to affect quite diverse spheres of collective life" (*Empire of Fashion*, 16). It does not take much of a leap to see something familiar in this "form of social change." Fashion is not only a kind of prearticulate anthropology, but the temporal institution prerequisite to the genre of the novel: the genre that narrates "social change" taking place over a "brief time span" by means of "fanciful shifts," from the subject position of the inhabitant (however resistant) of fashion's accelerated temporality. In such a time span, the details and events of the trivial can become narratable (while the unnarrated alien life in which new fashions often have their design sources is objectified as material); thus the life of an ordinary person in the "fashion system" acquires what can be recognized as content. An other world is eaten up, incorporated, to become visible as a set of self-defining signs in *the* "World."

Furs, silks and fine cottons, stimulants—tea, coffee, sugar, rum, gin, tobacco and spices of all kinds—scrimshaw and curios for cabinets, travel books and atlases, topazes, feathers, orientalizing and Americanizing changes in clothing and ornament: these things did not simply "improve the quality of life" in the metropole, they altered it, and altered the people who

---

2. See Laura Brown, *Ends of Empire*, and Neil McKendrick, John Brewer, and J. H. Plumb, *Birth of a Consumer Society*, in particular McKendrick's article, "The Commercialization of Fashion," 34–99.

3. Recall from Chapter 2 the story James Axtell recounts, of the "Englishmen's implausible [to the Iroquois] preference for the greasy beaver robes they had worn for a year or more in their smoky lodges. . . . [The Indians] could not appreciate that prolonged wear removed the long guard hairs from the downy, barbed underfur used in felting" ("English Colonial Impact," 253).

wore, ate, owned, contemplated, and changed their moods with them. "You are what you eat," and Europe was cannibalizing the places and peoples that eventually made up its empires.[4] No wonder some commentators saw monstrosity in fashion's "logic of inconstancy, its great organizational and aesthetic mutations" (Lipovetsky, 4).

Obviously, the speeding up of fashion's metamorphoses in the centuries immediately following the securing of routes to America and to eastern Asia round Africa's cape has many determinants — social mobility was spurred by the large projects of exploration and colonization but also by the "print revolution," the new money economies, and the emergence (and urbanization) of new population growth after the long stagnation brought on by the Black Death.[5] Like the reciprocities of colonization, transgression of class lines bothered some people, as did transgression of gender lines by the "masculine-feminines" and fops. We might now consign some of these lines of demarcation and transgression to sociology and others to anthropology; in the seventeenth century they are all fault lines that manifest the new salience of "manners and customs" as variable matters — thus, interesting problems. Culture begins to be considered, systematically, as its variants begin to impinge — through trading partners, threatening Ottoman "infidels," new-made knights, visiting "Indians," "experts," and what Anne Middleton has termed (for England) "the new men" of the bureaucratic and centralized state — on the experience of members of the once relatively stable, hegemonic, and isolated feudal chessboard of Latin Christendom.[6] Access to print was open mostly to those likely to fear the relativizing of European Christian culture and its structures of difference. Thus our most frequent attestations to the colonialist anxiety of influence appear in works designed either to disparage or to mediate the changes wrought by La Mode in *heimlich* manners and customs.

A few works or genres have been mentioned in earlier chapters which share qualities with collections and catalogues of collections — anthologies of travel accounts, encyclopedic works of cosmography, Bacon's lists of natural histories to be written, the series of advertising portraits of courtesans, the

---

4. See Jack Weatherford, *Indian Givers*, for a vivid account of the American branch of this process; other sources abound. Benedict Anderson's narrative of the European importation of revolutionary politics from its own colonies (*Imagined Communities*) is relevant, but largely involves European imitation of transplanted Europeans. On European cannibalism, see Peter Hulme, chaps. 1 and 2 in *Colonial Encounters*, and Jonathan Goldberg, chap. 6 in *Sodometries*.

5. See Daniel Roche, *Culture of Clothing*, or René Colas, *Bibliographie générale de costume et de la mode*.

6. Middleton, "'New Men' and the Good of Literature."

medallions illustrating ethnic costumes surrounding maps in the big atlases. As I have pointed out already, western and especially northern Europe's was a list-making, "anatomy"-composing culture. In addition to actual works of physical anatomy like that of Vesalius (or such systematically metaphysical extensions as Burton's *Anatomy of Melancholy*) a number of rather different texts also arranged themselves anatomically, titling parts and chapters with the names of body parts: Thomas Jeamson's *Artificiall Embellishments or Arts Best Directions How to Preserve Beauty or Procure It* (1665) divides numerous chapters (e.g., "To whiten a tan'd visage") into four main sections: "Of the Whole Body and Beautifying Thereof," "Of the Head, Neck and Breasts," "How to Beautifie the Arms, Hands, Leggs and Feet," and "Sents and Perfumes fitted for severall occasions." The anonymous *England's Vanity: or the Voice of God Against the Monstrous Sin of Pride in Dress and Apparel* (1683), on the opposite side of the moral argument, similarly itemizes body parts, though the body has by now subsumed a number of "artificiall embellishments." The title continues: *Wherein Naked Breasts and Shoulders, Antick and Fantastick Garbs, Patches, and Painting, long Periwigs, Towers, Bulls, Shades, Curlings, and Crispings, with an Hundred more Fooleries of both Sexes, are Condemned as Notoriously Unlawful.* John Bulwer's astonishing and encyclopedic *Anthropometamorphosis* (1650), which I will examine here at length, is divided into chapters (called "scenes") titled by the name of the body part whose many "embellishments" and alterations in various nations a given chapter collects ("Neck," "Lips," "Privy-Parts").

Like Bulwer's text in encompassing many cultures, including non-European ones, but unlike his in emphasizing the clothed, and unified, body, earlier works such as the *Receuil de la diversité des habits* (1562), with quatrains by François Desprez, or Boissard's *Habitus variarum orbis gentium* (Mechlinburg and Cologne, 1581) consist almost entirely of "fashion plates," crudely dividing the people of various locations into such taxonomically incoherent sets of horizontal categories as *matrona, virgo,* and *femina* (see figure 21).[7] These books of fashion plates actually have the most obvious bearing on the

---

7. Before 1610 in France, according to Jacqueline Tuffal, more than two hundred *recueils de costumes* were printed, amounting to 5 percent of all printed books. The number roughly doubled in the seventeenth century, and skyrocketed at the end of the period with which the present book is concerned. For a very brief overview of this period of production (1520–1799), see chap. 1 of Roche, *Culture of Clothing*. Roche says these texts were first modeled on the Latin texts of Lazarus Baïf, Bertellius, and Abraham Bruyn, but in the case of "Bertellius" at least (presumably he means Bellerius, or Ionni Bellero, who published the Latin version), the Latin *Omnium fere gentium* (Venice, 1563) was in turn a translation of the French *Recueil de la diversité des habits* of the previous year. My quotations and illustrations are drawn from the bilingual edition (Antwerp, 1572).

Fig. 21. Virgo, matron, femina. Plate from Jean-Jacques Boissard's *Habitus variarum orbis gentium* (Mechlinburg, 1581). By permission of the Houghton Library, Harvard University.

emerging organization of ethnological information: their concern is to display not only the clothing styles of various nations but the differences in habit *within* each nation; thus they must develop rudimentary sets of social functions or types. Predictably, but tellingly, these are not standard for each nation or culture depicted: *Receuil de la diversité des habits*, for instance, divides the residents of six cities or regions of modern France into twenty-eight categories, while the Moors or the natives of Brazil are classed into only two groups, male and female. A number of groups are not even subdivided that far: for instance *Asiana mulier*, *Polanus*, and the very singular *Cyclops* (see figure 23, below, and for more on this influential text, see the final section of this chapter).

The relationship of body to "culture" and of foreign culture to rhetorical display was most literally spelled out in the medium of dance (sometimes "native" dances performed by imported "natives"), especially the emerging form of ballet, featured between the acts or at the end of plays or operas and crucial to the masque.[8] These early versions of "ballet" intensify the erotically

---

8. On "native" dancers and mimicry of them in court and public fêtes, see Boorsch, "America in Festival Presentations." As much history of ballet is written with a telos in the high theater art of the late nineteenth and twentieth centuries, it is not easy to find secondary sources focused on these ethnographic origins. Miriam K. Whaples, "Exoticism in Dramatic Music, 1600–1800," is the most useful; see also Elise-Noël McMahon, "'Le corps sans frontiers,'" as well as

visual emphasis of the fashion plates' version of foreign "manners and customs," in an art where actual human bodies model the costumes and enact stylized "manners" against painted landscapes.⁹ Perhaps the most famous example today is that of the "Indian" dance in Dryden and Purcell's *Indian Queen*, in which the dancers wore feather headdresses acquired by Aphra Behn during her stay in Surinam.¹⁰ It is suggestive that the modern art of ballet as we know it developed in a direct parallel with modern fiction, from an eroticized quasi-informational representation inspired in part by exploration and discovery into a fictional medium focused on the secrets of female emotional experience (the dominance of female dancers can be dated to the early nineteenth century via the invention of the toe shoe). As theatrical display, it also developed into the fashion show—which still displays a "collection."

---

the suggestive article on "ballet" in the *Oxford English Dictionary*. Nonexotic ethnographic links are obvious from looking at any dance manual of the period: Thoinot Arbeau's *Orchésographie* (1588) is easily available. Most dances are identified by country or region, or have one imputed to them. Of the dance called *Canaries*, Arbeau (Jehan Tabourot) explains: "Some say that this dance comes from the Canary Isles, and that it is regularly practised there. Others . . . hold that it is derived from a ballet composed for a masquerade in which the dancers were dressed as kings and queens of Mauretania" (150).

9. Of the subgenre of the *mascarade*, the seventeenth-century choreographer Saint-Hubert says "The *mascarade* is improvised by people who disguise themselves. . . . It is just a pretext for wearing imaginative costumes" (Cohen, ed., *Dance as a Theatre Art*, 32). On early ballet generally see (in addition to Whaples) M. de Saint-Hubert, *La Manière de composer et faire reussir les ballets* (1641). The earliest human ballets (horse ballet was already familiar in military spectacle and triumph) are usually said to have taken place in Italian courts in the 1580s. Ethnological missionary Joseph Lafitau (see Chapter 9) had a different aetiology: the ballet was descended from the Cretan pantomimes so popular in imperial Rome: "The gestures were so marked and mimicked men's customs, affectations and actions so vividly that (for this reason) authors compare them to painting and poetry. . . . These dances were the forerunners of our ballets out of which our plays have developed" (*Moeurs* 1:321). The new "discipline" was institutionalized (and nationalized) when schools of ballet were chartered by the French and English crowns in 1661 and c. 1672. On the issues of discipline and "nation" in the ethnographic dances of Molière's *Bourgeois gentilhomme* (1670), see McMahon, "'Les corps sans frontiers.'"

10. The "Turkish Ceremony" in the last act of Molière's *Bourgeois gentilhomme* (1670) is perhaps an even more famous example, although in fact it parodies the exoticism of the comédie-ballet, rather than simply offering its Turkish dances and costumes up for delectation. The play is in some ways a theatrical presentation of my argument in this chapter that "clothes make the man," and Europeans of fashion and new money like the virtuoso M. Jourdaine appropriated foreign lifeways through costume (as did the rebellious and demimondaine). For closer analysis of the play in a similar context, see chap. 3 of Michèle Longino's forthcoming *Staging of Exoticism* ("Acculturating the Audience"). She notes, in the closing "Ballet des Nations" (which follows the "Ceremony" like an answer or revenge), a version of what we will later see in the fashion book, *Receuil de la diversité des habits*: the dance is made up of representatives of "provinces différentes," but the categorizing is utterly incoherent. "Differences of class, country and locale combine under the rubric 'province' to make a mockery of the very grouping they are supposed to be constituting."

La Bruyére's *Caractères* (1688), the international bestseller that grew from edition to edition until it had doubled its size, is another kind of anatomy of fashion. Its organization is based on demographic or social worlds—"The Court," "The Great" (though also "Mankind"); its chapter on "Fashion" begins by delineating, not characters especially conscious of style in clothing, but collectors and virtuosi. La Bruyère's satire shares with the fashion books, and amplifies, the tendency towards presentation of what we might call microcosmoi (taking off from John Earle's 1628 collection of characters, the *Microcosmographie*). In these stereotypes we see large numbers of people represented metonymically, sometimes numbers constituting a subculture, or "sub-World," like those of the virtuosi or the fops. The possibility is emerging here (and elsewhere) of a fictional character who can be representative horizontally (metonymically), rather than vertically, as are the aristocratic heroes of epic, romance, and tragedy. But the *Caractères* lacks metaphoric resonance with the sorts of curiosity and representation leading up to "anthropology," which one can see in the works entangling "habit" with nation and deriving fashion from foreign trade.

Louis Von Delft has in fact examined "characters" as ethnological prototypes, but it is mostly a rhetorical likeness: the characters are elaborate, satirical pieces of *descriptio* written in the "ethnographic present," certainly, and clothed in a single archetypal habit or mannerism.[11] But rather than providing exotica, they are offered as immediately recognizable portraits, and not merely of one's neighbors typically but sometimes of *specific*, individual neighbors—the English edition of 1698 offers one of many "keys" identifying by name almost every "character" in the work. La Bruyère is doing in a satiric vein what Thevet did earnestly almost a century before in his *Pourtraicts*; both works have didactic designs on the reader's own character (as, in a way, would anthropology on the collective scale: "We study customs in order to form customs," said Lafitau in 1724).

Closer to Bulwer's literally anatomical anatomy is M. de Fitelieu's *La contre-mode* (1642), in which chapters with titles such as "Teste à la mode," "Yeux à la mode," "Bouche à la mode," and "Piez à la mode" advance the argument that fashion "corrompait les functions qui avaient été originallement attribuées a chaque partie du corps."[12] In an echo of earlier literature on the putatively foreign invader, syphilis, fashion is represented as a *mal funeste*, a

---

11. See "Caractèrologie et cartographie à l'âge classique" and/or "Moralistique et topographie: *Caractères* et *lieux* dans l'anthropologie classique."

12. See Françise Waquet, "La Mode," 92. The essay discusses a number of seventeenth-century French works attacking or defending "la mode."

terrible maladie ravageant le monde, cette peste. (Elsewhere, in another association to New World literature, it is denigrated for making the world resemble "l'Eutopie de Thomas Morus où l'on ne distinguoit personne."[13])

These figures betray an anxiety about identity we will see blossom extravagantly in Bulwer. Fashion, the quickened temporality of a world under the pressure of new discoveries and information overload, is feared as an "oeuvre de dénaturation" (Waquet, 93) and a dissolver of class, gender, and ethnic distinctions: de Fitelieu describes the "corps à la mode" as "espagnol jusqu'à la ceinture" and "dès la ceinture en bas italien" (15–16).[14] La Bruyère's Characters, on the other hand, suggest an essential stability, however foolish, in human custom and personality; their *heimlich* recognizability and putative generality must have allayed the kinds of anxiety Bulwer's work bespeaks so loudly. The publication of Theophrastus's original "Characters" in translation as an appendix would have underlined the consoling assumption of "timelessness" in the genre of the Character—though ironically it would also have followed (in reverse) the precedent in seventeenth-century commentaries on Homer and other ancient works of emphasizing contemporary, presumably analogous, ethnological information from the very *unheimlich* places that seemed to be transforming European manners.

Fashion's assumption of the mutability of the individual body and countenance, clothes and habits—"habitus"—has private satisfactions to offer. The eventual realistic psychological novel will need this sense of individual/personal mutability—its drama is not development (as in romance) or historical transformation (as in epic) but personal *change* (like the later drama of psychoanalysis). But as we will see, and as Spenser had already vividly shown, Mutabilitie can be frightful, when "this *Titanesse* aspire[s], / Rule and dominion to her selfe to gaine" (*Faerie Queene* 7.6.4.1–2). For she not only broke the laws of Nature, but "death for life exchanged foolishly" (7.6.6.1–

---

13. Jean Paul Marana, *Lettre d'un Sicilien à un de ses amis* (quoted, from the author's seven-volume seventeenth-century collection, in Waquet, 93). In the utopian New England of the New World, we hear the Puritan minister Nicholas Noyes inveighing against periwigs, especially those made of women's hair, because they collapse distinctions between sexes, ages, and individual identities: "It removeth one notable distinction, or means of distinguishing one man from another" (Nicholas Noyes, "Reasons against the Wearing of Periwigs," 213). See Lipovetsky for a fully theorized account of Marana's perception. As Richard Sennett's introduction puts it, Lipovetsky "explains how the phenomena of 'difference' in modern society have shifted to the realm of fantasy rather than direct confrontation; in the erotic-tinged realm of consumer fantasy, all differences can be overcome" (Lipovetsky, ix).

14. La Bruyère: "We blame a fashion that divides the shape of a man into two equal parts, and takes one of it for the waist, whilst leaving the other for the rest of the body" (*"Characters,"* 389). See also Bulwer's legless ethnic dancer, Catherine Mazzina.

4). "Cette peste" may confer narrative suspense and a meaning-effect on private life (which indeed it helps to create), but there is no denying its corrosive effect on the grounds of external identity, which it wants to replace with commodities.

ANTHROPOMETAMORPHOSIS

Our English Ladies, who seeme to have borrowed many of their Cosmetical conceits from barbarous Nations, are seldome knowne to be contented with a Face of Gods making.

—John Bulwer, *Anthropometamorphosis* (1654)

John Bulwer was a seventeenth-century London physician and author of several books on sign language and lip reading, based at least in part on his successful experience in teaching the deaf.[15] His ethnological book, quoted above, is of interest here in constituting so full an expression of the "entangled" matters of racial identity and consumer culture at such an early point in the development of international capitalism.[16] Although racial science did not exist yet (and artists had a hard time conveying the "racial" features that are now visual conventions, and seem transparently visible), Bulwer's book conveys an anatomical sense of ethnic identity and understands the coexistence of different racial or ethnic physical features as mimetically unstable: one physical type is likely to alter itself in the direction of the other. His book does not concentrate on miscegenation as a threat to ethnic identity, though doubtless Bulwer feared it, because it is preoccupied with the problem of what he calls in his opening poem the "self-made man" (as the author of *England's Vanity* put it, "each pittiful fellow check-by-joleing it [*sic*] with your *Lordships*, and every *Mechanicks* wife *Apeing* your high-born Ladies"

---

15. Besides the work under discussion here, Bulwer's published works (which earned him the epithet "Chirosoph") include *Chirologia: or, The Natural Language of the Hand* (1644), *Philocophus; or, The Deafe and Dumbe Man's Friend* (1648), and *Pathomyotomia, or a Dissection of the significative Muscles of the Affections of the Minde* (1649). *Chirologia* and its companion treatise, *Chironomia*, have been published in a modern diplomatic edition by James Cleary. The British *Dictionary of National Biography* notes the oddness of Bulwer's failure, given his interests, to invent a sign language for the deaf and also points out that his "discovery of methods for communicating knowledge to the deaf and dumb" preceded John Wallis's celebrated 1662 presentation of his deaf pupil to the Royal Society by fourteen years. Work in this area had been done previous to Bulwer's in Spain (and reported in England) by two Benedictine monks, Pedro Ponce and Juan Paulo Bonet. See also Dilwyn Knox, "John Bulwer," 20.

16. I have borrowed my use of the word "entangled" from Nicholas Thomas, *Entangled Objects*; see esp. chap. 4, "The European Appropriation of Indigenous Things." But see also the essays collected in Arjun Appadurai, *Social Life of Things*, and Mary Helms, "Essay on Objects."

[31]).¹⁷ The fear is above all of the "fake," that is the *fait, factus*, fabricated, the culturally produced, the constructed.¹⁸ For all the intervening centuries of state-supported ethnology, that fear persists, along with the compensatory belief in the God-givenness of one's "natural and lively Image, Forme, and Beautie" (and power and wealth).¹⁹

The first of three editions of the many-titled *Anthropometamorphosis* appeared in 1650; its dedicatory letter refers to Bulwer's previous books as "public paroxysms."²⁰ The label seems at first both appropriate and weird for printed books about gesture, especially one's *own* books. But Bulwer is in fact a paroxysmal writer, and, as such, a good registrar of crisis. In *Anthropometamorphosis* we find a tantrum marking the crossroads of a number of emergent social and intellectual structures, a text that might serve as magnetic center of several analyses: not only the histories of fashion, the body, and anthropology, but those of wonderbooks, monstrosity, abjection, semiosis, plastic surgery, nationalism, commodity capitalism, and subjectivity meet here, as well as histories of the concepts, central to all of these topics, of "nature" and

---

17. A really remarkable tract on this popular subject, often considered by way of the face-patch, is that of "Misospillus" ("Spot-hater"), *Wonder of Wonders, or, A Metamorphosis of Fair Faces Voluntarily Turned into Foul, or, An Invective against Black-Spotted Faces* (1662). It reaches its climax in some lines of verse that explicitly invoke miscegenation: "Complexion speaks you Mungrels, and your Blood / Part Europe, part America, mixt brood; / From Britains and from Negroes sprung, your cheeks / Display both colours, each their own there seeks." See also the anonymous *England's Vanity* ("Jude informs us what a plague the coming in of some black Sheep (that were all *Spots*) proved to the poor flock of Christ that fed among them" [97]), and the later, more simply amused, poem "The Patch. An Heroi-Comical Poem... in Three Cantos... By a Gentleman of Oxford" (1724).

18. The penultimate couplet of the opening poem in the first edition, "The Full intent of the Frontispiece unfolded, Or, A through-description of the National Gallant": "Thus *capa peia* is that *Gallant great, / Horrid, Transformed self-made Man*, Compleat" (1650, A4ᵛ). Compare Edward Brathwaite's conduct books, "The English Gentleman" and "English Gentlewoman" in *Times Treasury, or, Academy for Gentry* (1652). I thank Helaine Razovsky for instruction concerning early modern conduct books; her book in progress is "'The Right True Way to Happiness': English Reformation Spiritual Conduct Books."

19. William Prynne, *Unloveliness of Love-Lockes* (1628). For a resonant contemporary expression of this repugnance, in a reversed context, see the epigraph to my article, "*Anthropometamorphosis*," quoting the words of Jonathan Haynes, a Chicago man who had murdered a plastic surgeon and a hairdresser: "I condemn fake Aryan cosmetics. I condemn bleached blond hair, tinted blue eyes and fake facial features brought by plastic surgery. This is the time that we face up to it [!], and stop feeding off Aryan beauty like a horde of locusts in a field of wheat" (202).

20. For the most part I will be quoting from Bulwer's expanded edition of 1654 (*A View of the People of the Whole World*). Where there is reason to cite from the first edition (1650) I will point out that I am doing so. The 1653 edition, which I have looked at in the Houghton Library at Harvard, has considerably more front matter (including the amazing allegorical frontispiece, shown here at the opening of Part 3 in figure 20) than does the microfilm of the 1654 version, but it is not widely available.

"culture." It is of serious interest to a student of the history of "culture" that these areas of experience and expressivity come together in one text, and that the text in question *is* a paroxysm.

As will become obvious, Bulwer's hysterical categorizing offers concrete and literal support to some of Julia Kristeva's elusive psychoanalytic articulation of the structure, personal *and* social, of abjection: "the twisted braid of affects and thoughts I call by such a name does not have, properly speaking, a definable *object*. . . . The abject has only one quality of the object—that of being opposed to *I*" (*Powers of Horror*, 1). The "abject" is then nearly synonymous with the Other conceived as intolerable, impossible to identify with (as in Silverman's account of the non-ideal body, discussed in the last chapter). The "deject" is the one who cannot tolerate, or incorporate: "a deviser of territories, languages, works, the *deject* never stops demarcating his universe, whose fluid confines—for they are constituted by a non-object, the abject— constantly question his solidity and impel him to start afresh" (8).[21] The subtitle alone (of the first two editions) can make the point: *Historically Presented, In the mad and cruel Gallantry, Foolish Bravery, ridiculous Beauty, Filthy Finenesse, and loathsome Lovelinesse of most NATIONS, Fashioning and altering their Bodies from the Mould intended by NATURE*. But this book should not be read as a case of personal anxiety, though it is that. Bulwer's strange grab bag of ethnographic shudders seems to have spoken well for some of the rage and bewilderment of its time. He published an expanded, illustrated edition in 1653 and again in 1654, under a significantly different title: *A View of the People of the Whole World, or, A Short Survey of their Policies, Dispositions, Naturall Deportments, Complexions, Ancient and Moderne Customes, Manners, Habits and Fashions*. The personal psyche has been projected here onto the widest possible screen.[22]

Although this title page boasts proudly that the work is "everywhere *adorned*" (with "Philosophical *and* Morall . . . Observations") and that the "Figures are annexed" "for the Readers greater *delight*" (emphases mine), it

---

21. See especially Kristeva, chap. 4, "From Filth to Defilement," which attempts to find "correspondences" between the abjection of individual subjects and the collective abjection of the symbolic systems they inhabit and maintain (67). The concept is not unconnected to Mary Douglas's work on "purity," or Latour's notion of the hybrid: "it is . . . not lack of cleanliness, or health that causes abjection, but what disturbs identity, system, order. What does not respect borders, positions, rules. The in-between, the ambiguous, the composite" (Kristeva, 4). This is what constitutes impurity for Mary Douglas and has been defined for centuries by Kristeva's "deject" (for instance Leviticus) as "the unclean."

22. According to William Oldys's *British Librarian* (1738), the new title was "seemingly added by the Printer to advance the universal reading of the Author" (365). For an extended reading of the kind of gaze (and its growing popularity) implied in such a title—"A View of the People of the Whole World"—see Mary Louise Pratt, *Imperial Eyes*, chaps. 1 and 2.

is in fact adornment and its delight—fashion in its most ordinary modern sense—that focuses the anger of Bulwer's text. The work participates in a discourse ongoing throughout the century and beyond, "le discourse de la Mode" (as so many French treatises were titled).[23] But the usual pair of polarized terms at the heart of the discourse—substance and ornament would be one version of this pair—do not hold when the matter to be stylized is the human body itself. I want to describe this work, not as it presents itself—at once a wonder book and a satiric fulmination against "gallantry," an entry in the seventeenth-century European resistance to "newfangledness"—but as part of an answer to the question that informs all of Part 3, Where did we come by the modern notion of "culture" (and what did it look like in its youth)?

The twenty-four chapters of Bulwer's encyclopedic book, each focused on a specific body part, offer us a long, rabid, but usefully syncretic (or hybrid) example of the forces of "Nature" marshaled rhetorically against the inherent monstrosity of "Culture." It does not *assume*, as later writers of the Enlightenment so often would, a virtue in Nature superior to those on parade in the Culture of European cities: it *declares* it, over and over again. Even less like the later proponents of the Noble Savage (and more like modernist anthropologists), it locates Culture characteristically in the islands and jungles of the lands newly discovered or rediscovered to European commerce and cosmography. A grossly oversimplified summary of *Anthropometamorphosis* would claim that it represents England as Nature; Asia, Africa, and America (and occasionally southern Europe) as Culture; and the recent discovery of Fashion by "our English Gallants" as the corrupting effect on English Nature of its new commerce with foreign Culture(s).

Such a summary would be partly true. In the chapter on "Face Moulders, Stigmatizers and Painters," Bulwer explicitly chastises "Our English Ladies, who seem to have borrowed many of their Cosmetical conceits from barbarous Nations" (260–61). (For a visual comparison, see figure 22, an illustration from the appendix.) But no hysterical text is simple. Bulwer's loathing of (and incompletely repressed fancy for) everything but himself

---

23. The topic is increasingly interesting to cultural historians and the literature is large. In addition to works already cited, I have benefited from Sylvain Menant, "Les Modernes et le 'style à la Mode'"; Louise Godard de Donville, "La Femme dans la discours sur la Mode au XVIIe siècle"; and Philip Berk, "De la Mode: La Bruyère and the Myth of Order." For fashion and animosity toward it in early eighteenth-century England, see Erin Mackie's *Market à la Mode*. A translation of Roland Barthes's 1967 structuralist treatment, *Système de la Mode*, came out in 1983.

The Artificiall Changling. 261 Spotted Faces affected.

Barbarous Nations, are seldome known to be contented with a Face of Gods making; for they are either adding, detracting, or altering continually, having many Fucusses in readinesse for the same purpose. Sometimes they think they have too much colour, then they use Art to make them look pale and faire. Now they have too little colour, then Spanish paper, Red Leather, or other Cosmeticall Rubriques must be had. Yet for all this, it may be, the skins of their Faces do not please them; off they go with Mercury water, and so they remaine like peeld Ewes, untill their Faces have recovered a new *Epidermis*.

Our Ladies here have lately entertained a vaine Custome of spotting their Faces, out of an affectation of a Mole to setoff their beauty, such as *Venus* had, and it is well if one black patch will serve to make their Faces remarkable; for some fill their Visages full of them, varied into all manner of shapes and figures.

This is as odious, and as senselesse an affectation as ever was used by any barbarous Nation in the World; And I doubt our Ladies that use them are not well advised of the effect they worke: for these spots in
Faire

Fig. 22. Cross-cultural cosmetics. From the appendix to John Bulwer's *Anthropometamorphosis* (1653 edition). By permission of the Houghton Library, Harvard University.

(whatever that might be) requires him to castigate as monstrous morphologies both congenital and "artificial," native and foreign, attractive and repulsive to him and to others. To extrapolate from this "View of the People of the Whole World" a coherent or even stable definition of monstrosity or nature would be beside the point. Instead we will examine a tempest, and the disastrous path of its associations.

Two general associations I have already made are those (1) between the early European study of foreign cultures and a contemporary, mostly satiric discourse on fashion and (2) between fashion and monstrosity—especially monstrosity as conceived in Julia Kristeva's analysis of horror. Abjection and absorption of other cultures must have been important functions in a period that saw established the economic importance of the consumer to early capitalism; what bourgeois Europe could and would not swallow morally (culturally) in the early days of colonial empire is interestingly nonidentical with the geography of its material (and human) consumption. In Bulwer's book one sees, most saliently of all for the investigator of early anthropology, the mechanics of abjection applied to European and homegrown British cultural practices. From the beginning of empire to the end, we find books about British "style" that start with descriptions of the lifeways of its colonies; British mores are Bulwer's moral destination.[24] His xenophobia ranges appropriately far, but it comes home to roost in his closing appendix, "The Pedigree of the English Gallant."[25]

Bulwer states his aim this way in the dedicatory letter of the first edition of 1650: "What I here present you with, is an Enditement framed against most of the Nations under the Sun; whereby they are *arraigned* at the Tribunal of Nature, as guilty of High-treason, in Abasing, Counterfeiting, Defacing and Clipping her coin instampt with her Image and Superscription on the Body of Man." A Nature that mints coins and presides over law cases is certainly English, but whether its behavior is strictly natural is a question Bulwer avoids. His Nature is whatever he wants her to be, mainly, and he knows her secret aims and wishes intimately: using one's hand as a dish (like Diogenes) is "no way contradictory to the intention of Nature," however, "it is plain, by the full length and position of the Hand . . . that nature never intended the Hand to be as a Fork" (*A View of the People*, 184–85). "For a woman to be shorne, is clearly against the intention of Nature" (58), but Nature

---

24. At the other end of the story, see, e.g., Dick Hebdige, *Subculture: The Meaning of Style*.

25. "If it be true that the abject simultaneously beseeches and pulverizes the subject, one can understand that it is experienced at the peak of its strength when that subject, weary of fruitless attempts to identify with something on the outside, finds the impossible within" (Kristeva, *Powers of Horror*, 5).

"made it lawfull for us [men] to cut [our hair]" (59). The true "office of Cosmetickall Physick" is "to conform [the Bodies of Infants] most to the advantage of Nature" (Introd., B2ʳ), though all such shapings of skull and frame as are practiced by Russian, Tartar, and Native American "Nurses and Midwives" are abasements of that coin "instampt" with Nature's image. Bulwer's touchstone in making determinations about Nature's design, clearly enough, is how he and his friends dress, wear their hair, and raise their children. He has company in this attitude, but few confessedly reactionary provincials have composed an ethnological encyclopedia around their intuition of the Natural.

Given Bulwer's self-centered conception of Nature, it is little wonder that his implicit definition of the monstrous overlaps almost completely with the category of "foreign." He could have left it at that—most people do—but in fact he was an intellectual; the semantically functional dichotomy was and is "natural/artificial," not "natural/foreign." The latter pair is merely circumstantial. It *so happens* that foreigners are mostly unnatural and monstrous. Since it sometimes happens that the English ("our English Gallants") are unnatural, too, especially fops and women, there must be a metaphysical term transcending geographical location in which to frame the charges brought before Nature's "Tribunal." The term of course is "artificial," though it does not apply to all cases we might imagine: "Every part of the new-born Infants Body is to be formed, and those parts that ought to be concave, must be pressed in; those which should be slender, constrained and repressed; and those which are naturally prominent, rightly drawn out: The Head also is diligently to be made round" (Introd., B2ʳ). This manipulation is not artificial, because the result will be a head of the sort Bulwer knows is "intended by Nature." So that you will know it too, he details it geometrically in the chapter on Skull-fashions. What *is* artificial is the distinctly un-English (i.e., unnatural) skull-fashion of, for instance, the Russians, "who love a broad forehead, and use Art to make it so" (78). "To vindicate the Regular beauty and honesty of Nature from those Plastique imposters, we say, that a forehead which keeps its natural magnitude is one of the unisons of the face, whose longitude . . . is the third part of the face, and ought to answer the length of the Nose, so that if we compare it to the rest of the face, it ought to have the proportion of a half part to a duple" (81), et cetera.

This and other clear geometries of the natural face make it obvious why Africans of Guinea or Aethiopia must have artificial, i.e., monstrous or cultural faces, for "how can [a flat] Nose beautifie a round Face, such as the Guineans, and they of Caffarain the lower part of Aethiopia are said to have, unlesse we will imagine such a rotundity, as makes a Concave or hollow

Face, with which a Camoise Nose may have some indifferent correspondency. To speak the truth, this Nose being gentilitious and native to an Ape, can never become a Man's face" (1650 ed., 86).[26] (Bulwer is humorously aware that bodily norms are contested; he delights in such ironies as the absurd ethnocentricity of "they of Bolanter," who "have Eares of a Span long, and it is held such a note of gallantry among them, that those that have not their Eares long, they call them Apes" [1654 ed., 145]. It's a crazy mixed-up world.)[27]

To support his category of the "artificial" Bulwer depends heavily on the notion of acquired traits. The usual aetiology of monstrous differences from the English norm he assumes to be the gradual pseudo-naturalizing of one or another people's "fashionable elegancie," his most spectacular example, borrowed from Browne's *Pseudodoxia Epidemica*, being the "Blackness of Negroes," with which the final ethnographic chapter of the first edition ends. (The section is shifted in the later editions to the middle of the final chapter.) This is an extreme; often enough Bulwer is perfectly accurate in his perceptions of the "constructedness" of the traits and practices he details — tattooing, for instance, or femininity. His response to these constructions is axiomatic and unmodulated: "considering these strange attempts made upon the naturall endowments of the Face, one would think that some men felt within themselves an instinct of opposing Nature, . . . whereas they should strive against their own inward, they oppose their outward Nature" (241).[28]

One would certainly think that about "some men," and not only in respect to the care and adornment of the person. The passage expands its focus at this point and begins to detail a long and elaborate topographical and even cosmographical metaphor, adumbrated earlier and supported in later figures: the face, already defined in terms of its most proper longitudes and lat-

---

26. I have quoted from the first edition here; in the expanded 1654 edition the passage appears on pages 130–31, except for the last sentence, which has been moved to an earlier location on page 128. Some of the most excruciating sentences and passages about African physiologies in the 1650 edition are moved or broken up in the later editions.

27. Bulwer fears the unlike likeness of apes as well as of foreigners, and makes a point of their threatening nearness to us in his introduction, where he anticipates Darwin, or rather Lamarck: "in discourse I have heard to fall, somewhat in earnest, from the mouth of a Philosopher . . . That man was a meer artificiall creature, and was at first but a kind of Ape or Baboon, who through his industry (by degrees) in time had improved his Figure and his Reason up to the perfection of man" (B3ʳ).

28. The introduction here of an interior to the monstrous helps inaugurate another long history, the history of abnormal psychology, which is to say psychology. (See Dennis Todd, *Imagining Monsters*, for the eighteenth-century English strand of this story.)

itudes, becomes first a natural landscape and then a political territory.²⁹ Although the lesson seems to be lost on Bulwer (that "men . . . should strive" inside and *not* outside their own proper political borders), it is hard to miss either the avowed or the unconscious dynamics of colonial and imperial expansion in this account of "the little world of beauty in the face" (242).³⁰ Where "man transported with vaine imaginations . . . findes Hils, he sets himself to make Plains; where Plains, he raseth Hils; in pleasant places he seekes horrid ones, and brings pleasantnesse into places of horrour and shameful obscurity." If the body is the Other (World), Bulwer's allegory tells us the moral price of conquest: "when [Man] thinks he triumphs over his subdued and depraved Body, his own corrupt Nature triumphs over him" (241). This victory of corruption is Satanic, naturally, introducing yet another character not to be confused with the self: it is "a stratagem of the enemy of our Nature," who it turns out is a "cuning politique Tyrant" sending domestic opposition out of the city (of the self) "to fight with the enemy, to the end that venting his violence and fantasticalnesse abroad, [the Tyrant] may have plenary power to tyrannize at home at his pleasure" (241).³¹

29. "Through frustrations and prohibitions, this [maternal] authority shapes the body into a *territory* having areas, orifices, points and lines, surfaces and hollows, where the archaic power of mastery and neglect, of the differentiation of proper-clean and improper-dirty, possible and impossible, is impressed and exerted. . . . Maternal authority is the trustee of that mapping of the self's clean and proper body" (Kristeva, 72). In Kristeva's description of abjection (that which creates the symbolic forms of the monstrous, the sacred, the Other), especially in its public forms (e.g., religious ritual), the threat posed by the maternal is "that of being swamped by the dual relationship, thereby risking the loss not of a part (castration) but of the totality of [one's] living being. The function of these religious rituals is to ward off the subject's fear of his very own identity sinking irretrievably into the mother" (64). It is easy to see the mapping and itemizing proclivities in Bulwer's text as, in part, magical hedges against such loss, and they would help to explain his frequent reference to "Woman" as "that impotent Sexe"—an otherwise somewhat surprising epithet in a work that emphasizes "that Sexe's" power to create human (and monstrous) shapes.

30. This collapse of body and particularly face with landscape is not Bulwer's alone. *England's Vanity* similarly reserves to powers higher than one's own the right to mold either prospect: "Though the Face of the Creation hath its variations of Prospect and Beauty, by the alternate intermixtures of Land and Waters, of Woods and Fields . . . God here mounting an Hill, and there sinking a Vale, and yonder levelling a pleasant valley; Designedly to render the whole more delectable. . . . Yet hath he nowhere given us more admirable expression of his Infinite Power and Wisdom than in the little Fabrick of mans Body. . . . Nor is it possible for the heart of man . . . to adore enough the Transcendences of his Divine Hand. . . . But amongst them all . . . to survey onely the Glories of the Face" (81–82).

31. The "tyrant" of Bulwer's metaphor sounds more like Cromwell than Charles. Though at times Bulwer sounds like a Dissenter, Jeffrey Wollock's scholarship points to a High Anglican identity; see "John Bulwer's (1606–1656) Place in the History of the Deaf." Thanks to Katherine Rowe for finding this source.

It is in this extended figure that Bulwer defines the matter of his book, the intersection of the cosmos and the cosmetic: for "as the greater World is called Cosmus from the beauty thereof, the inequality of the centre thereof contributing much to the beauty and delightsomenesse of it; so in this Map or little world of beauty in the face, the inequality affords the prospect and delight" (242).[32] The bodies of those "barbarous nations" afford the perfect synecdoche both for the nations, with their competing *cosmoi* and, in their obviously decorated state, for the anti-natural business of Culture per se. The Aethiopian body, generated at first by the "prevarication of Art" (174, quoting Cardanus) both stands for and is a part of the Aethiopian cosmos, "subdued and depraved" but potentially triumphant over the cosmetic styles of England, "our Nature."

Although the book attacks foreign fashions, the true target of its didactic wit is the "English Gallant"; its tough love is designed as a kind of prophylactic against the "vaine imaginations," "foolish bravery" and "filthy finesse" discoverable in the "People of the Whole World" here set before us for our undeniable but dangerous delectation. The multiculturalism of seventeenth-century English consumer culture has already begun to break down the sovereign borders of gender identity, apparently, and God alone knows what further dissolutions might stem from this: the end of the human species is a possibility, since sex change on the part of women and sodomy on the part of men are consequential expressions of the gender inequality at "the centre of" English society. (In fact Parcelsus offers good information on how to propagate homunculi "without the conjunction of women" (492), but such "non-Adamiticall men" are not strictly speaking human.)[33]

Gender is the snarl in Bulwer's argument. He does not want to argue, with

---

32. On cosmos and cosmetics, see Angus Fletcher's superb and still startling chapter in *Allegory*, "The Cosmic Image," esp. 108–20. The whole chapter richly repays reading, making suggestive links between ornament, sphere, world (as in Kepler's earth-bijoux, seen from the moon), seventeenth-century poetics, and even taxonomy—which has an etymology related to that of cosmos, as *taxis* and *kosmos* were synonyms. Costume and custom are also rooted etymologically in *kosmos*. See Kim Hall, *Things of Darkness*, chap. 2, for a discussion of cosmetics, colonialism and race in seventeenth-century Britain, esp. 85–92.

33. Bulwer's discomfort with male asexual reproduction would seem to challenge Marie-Hélène Huet's claim that "what made monstrosity monstrous was that it served as a public reminder that, short of relying on visible resemblance, paternity could never be proven" (*Monstrous Imagination*, 33–34). She is partly right, but monstrosity seems to involve several different categories of identity and their interrelations. A child created without the assistance, however theoretically passive, of a female partner, though it might stand as unusually clear evidence of paternity, would not stand as evidence of female submission to male sexual domination—would not then be a sign of the difference (only conceivable hierarchically) between male and female.

Aristotle, that women are technically monstrous in their dissimilarity from the norm (not that he is friendly toward women, but his differentia are ethnic or "national");[34] on the other hand, many of the differences women exhibit seem to be cultural, that is constructed, *ergo* monstrous, or at least supported and augmented by cultural means. I have already mentioned the difficulty with maintaining the naturalness of gender differences in hair style: if, as Bulwer declares in the chapter on "Eyes," "true beauty is referred to the successe and goodnesse of utility," then one or the other of the genders does not exhibit true beauty in the trimming of its hair, since the fundamental uses of head hair cannot be different between the sexes. ("The prime end of the Haire of the Head is to defend the skin, the second use is to defend the Braine from injuries from without, or from within" [50].) I have also alluded to the difficulty of maintaining the dichotomy natural/foreign in the face of that gender whose features and adornments are both English and unmasculine. Shaving the chin must be considered "piacular and monstrous" in men; "to be seen with a smooth skin like a woman, a shameful metamorphosis" (199). Yet it cannot be denied that, in general, "men of the New world . . . have store of milk in their breasts" (319). Are the spontaneous beardlessness of English women or breast milk of Brazilian men natural *and* monstrous at once? Are the English customs that discourage facial hair in women and breast milk in men artificial? But English customs are the very definition of Nature; except for the foppish manners of Gallants there should be no need to speak of English customs, there should be none to speak of.

Sadly, custom must be called in: "to what ends should we either mingle or change the custome, or the sequestering variance of virile nature with feminine, that one sex cannot be known and distinguished from another?" (60). As with the production of Nature-shaped infants, Nature needs a little help from art in realizing some of her intentions. (The "tip of the Ear," for instance, "seems in a manner to be perforated with an invisible hole . . . wherein the Athenians were wont to hang their golden Grass-hoppers." So pierced ears are natural. And yet, "this is no warrant for the monstrous practices of these men . . . who so shamefully load it with Jewels . . . , and use such force of Art to tear and delacerate the most tender particle thereof, stretching it to so prodigious a magnitude" [156].)

In an unsurprising dodge, women become precisely what more recent

---

34. See Aristotle's *Generation of Animals*: "The female is as it were a deformed male" (2.3.175); "The first beginning of this deviation [from the generic type] is when a female is formed instead of a male" (4.3.401). Quoted in Huet, 3 and 93.

misogynists have so often and oddly called them, the guardians and transmitters of culture. Eluding the logical conundrums they generate as phenotypes themselves for his matrix of assumptions about the natural, Bulwer concentrates on them especially as the nurses and midwives who introduce so many artificial shapes in the bodies of the infants under their care, or as the dangerously imaginative maternal vessels of infants in their most formative stage. Midwives not only swaddle and deform us as children, snipping septums and foreskins and pressing down protuberant "asses ears," but their interferences instill in us from the start a taste for the altered and the perverse; Gallants are the logical outcome of meddling midwives and nurses, as are the less emphasized ladies of fashion. Bulwer also supports the thesis, recently analyzed by Marie-Hélène Huet in *Monstrous Imagination* (see note 39), that monsters are produced by the vivid fantasies or perceptual experience of their mothers during pregnancy; such a thesis supports his notion of women as female agents of interruption in the male continuity of natural process.

Women are in some ways represented here as the missing link—ontogenically intermediate as well as intermediatory. Even the normative geometrical descriptions of "absolute" women and their body parts produce an effect similar to the static ethnographic writing quoted on every page of Bulwer's book. The slide is easy from a description of a woman's "natural" lips— "somewhat full . . . , coral, imitating Vermilion, a little disjoyn'd, yet so as the teeth are scarce discovered, while she holds her peace or laughs not, unmoved" (132)—to the more fully alien *effictio*, which characteristically reads as if it describes a statue or an engraving rather than a living agent.[35] Women are intermediate between men and the beasts they continually threaten to revert to, as well as between Englishmen and foreigners: some men, as Bulwer sternly records, have intercourse with beasts *instead of* women (493). And the figments of their imaginations are intermediate between bodily and Platonic realities: the hairy girl of Pisa was conceived as an effect of "the Picture of

---

35. Here, for instance, are the Giachas: "The Giachas or Agagi of the Ethiopian Countreys beyond Congo, have a custome to turn their Eyelids backwards towards the Forehead and round about; so that their skin being all black, and in that blacknesse shewing the white of their Eyes, it is a very dreadfull, and divilish sight to behold; for they thereby cast upon the beholders a most dreadfull astonishing aspect" (93). That the slide to the generic barbarian, male and/or female, is not blocked by gender seems partly due to the sense of gender's disappearance into monstrously blurred "variance" in those distant landscapes and facescapes. Female gender was close to no gender. (For recent contrasting views of the ontological status of the categories "male" and "female" in early modern Europe, see Thomas Laqueur, *Making Sex*, and the review essay by the historians Katharine Park and Robert A. Nye, "Destiny Is Anatomy.")

St. John the Baptist, painted after the usual manner clothed in Camels haire, whose image hanging in her Chamber the mother had wishtly beheld" (475).

It is not, I think, beside the point to wonder what Bulwer considered his illustrated book's effect was likely to be on the unborn children of its owners' wives. The imagination, and the cultural intelligence that designs and stores conceptions of decor—the cosmetic intelligence—are dangerous shifters in the multiplicitous text of the "greater World." As long as Images and Superscriptions can be "instampt" in books and broadsides they can be disseminated in the frighteningly manipulable flesh of our progeny, as well as the less intimate fashions of clothing, adornment, and sexual taste adopted in adulthood by gallants and ladies of fashion. Instead of (or in addition to) "Nature's coin," a considerably increased quantity of commercial coin was circulating in the 1650s between England and places very different from England, between the English and the very different people whom and from whom the English bought. Even the innocent commodities of this trade could smuggle in notions of physiological difference dangerous to the national stock: "the shape of Spanishe Stockings sold upon our Exchange, whose shortnesse speaks them to have been made for women, seems to intimate that the women there, have great Legs and very little Feet" (426).

In this formative stage of British use of knowledge about Others, monstrosity remains as important as ever in the popular consciousness. The difference is that it is now seen as transmissible to home—thus Bulwer's appendix on "the Pedigree of the English Gallant." Commodity capitalism and print have disseminated monstrosity widely and provocatively: biological and mechanical reproduction are intimately linked, as is made explicit in Bulwer's use of printing and engraving terms in discussing Nature's minting of the normal.

Culture (that is, "fashions" or "manners and customs," "moeurs," "habites"), at this stage, seems associated positively with barbarians, women (especially loose women) and "effeminate men"—as it still is, in its aspects of the artificial and the unnatural. It has not become a universal category yet by Bulwer's time; it is itself Queer, because the people who carry and transmit it are Queer.[36] Contact with such people may be necessary, to reproduce the race and/or grow the economy, but it is also one of the greatest dangers fa-

---

36. "Culture" in the contemporary sense is not found in English until the early twentieth century (even Tylor's *Primitive Culture* understands the term in a restricted sense, as having more to do with the arts and crafts of daily life than with government or politics). In the period we are talking about it is still close to its etymological roots; its chief areas of reference are agriculture, husbandry, and worship.

cilitated by the age of print. Bulwer's own book, patched together with chaotic amplitude from countless and diverse prior texts, organized by a blazon of body parts, stuffed with dangerous images of alien cultures and suggestive comparisons, raging in affect and unrestrained by conventions of genre or discipline, is as monstrous as any text "instampt" and engraved in its moment.

Perhaps the most uncanny of Bulwer's monsters is the multicultural female folk-dancing monster introduced in the penultimate chapter ("On Leg and Foot-fashions"), Catherine Mazzina, born in Avignon in 1594,

> of a comely forme, and 27 inches and a Palme over in heighth, but wanting Hips and Legs, and consequently Feet, her Armes were perfectly formed, being longer than her breast and trunke, the lower part of her body did in a manner appear bifid, emulating the bottom of a Harpe;[37] She spake to purpose, sung, plaid on a Lute, danced with her hands Spanish, Mauritanian, Italian and French dances, in like manner to the sound of Musique she so composed the Gestures of her imperfect body, that they who had seene her afar off, would doubtelessly have said, she had danced with her Feet. And as to the endowments of the mind, there was nothing wanting to her which is granted by Nature to other men. Moreover she was endowed with both Sexes, yet she drew nearer to a woman, and was more vigorous in that Sex, and therefore was rather called a woman than a man. (453)

Catherine is several kinds of hybrid at once, including a representative of mixed notions of monstrosity; although unmoralized, she sounds like a typical broadside monster in physical shape, while at the same time coming under Bulwer's new more "artificial" and cultural paradigm.[38] Her bifid, "Harpe"-like "lower part" could be the result of her mother's having stared at printed broadsides of monsters — or perhaps of her mother's desire to have a Harpe?[39] The lute playing and dancing of the "comely" Catherine show her a typically feminine and aspiring bourgeoise — a familiar, homely character

---

37. Recall the remarks of de Fitelieu and (in n. 14) La Bruyère, on the split bodies of the fashionable.

38. For easily available reproductions of broadside monsters consult the many collections of Hyder Rollins. Several of the monstrous broadside ballads are discussed and located specifically in Helaine Razovsky, "Popular Hermeneutics: Monstrous Children in Renaissance Broadside Ballads." Janis Pallister's English translation of Ambroise Paré's *Des monstres et prodiges* (1573) contains a multitude of French woodcuts of human or humanoid monsters (including the Monkfish and Bishopfish discussed later).

39. The relation of female desire to monstrosity was a long-lived one (not deceased even now). Huet quotes Pietro Pomponazzi: "If a pregnant woman greatly desires a chickpea, she will deliver a child bearing the image of a chickpea. That is how Cicero's family got its name" (*De naturalium effectuum admirandorum causis* [Basel, 1556], quoted in Huet, 17). More than 150 years later James Blondel sees the need to satirize those who believe that "the mere Longing for *Muscles* [mussels] is sufficient to *transubstantiate* the true and original Head of the Child into a

to Bulwer's readers; but she sounds a little too bright, her knowledge of dancing a little too cosmopolitan, and of course she has a penis. As a hermaphrodite, she is a regular, unsurprising (if still rather risqué) monster—her shock value in the context of Bulwer's previous 452 parasitically ethnographic pages comes from her knowledge of so many cultures, and her ability to translate the language of dance from her feet to her hands. In other words, she is a hermaphrodite of *culture* as well as of sex and gender.

She *is* moralized, though—not explicitly, but by the tone of Bulwer's opus generally. Bulwer seems to intend her as the emblem of what is ridiculous in "fashion": the cosmopolitan quality of her knowledge combines with its uselessness to make her the perfect creature of empty fashion, and her leglessness and consequent foot-lessness both mark the emptiness of the attainments and punish her for them. Bulwer has elsewhere explained why women are not inherently monstrous, and how we know that; quoting another Italian (the "Marquesse of Malvezzi") he stresses that their differences have a *use*: "they who believe that Woman . . . is not an Errour or a Monster, must confesse she is made for Generation, and if she be made for this end . . . it is necessary she be endued with parts that move unto that end." Nature informs us directly about this telos of womankind, for "so soone as she is represented unto us . . . man doth by Nature hasten to contemplate her for the end to which she was made by Nature" (493). A woman without a "lower part" (without an "end") has taken the sterility of fashion to its literal extreme—as is demonstrated, at least in Bulwer's terms, by her veritably ethnographic knowledge of dances, of foreign body languages.

The discerning reader may already have noticed that Catherine Mazzina sounds a bit like Bulwer—a specialist in body languages and gestures, an armchair dilettante of ethnography, a multilingual European with a penis. Both seem avid, hysterical, filled with a weird *jouissance*. Their differences, then, should tell us something about Bulwer: one is a "monster" and the other a potentially monster-genic analyst of monstrosity; one is a dancer and the other a student and teacher of body language; one is a set of texts (by Aldrovandi, Hoffman, and others), the other an author whose texts are all patchworked together from other texts. One is an Italian born in France who performs African dances, the other an Englishman from England who writes in English.[40] One is a "woman," and the other is a "man." It's a good bet that

---

Shell-Fish" (*Power of the Mother's Imagination over the Foetus* [1729]). See Huet, 14–16, and (for Blondel's pamphlet war) 64–67.

40. Bulwer includes at least one other European monster who is not living in the "right" country: "Scaliger remembers a certaine little Spaniard covered with white haires, which he reports to have been brought out of India, or to have been borne of Indian parents in Spaine" (475).

Catherine would not score well on the "intended by nature" test, despite her comeliness, her vigor, her intelligence, and her penis. She seems easily read as the precise shape of Bulwer's nightmare self, doomed to live out, as subject, the role of all those abjected objects in the twenty-four "Scenes" of his ethnoballetic encyclopedia.

Michel-Rolph Trouillot in his essay "Anthropology and the Savage Slot" reminds his fellow anthropologists that, "like all academic disciplines, [anthropology] inherited a field of significance that preceded its formalization" (18). He is interested in the construction of the phantasm "savage," rather than in the "monster" that preoccupies Huet, and insists, "Anthropology did not create the savage. Rather, the savage was the raison d'être of anthropology" (40). His essay complicates the by now received truth that "the savage or the primitive was the alter ego the West constructed for itself" by pointing out that "this Other was a Janus, of whom the savage was only the second face. The first face was the West itself, but the West fancifully constructed as a utopian projection" (28). A reading of Bulwer's *Anthropometamorphosis* might complicate that complication further. The "first face" need not only be the paradisal utopian one of Montaigne or the orderly one of More, it could also represent the simply normal utopia of home—or rather home reimagined *as* "normal" (Dorothy's Kansas). And the second face need not only be "the savage," constitutively distant. Monsters can live next door, especially if monstrosity is construed as meaning altered, "Transform'd," "ab-normal."[41]

The concept of the savage, then, was not the *only* "raison d'être of anthropology." Bulwer's book pieces passages of what Trouillot refers to as "paraethnography" (as in "parascience") into a satirical treatise on cosmetics and points it at local as well as exotic targets, home-grown deviations from a bodily standard imagined as "natural," if rarely encountered. The "self-made man," that is, the monster/fashion plate, seems to have been another slot that "preceded the formalization of anthropology." Anthropology has from al-

41. Medieval monsters were mostly imagined as species, and species on the world's margins. But by the seventeenth century they had become individual and local, even intimate: Pepys's servant James Paris, in his manuscript history of monsters (c. 1680) begins literally at home, with the birth of a monster in his childhood home to a friend of his mother's, an inveterate reader of illustrated "Almanacks": "this Accident was Kept very Secret, and the Child being a Monster and not having been Cristened was wrapped in a Clean Linnen Cloth and put in a littel wooden Box and Buried very Privately, in a part of our Garden which I Caled my Garden.... A few Dayes After being Buisy in my little Garden, I Discovered a little Box, in which I found this Little Mounster, which I Buried Again" (*Prodigies and Monstrous Births of Dwarfs, Sleepers, Giants, Strong Men, Hermaphrodites, Numerous Births and Extreme Old Age etc.* [British Library, MS Sloane 5246], qtd. in Dudley Wilson, *Signs and Portents*, 94).

most before its beginning had antiquarian and folklore studies as (paradoxically) marginal poor relations—middle-class European and later "Western" study of the peasantry and the working classes has always seemed parallel to the study of cultures geographically exotic. Early sociology, as well, concerned itself with "deviants" and the working class. In Bulwer one can observe all these abjections unified under the rubric of the "artificial," and find a hysterical near-admission that one can barely escape the rubric oneself. Between the postfeudal social mobility that permits and even requires men and women to "make themselves,"[42] and the circulation of exotic imagery that encourages them to make themselves strange, it is hard to imagine how poor Nature will communicate her intentions to an increasingly imperial and self-conscious Europe.

"Manners," "customs," and "fashions" are the early modern terms that covered the terrain roughly equivalent to the "culture" of modern anthropology. Clothes and dances, jewelry and body language, erotic self-presentation and carriage were observed at home and abroad, by both the fashion conscious and the protoanthropologists. This makes sense, and seems obvious. But it also seems somewhat neglected as a way of more deeply imagining "the larger thematic field" (Trouillot) on which anthropology set up its tents. The "discourse de la Mode," with its demonic women, fops, virtuosi and social climbers, its innovations and superficialities, its abandonment of all that is sturdy, deep, and "natural" about the suddenly self-conscious culture of Europeans at home, is closely tied to the sudden appearance of a (distant) *mirror* in the sixteenth century. Michael Taussig is talking about that self-consciousness and theatricality in "The Report to the Academy," which opens *Mimesis and Alterity*:

> Now the strange thing about this silly if not desperate place between the real and the really made-up is that it appears to be where most of us spend most of our time as epistemically correct, socially created, and occasionally creative beings. We dissimulate. We act and have to act as if mischief were not afoot in the kingdom of the real and that all around the ground lay firm. That is what the public secret, the facticity of the social fact, being a social being, is all about. . . . Try to imagine what would happen if we didn't in daily practice thus conspire to actively forget what Saussure called "the arbitrariness of the sign"? Or try the opposite experiment. Try to imagine living in a world whose signs were "natural." (xvii–xviii)

Bulwer seems to have felt himself, as many did, confronted with a choice between these two experiments, and wishing earnestly he could "return" to

---

42. See for instance Brathwaite's conduct books, "The English Gentleman" and "English Gentlewoman."

some world before the idea of natural signs had to be considered a thought experiment. He has Taussig's *jouissance* but not his serenity, so his book is silly if not desperate. Still, it is unsettling to imagine Taussig's "unstoppable merging of the object of perception with the body of the perceiver" (*Mimesis and Alterity*, 25) from the point of view of someone like Bulwer who had never heard of genes, whose country had just been through a civil war, and whose civilization had recently undergone the discoveries of such image-cornucopias as print, America, engraving, and the microscope. The idea that the social fact was "factitious" and that the customs preserving gender identity were malleable might well make someone anxious, especially someone atop the gender hierarchy; the prevalence of alien or fictional representations might seem to threaten (or excite) all who perceived them with reproductive chaos.[43] Bulwer sees in fashion a bone-deep penetration of the body politic, a self-consciousness that could give his people fatal vertigo. The work of nature is hard to preserve in an age of mechanical reproduction. Huet is too narrow in her reading of what the "monster" is a sign of—it is not a sign of the fact that we can never be sure of paternity but, perhaps, of the fact that we can never be sure of any kind of identity, at least not through the means of a stable visual signifier.[44] After the deconstructions of Hooke's microscope and the illusive scintillations of Cavendish's Blazing Empress, we can hardly imagine such a thing.

CATEGORIES AND MISFITS, OR, WHAT IS THAT NAKED MONSTER DOING IN A BOOK ABOUT CLOTHES?

It may be instructive to go back to the influential French and Latin picture book of the 1560s and 1570s, the *Receuil de la diversité des habits* (in Latin and bilingual editions, *Omnium fere gentium, nostraeque aetatis Nationum, Habitus & Effigies*). For there was a wonderful as well as a hysterical approach to the category challenges represented by so much and such disseminated difference, and Bulwer alone cannot represent the effervescence of a time in which, at least for some (to quote my epigraph from Sedgwick), "because there can be terrible surprises . . . there can also be good ones." The *Receuil*

---

43. For a science fictional population management technique that also provoked fears of lost distinction and reproductive chaos (free love, or "the community of women"), see Campanella's utopia, *De civitate Solis*, and the discussion in John Headley, *Tommaso Campanella*, 298–314.

44. See Marjorie Garber's "Spare Parts: The Surgical Construction of Gender," a discussion of the twentieth-century discourse of the "transsexual" which resonates everywhere with Bulwer's terror (and joy) in the face of the "self-made man."

contradicts its apparently orderly categoricity so flamboyantly that we can see in it at once the moment of transition to intellectual modernity (Reiss's "analytico-referential discourse") and the moment of joy or excitement in otherness and confusion. It recalls Kepler's marvelous fence-sitting in the *Somnium*, or Bacon's beauty-laden manifesto against the pleasure that distracts in the search for Truth. It is anthropology without mastery, and without the wish for it.

The *Receuil* organizes an entire world of "habit" (another double-entendre like "custome") — each illustrated habit corresponding to a single type of human wearer. Although, as I have already pointed out, it does not achieve a taxonomy, it does manage to comprehend the basic categories by which we tend to distinguish human social and reproductive functions to this day: nationality, sex, occupation, marital status, and (implicitly) social status. Like the maps of the period, it is unsurprisingly more data-rich closer to home (it was printed in Paris first, in French), but it spreads a wide net: the little octavo seems to aspire to a global coverage of ethnic costumes. I have noted one pole of this globe — familiar France, with its twenty-eight types; the other, the exotic singular, remains to be contemplated.

In fact the work offers a number of antitheses to home and its diapason of internal distinctions: there are nations that produce only one representable type (for instance, *Asiana mulier*— eleven out of the seventeen explicitly gendered one-type nations are represented by a female); there are types with no nation (*Mulier sylvestris*), nonhuman types (*Simia erecta*), and monsters (*Episcopus marinus*, "Cyclops"). What is perhaps most intriguing in a book about *habits* (of *omnium gentium*, all nations) is then the naked character with no nation, or gender; this character ought to function as the fundamental Other of such a text, the undistinguished ground, Kristeva's "not-I." And yet we are invited to think of the Cyclops as in fact clothed. S/he does not in any other way interrupt the monotony of the book's presentations: French and/or Latin verses on the left, inviting the reader to look at the plate on the right. The Wild Woman "au naturel vous est icy depeinte, / Comme voyez qu'il appert a vostre oeil" (is depicted for you here *au naturel*, as you see it appear to your eye); as for the Cyclops, "Hic vertas oculos amice Lector, / Si te vera juvat videre monstra" (Please turn your eyes here, dear Reader, if it pleases you to see true monsters). By page forty-four we are used to looking at costume on the right, so the Cyclops's extremely complicated naked body looks like a variation on this theme: what then *is* costume? And why does it so much absorb the focus of Europe's initial ethnological attention?

Some answers spring to mind: the "nakedness" of the New World's Americans had obsessed European letters since Columbus first combined the

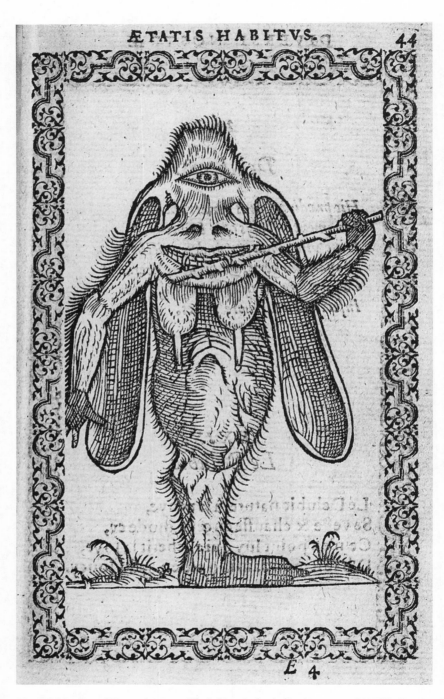

Fig. 23. "Cyclops." From the *Recueil de la diversité des habits* (Antwerp, [1572]). By permission of the Houghton Library, Harvard University.

claim of nakedness with a description of their clothing. In the Edenic settings of the pristine Bahamas, Genesis 1–2 was an easily available frame of reference, and it was politically convenient as well as more mysteriously appealing to play with the idea of the Americans as nonmembers of the biblical human family (constitutively clothed). Against that background, every representation of costume was a nonabsurd predication of the wearer's full humanity: one might imagine that the *Simia erecta* is depicted in this collection to remind us that not all featherless bipeds are human, that cultural expression (of which the metonym is clothing) is the sine qua non of species membership in the human.

But the *Simia erecta* turns out to be clothed, unlike the Cyclops or the *Vir* and *Mulier sylvestris*. And so are the other two monstrosities, the Episcopus and the Monachus piscus (Bishop- and Monk-fish). Or so they seem — in fact their monstrosity lies precisely in the fact that the apparent mitre, cowl, and so on that mark the habits of bishop and monk are actually flesh (even as, for Bulwer, the altered bodies of some people are "actually" habits). This group of categorical surprises (none of which is designated by nation or region) really forces the question of the meaning of costume and its relation to the body and to the borders of the human. If clothing signifies human membership, then are the wild man and the wild woman not human? What is a nonhuman *vir* or *mulier*? Does the thick mantle of the *mulier*'s hair count as covering the body? If flesh imitates particular styles of clothing as perfectly as it does in the illustrations of the Bishop- and Monk-Fishes, does that mean that clothing need not be understood as counter to (or covering for) body? Because the book arranges its entries alphabetically (by Latin name), the "Cyclops" is the first unclothed or questionably human character to appear. What is the meaning of the moment in which we interpret his/her multiple morphology as a habit and his/her name as the name of a *gens*?

Works like this are one of the great pleasures for a modern researcher reading around in a past not yet recognizably organized; as Gabriel Chappuys included Utopia in his encyclopedia of *royaumes*, so the compiler of this *Omnium* adds the wild, the monstrous, and the multiple to his index of persons.[45] The implication is that "human" is a spectrum, within a larger spectrum, and that the borders are fuzzy: the Scottish *Mulier sylvestris is* clothed, the Brazilians wear headdresses and garters, and if this seems too little for proper civility, it is clear enough on the other hand that the various persons

---

45. Chappuys, *L'estat, description et gouvernement des royaumes et républiques du monde, tant anciennes que modernes* (1585).

in *dueils* (mourning), totally covered with fabric, including their eyes, are ominously hidden—are they automata? (See figure 24.) One reads, as perhaps the compiler compiled, in the interrogative.

This interrogative invokes a different mood from the anxious didactic one of the later Bulwer, and the hybridity of the Cyclops—composed not only of both sexes but of every major species among the "monstrous races"—is hilarious rather than abject. S/he brings, in fact, more structural features of the cosmographical project into question than the division between the human and the not—*all* categories of identity fail to obtain in this image. Human or not? Male or female (or both)? Of what *kind*? Singular or plural? Of whose making? Clothed or naked? None of the above? These collapsed categories add up to a kind of whacky sublime to which later ethnological surveys are strangers. Even the disapproving Bulwer is drawn to hoard and publish descriptions of human strangeness *as* strangeness—he takes an illicit (for him) pleasure in transgressions of the Normal Body, so painstakingly measured. But Bulwer's pleasure is ethically modern, grounded in the very disapproval through which he hopes to escape responsibility for it. The pleasure offered by the calmly, democratically alphabetical parade of human, subhuman, semihuman, and nonhuman fashion plates in the 1562 *Receuil de la diversité des habits* is something at home in a world still short of knowability. Instead of reverting to the ethical realm, where foreign cultural puzzles can be turned into homegrown moral certainties, it simply follows the random logic of the alphabet through whatever the alphabet presents, leaving the reader to extrapolate a structured cosmos on her own.

Fashion and the awareness of fashion are aspects of cultural self-consciousness and a prelude to cultural analysis; they are also sources of pleasure—of a narrative system of desire and satisfaction so potent it has blurred (as some writers feared) the visibility of class and ethnic distinctions studied by the social sciences that followed in its wake. In the *Receuil de la diversité des habits* we can see the contemplation of basic anthropological categories mixed with the pleasures of fashion and, more deeply, of carnival comedy. The categories could be transgressed comically in a context that shared so much with forms of visual pleasure and exercised so immediately the mechanisms of recognition and identification. The fashion show maintains this capacity for whimsy and the challenging of assumed social structure (as in the mid-1980s' skirts for men). But ethnology lost its sense of humor, because knowing—or at any rate Bacon's version, the kind that "enlarge[s] the bounds of human empire"—precludes the ambiguities and shifting or hybrid identities that make us laugh. Almost a century passed between the *Receuil* and Bulwer's *Anthropometamorphosis*, and in that century Quebec City

Fig. 24. Woman from Brabant in mourning. From the *Recueil de la diversité des habits* (Antwerp, [1572]). By permission of the Houghton Library, Harvard University.

was built, Harvard University established, the East India Company formed, and the bounds of empire enlarged so far that the empire was already striking back:

> Infinite and many are the sinfull, strange, and monstrous vanities, which this unconstant, vaine, fantastique, idle, proud, effeminate, and wanton age of ours, hath hatched and produced in all the parts, and corners of the World; but especially in this our *English* Climate; which like another Affrique, is always bringing foorth *some new, some strange, misshapen, or prodigious formes*, and fashions, every moment. (William Prynne, *Unloveliness of Love-Lockes*)

VIII • "MY TRAVELS TO THE OTHER WORLD"
Aphra Behn and Surinam

[The Indians] had no sooner spy'd us, but they set up a loud Cry, that frighted us at first; we thought it had been for those that should kill us, but it seems it was of Wonder and Amazement. They were all naked, and we were dress'd, so as is most commode for the hot Countries, very glittering and rich; . . . my own Hair was cut short, and I had a taffety Cap, with black Feathers on my Head; my Brother was in a Stuff-Suit, with silver Loops and Buttons, and abundance of green Ribbon.

    I had forgot to tell you, that those who are nobly born of that Country, are so delicately cut and raised all over the Fore-part of the Trunk of their Bodies, that it looks as if it were japan'd, the Works being raised like high Point around the edges of the Flowers. Some are only carved with a little Flower, or Bird, at the sides of the Temples, as was *Caesar*; and those who are so carved over the Body, resemble our ancient *Picts* that are figur'd in the Chronicles, but these Carvings are more delicate.

    The Fame of *Oroonoko* was gone before him, and all People were in admiration of his Beauty. Besides, he had a rich Habit on, in which he was taken, so different from the rest, and which the Captain cou'd not strip him of, because he was forc'd to surprize his person in the minute he sold him.

—Aphra Behn, *Oroonoko* (1688)

THE "SELF-MADE" RESTORATION playwright Aphra Behn seems to have combined in one person interests in all the issues and materials and often even the very texts that have occupied the previous seven chapters of the present book. Herself a traveler, like Thevet and Hariot, to the New World, she was also clearly a reader of cosmographia and colonial propaganda. She was a fan of the new science (see her Copernican preface to her translation of Fontenelle's *Entretiens sur la pluralité des mondes*— and note her apparent gifts of insects and an Indian feather "Peruque" to the Royal Society), and dismissive of the occultist experience of wonder (if her translation of Fontenelle's rationalist *History of Oracles* was her own idea—see also the poignantly hilarious character of the would-be cabbalist and moon-voyaging doctor in *The Emperor of the Moon*.¹ Wonder, however, of the kind we have seen trapped in the sensuous amber of micrographic and pornographic detail is featured with exemplary success in Behn's *Oroonoko*, often (until Ian Watt's *Rise of the Novel*, and now again) treated as a candidate for the "first" novel in English.²

Oroonoko begins as a romance, set in Africa, about the eponymous prince and his thwarted love for Imoinda, a young woman in his grandfather's harem. Both lovers separately are kidnapped and sold into American slavery, ending up on the same plantation in Surinam where the narrator, Behn apparently speaking as Behn, is a frequent visitor. Here the work's mode changes into that of an ethnographic novel, in which the African protagonist starts an unsuccessful rebellion among the slaves, kills his wife before he is recaptured (to prevent their separation and the birth of their child into captivity), is tortured and finally dismembered alive as a lesson for the other slaves, by the colonial administrators who are the narrator's peers and hosts.

The epigraphs above suggest some of the salience of fashion, including body marking, in *Oroonoko*. Here its valence with bodily transformation is made pornographically identical with the drama of colonial domination and subjection in the two scenes of mutilation that kill off the novel's African protagonists. This chapter, concentrating on *Oroonoko* as an early and brilliant illustration of the cloven dialectic of fiction and anthropology, will shift attention from fashion to punitive mutilation via the intermediate realm of tattoo and ritual mutilation, and from the terror of the torture-prone social world of the slaveholding colony to the threats to identity and knowledge it produces in victim and victimizer alike. Fiction and science now offer differing means of managing such terror and confusion: in Behn's early novel one sees fiction appropriating the voice and focus of natural history, for ends related to the lived realities of colonial existence as much as to any formal logic of fictional realism.

Oroonoko, long read as marking an early moment in the development of literary realism, is indeed richly relevant to this book's web of concerns with

---

1. All of Behn's works will be quoted from *The Works of Aphra Behn* edited by Janet Todd; *Oroonoko* is in vol. 3, *The Fair Jilt and Other Short Stories*.

2. To say "until Watt" here is clearly to oversimplify. But the way in which this influential critic excluded Behn from the genealogy is interesting in light of this book's insistence on the mutual enfolding of new or "foreign" worlds and the "real" one: "Earlier types of prose fiction have also tended to use proper names that were . . . non-particular and unrealistic in some way; names that . . . like those of Lyly, Aphra Behn or Mrs. Manley, carried foreign, archaic or literary connotations which excluded any suggestion of real and contemporary life" (*Rise of the Novel*, 19).

developments in representation: in its choice of New World setting, its earnest claim to historicity, its sensationalism and accuracy of detail, its narrative narcissism and the fragmentation of the protagonist, its assumption of female subjectivity as the normative vehicle of narrated experience, its reenactment (for the narrator) and reversal (for the main characters) of Gonsales's and Cyrano's spatialized escapes to an alternative world from a reality in which they are marginal. In all these features it will seem to resonate with one or another of the narratives I have been discussing for several chapters. It also maintains relations with the values espoused by Bacon and Browne: natural history, as we will see, plays an important role here both as intermittent stylistic model and as exorcism or denial.

I have written at length about this single short fiction because, though no one any longer believes in such conceptions as "the first novel," I think *Oroonoko* is something more than representative.[3] Its formal innovations and evasions, its struggles with its disturbing material, constitute at least in part what the genre exists or existed to *do*. I would consider tying the early novel's mediation of what Michael McKeon (analyzing the "categorial instability ... the novel ... originate[s] to resolve" [*Origins*, 20]) calls "questions of truth, questions of virtue" to a quite specific historical cluster of events and developments. In what he refers to as "the old problem of mediation," "means become ends, . . . vehicles become tenors" (411). I do want to read a significant content into the vehicles — plots, characters, narrative situations — and into their various formal characteristics, but I am not sure that formal vehicles *become* tenors. Might it not be the case that they already are, that the material of representation in part provokes its fundamental means? Might the formal properties and tendencies of a literary kind express, more than any particular *plot* does, the exhilarations, conceptual impossibilities, and catastrophes of a large-scale cultural situation?

Let me start near the beginning of the chain of transformations with the clothes which make their culture-bearing men and women, for above all

---

3. Although not as successful as Thomas Southerne's dramatization of *Oroonoko* (issued in print at least twenty-four times in the century following Behn's death), Behn's novel had an afterlife, on its own as well as as part of its author's often-republished oeuvre. It came out twice in 1688, and her *Works* or at least her more or less complete plays and fiction were published eight times in the next forty years. *Oroonoko* appeared once by itself again in 1800, and in two or three editions of her collected works in the nineteenth century as well (not to mention its appearances in anthologies). Besides Southerne's dramatization, there were three more in England during the eighteenth century. (These estimates come from a preliminary study and include neither American nor Australian publication nor translations.) The record is richer than that of Madame de Lafayette's *Princesse de Clèves* in France and comparable to the English reception of *Gulliver's Travels*; *Robinson Crusoe* far outstrips them all in several European countries.

Oroonoko plays a part in the emerging European project that came to be known as anthropology—"a speaking or discoursing of men" (Thomas Blount, *Glossographia*, 1656). As a realistic or at least historicist fiction set in the *locus classicus* of the infant science and attending to its central concerns with human social relations, especially the relations of power and affinity, it sits at the vanishing point where now-divergent kinds of "speaking and discoursing" seem to merge into a single inquiry—but don't. In an era when clothes or "habit" was a major focus of ethnological literature, including body ornamentation, as I have shown, the term "fiction" primarily meant "the act of fashioning" or "that which is fashioned" (especially "fabric," see definitions 1a and 1c in the *Oxford English Dictionary*). Fashion, meanwhile, often meant "manners and customs," the early modern phrase for what has become the disciplinary object of ethnology—its particular reference to clothes (*costume*) is a metonymic derivation from its larger application to customs (*coutûmes*, or something like what we *now* call habits). All this is very pleasantly circular, but it is important to bear in mind, as we begin to consider Behn's "true history," that the seventeenth century's second meaning of "fiction" is "feigning, counterfeit." Social science and novelistic fiction may tell the same stories, but their relations to truth are constitutively different. Our interest continues in the nature of the sacrifices, exclusions, and excisions required by the emergent conception of scientific truth.

FASHION PLATES IN SURINAM

The narrator's nameless brother makes only a single appearance in *Oroonoko*, as a fashion plate attracting the admiring wonder of a village of Indians her party has traveled into the jungle to visit (or perhaps more precisely to spy on, to ogle): "This Feud [between the Indians and the planters] began while I was there, so I lost half the Satisfaction I propos'd, in not seeing and visiting the Indian towns. But . . . *Caesar* [Oroonoko] told us, we need not fear, for if we had a mind to go, he would undertake to be our Guard . . . About eighteen of us resolv'd, and took Barge" (100). The narrator herself is looking good that day, if not exactly respectable in her short hair and feathered cap. How the Indians are looking, besides "naked," this passage does not say. "Naked," as we have seen already, is not a word that always means what it says—it seems mostly to be a word that introduces (or summarizes) descriptions of the strange "habits" of non-European people.[4] Here it must

---

4. See, for instance, *Oroonoko* itself, in a lengthy earlier passage on Indian clothing and "adornment" that ends: "And though they are all thus naked, if one lives for ever among 'em, there is

suffice for the whole description. The point of this strange passage is that, after a long and fearful journey into the wilderness in search of a village of Indians who had never seen a White person before, the little party of Europeans—or rather their clothing—become the spectacle, the occasion of wonder. Since one might rather have expected Indians as a source for one's own vicarious wonder, it puts the reader in the position (familiar to readers of *Gulliver's Travels*)[5] of ogling European fashions along with the Indians, rather than—as in the opening pages of *Oroonoko*—ogling Indians with the European narrator.

This shift of perspective is not unprecedented in New World writing. It is the selling point of Cabeza de Vaca's *Relación* (1542), for instance, and Walter Ralegh (seventy years earlier than Behn, and west of Surinam on the Oronoko River itself) offers the narrative of an encounter with a warrior chief named Topiawari which, though it does not make Ralegh into a specular object, invites identification with both parties.[6] It is however, unusual, a point that needs little defending at this stage of the critical revisitation of colonial writing. I do not point it out as morally benign or politically progressive but simply as a surprise, a rhetorical moment that, in the context of late seventeenth-century writing about Surinam, invites a moment's wonder itself.

Aphra Behn has at times been more remembered for the Indian headdress she supposedly donated to the King's Company for a revival of John Dryden and Robert Howard's *Indian Queen* (1664, and used again in *The Indian Emperour*, its sequel) than for her own plays (including her shrewd and interesting American play, *The Widow Ranter* [1689], in which the headdress was also used, set in the Virginia of Hariot's propaganda and White's ethnography, in the midst of an Indian rebellion).[7] While this is mainly a sexist relegation of the writer to a passive historicity analogous to the "nature" in which women

---

not to be seen an undecent Action, or Glance" (2–3). *Oxford English Dictionary* citations of "naked" go back to the year 850, and except with reference to weapons or the condition of being unarmed, mean "without clothing" in an absolute sense. Even metaphorical usages (for the most part limited to the period covered by this book) signify divestiture, serious lack, rather than simply "less" than I have (on).

5. See the "Inventory" in *Gulliver's Travels*, part 1: "We observed a Girdle about his Waist made of the Hyde of some prodigious Animal; from which, on the left Side, hung a sword of the Length of five Men, and on the right, a Bag or Pouch divided into two Cells," and so forth (17–19).

6. For readings of both men's texts in terms of writerly and readerly identification, see my *Witness and the Other World*, chap. 6 and the epilogue.

7. "I had a set of these [little short Habits . . . and glorious Wreaths for their Heads, Necks, Arms, and Legs] presented to me, and I gave 'em to the King's Theatre, and it was the Dress of the *Indian Queen*, infinitely admired by persons of quality: and it was unimitable"

are believed to be inextricably stuck, it is a demonstrable fact that clothing, fashion, costume, and disguise mattered to her, not only as public figure and playwright, but as writer and thinker.[8] That she represents the narrator of *Oroonoko* as having donated an American Queen's headdress to the English King's Theatre is significant, whether it is also a fact true of Behn herself or not, and significant especially in light of all the other passages in the text in which clothing performs important functions of organizing attention, transmitting information, or symbolizing social identities.[9]

The African slave Oroonoko is (to Europeans) recognizably an aristocrat because of his "rich Habit"; Imoinda, because of her intricate tattoos. The narrator and her brother are (to "Indians") questionably human because of their feathers and glitter ("all . . . ask'd, if we had Sense and Wit? If we could talk of Affairs and War, as they could do?" [101]). The visual clichés of de Bry's American engravings and others like them (see figure 25) are reversed in this verbal text; more importantly, everyone is represented as belonging to the fashion regime (even the narrator's otherwise functionless and featureless brother). Clothing and "adornment" specularize the wearers in *Oroonoko*; nakedness renders them invisible or in writerly terms "unimitable."

What brings the body and face into visibility besides adornment here is mutilation. The Indian warriors at the end of the episode that displays the narrator's outfit demonstrate this fact when "Caesar" (Oroonoko's slave name) pays them a visit: "some wanted their Noses, some their Lips, some both Noses and Lips, some their Ears, and others cut through each Cheek, with long Slashes, through which their Teeth appear'd" (102). Perhaps what links these "imitable" wounds to clothing and adornment is their status as visual absences of certain body parts—like clothing, amputations could be said to hide (or to distract from the idealized body of erotic attention). But these are a very particular kind of wound—as mutilations they share something technically with Oroonoko's and Imoinda's tattoos, and as *self*-mutilations (they are inflicted "in competition for the Generalship . . . to shew they are

---

(*Oroonoko*, 2). As Robert Chibka points out in his excellent article on *Oroonoko*'s play with the issues of falsehood and fictionality, H. A. Hargreaves's insistence on Mrs. Bracegirdle's having worn the costume on stage in 1664 is a logically "odd" means of defending *Oroonoko*'s "position of importance in the history of English realism" (Hargreaves, 444), but "not atypical in Behn criticism" (Chibka, 512).

8. "For Behn and for her friends, the theatre had been life and a metaphor for life. It would be a long time before any woman would again feel able to accept so thoroughly the theatricality of her demeanour or to remain so masked that nothing about her could be declared 'authentic'" (Janet Todd, *Secret Life of Aphra Behn*, 434).

9. It is significant in a different way, less salient here, that she may in fact have donated headdresses to both the King's Theatre *and* the Royal Society (see n. 13 on this in Todd, *Works* 3:447).

Fig. 25. The overdressed and the "naked": Vespucci meets America. Woodcut by Jan van der Straet, *Nova reperta*, in *Speculum diuersarum imaginem speculatiarum* (Antwerp, 1638). By permission of Smithsonian Institution Libraries.

worthy to lead an Army," [103]), they share the intentionality, the *communicative* status of those aristocratic adornments. After these literally sensational representations are described, suddenly it is possible to describe the clothing of the supposedly "naked" Indians—from whom, one must assume, the narrator gets her feather outfits to donate to the King's Theatre.

With self-inflicted wounds functioning (as Bulwer would concur) as clothing, the final scene of Oroonoko's dismemberment and death can be seen as the text's supreme fashion plate—a tragic equivalent to the comic multiple en-memberment of the Cyclops we saw in the *Receuil de la diversité des habits*. Had Bulwer lived to read it (he seems to have died in 1657),[10] the scene would surely have found a place in a new revision of the *Anthropometamorphosis*, as the climactic act of culture. Indeed, it shares more than one feature with what Christians were taught to consider the climactic event of supernatural (thus supercultural?) history. The "frightful Spectacle of a Mangled King" may be too much for the narrator's friend, plantation-owner Colonel Martin (who, on being offered part of "Caesar's" body to display,

---

10. I thank Katherine Rowe for this information.

"swore . . . that he could govern his *Negroes*, without terrifying and grieving them" with it [118]), but it is an image Anglican and Catholic Christians such as the royalists living in Surinam were expected to imagine (complete with fear and grief) routinely, and above all it is an image of the slaves' truth, the truth of colonial and imperial domination. Although the Indians were called "cannibals," almost the only dismemberments early modern European texts narrate are performed by Europeans on the bodies of their slaves or on indigenous people who just happen to be there at the wrong moment.[11]

This image does not need the fashion regime to acquire significance or readability — we do not have to register the mutilation, with its Christian resonance, as a garment, a "habit." What then can we learn from Behn's colonial (proto-)novel by doing so, or more generally by reading her text against the now normative novelistic grain, as a work of (proto-)anthropology? One obvious benefit would return us to the norm by exposing the ways in which novels are not in fact the same as anthropology, but acts of imaginative thinking somewhat less rule-bound and more particularistic. Another benefit would be to give us a wider view of anthropology, one that included novels in its canons. But the words "novel" and "anthropology" did not exist with their present ranges of meaning in 1688, so both these benefits are limited by the anachronism of the thought experiment's terms.[12] The richest possibility I can hold out is that we may be able to see part way into the "epistemic murk," in Michael Taussig's phrase (*Shamanism, Colonialism, and the Wild Man*), which over time precipitated more distinctly the forms of thinking — of managing knowledge — that are expressed by novels, ethnographies, and cultural taxonomies.

Poetry in the modern West has come to seem less sufficient to its ancient role, partly epistemological, of analyst and preserver of the human, and of specific human cultures. Indeed it is possible from a purely formal point of view to see the discourse and institution of anthropology as functionally replacing it. Enlightenment universalism has made "the human" a category

---

11. The excitement of Hans Staden's account of his Brazilian adventure is based on his having been abducted by cannibal Tupinamba, but in fact he is not killed and dismembered (his worst fear, and nothing compared to being dismembered and *then* killed) but only anticipates it during his captivity. He is given reasons to believe that this can happen to him and has happened to other captives, but does not provide any unambiguous eyewitness narration to support the belief.

12. "Novel," at the time, referred to short prose tales like those of Boccaccio or Marguerite de Navarre; "anthropology" was a rare and vague term, mostly referring to the historical and genealogical study of "Man," though by the turn of the century it could refer to a kind of anatomy-cum-psychology: the *Oxford English Dictionary* cites J. Drake, *Anthropologia Nova, or, A New System of Anatomy* (1706).

within which what are conceived as "specific cultures" are arrayed as variants; as a logical consequence, the language of analysis and preservation must be external to the language of any one object of anthropological representation, rather than hyperbolically internal, as poetry necessarily is. Taussig's murk is the atmospheric effect of terror, and the intuition surfaces that an increase in the level and a change in the nature of terror could share some responsibility for poetry's drift toward the margins of cultural significance.[13] In the passage we have been looking at, the Indians' cry "of Wonder and Amazement" at the sight of the narrator and her brother initially terrifies the Europeans: wonder and terror are branches of the same high-tension aesthetic, both "cognitive emotions," both invoked by Aristotle as among the proper ends of poetry.[14] But terror can overwhelm more completely than wonder, unless a form of cognitive defense is available to mediate. One feels the pressure of the search for containment both here in Behn's fiction and in Lafitau's openly ethnological account of southern Canada three decades later (see the next chapter). The sublime has too much to do with reality when the reality is colonial helter-skelter.

Behn (who was probably in Surinam as a spy, and not to make real money) wrestles with this difficulty in her intermittently poetic testimony about the colony. The passage detailing the final dismemberment of Oroonoko, from which I have not yet quoted, serves as the climax of a novel intended to sell, while a representation of that scene *internal* to the novel — the displaying of Caesar's quartered body to the other slaves of Surinam — is imagined as too terrifying and foregone by the slaveowner Colonel Martin. Whose terror is at stake in this novel — that of the kidnapped, enslaved, abused, and mutilated Africans, or that of their owners? Taussig invokes this confusion painfully well in the first part of his *Shamanism, Colonialism, and the Wild Man*, "Terror," a discussion of the emotional atmosphere and social organization of the English rubber plantations in the Putumayo region of Colombia at the turn of our own terrifying century. "All societies live by fictions taken as real. What distinguishes cultures of terror is that the epistemological, ontological, and otherwise philosophical problem of representa-

---

13. Allen Grossman's essay, "The Passion of Laocoön, or the Warfare of the Religious against the Poetic Institution," has been important to my thinking on this matter.

14. On the connections of wonder and dread in medieval culture, see Caroline Walker Bynum, "Wonder," esp. 16; on wonder and horror, see the introduction to Lorraine Daston and Katharine Park, *Wonders and the Order of Nature*, 1998. None of these historians takes up the relation of *terror* to wonder; these two emotions are perhaps less continuous. The "cognitive emotions" do not characteristically threaten our identity as the experience of the sublime is defined as doing, from Longinus to Kant.

tion — reality and illusion, certainty and doubt — becomes infinitely more than a 'merely' philosophical problem. . . . It becomes a high-powered medium of domination" (121).

Taussig's exemplary story is of a runaway criminal, Don Crisóstomo Hernández, "who sought refuge in the deep woods of the Putumayo where he reigned over whites and Indians with great cruelty": "On hearing of a group of Huitotos whose women and children as well as the men were said to be practicing cannibalism, Don Crisóstomo decided to kill them for this crime, decapitating them all, including babies sucking at the breast. . . . Here we have the tale of a man killing off that [desperately needed Indian] labor down to children at the breast because of their alleged cannibalism — mirroring, at least in fiction, the spectacular show of carving up human bodies that, again through fiction, occasioned the white man's furious 'reprisal'" (27). The answer to my question (Whose terror is at stake?) is that terror does not "belong" to one or another party in a situation such as that of the Putumayo rubber plantations, or the Surinam sugar plantations, although it originates in the Europeans. Terror is the light in which all things are seen, or hidden. In his poetic analysis of another site in the modern "culture of terror," Peter Dale Scott borrows the Indonesian shadow-play character Rangda ("the terrible witch"), quoting Margaret Mead and Gregory Bateson's description: "*she is Fear / afraid as well as frightening*" (*Coming to Jakarta: A Poem about Terror*, 3.12).[15]

Taussig points out that, in the "jungle" narratives (some by anthropologists) from which he quotes passages of panic or foreboding, what is usually missing is any account of the rubber plantations. The weird nightmarishness of the fear experienced and acted upon by the Europeans in the Putumayo and the Caquetá comes from its projective origin; the nightmarish quality of its representation seems attributable to the consequent repression of the fear's actual source. The "privilege of unknowing" here becomes its own retributive curse.[16]

Robert Chibka ("'Oh! Do not Fear'") rightly mocks the long debate among literary critics over the historicity of Behn's narrative (not properly an ingredient in discussions of verisimilitude or realism), and substitutes for it a thorough and chilling account of the text's labyrinthine fascinations with

---

15. Rangda is visually defined by her "long hanging breasts, realistically made of bags of white cloth filled with sawdust" (Miguel Covarrubias, *Island of Bali*; quoted in Scott, *Coming to Jakarta* 3.12); note the salience of this same figure in Theodor de Bry's New World engravings, especially in depictions of "Indian" cannibalism. See Bernadette Bucher, *Icon and Conquest*, for more on this figure in the New World context.

16. I refer to Eve Sedgwick's essay of that title, reprinted in *Tendencies*.

the problem of truth in representation. But it seems to me that we do have evidence, however we might weigh its importance, of Behn's having lived in Surinam in the telltale presence of terror's "epistemic murk" at the heart of her representation of the place. This phenomenon also marks Conrad's *Heart of Darkness*, similarly presented as an eyewitness tale of other people's troubles in a rain forest burdened with and partially "exterminated" by European plantations, its narrative fraught with instances of mendacity and the unrepresentable.

I cannot improve on Chibka's blow-by-blow account of *Oroonoko*'s thematized mendacity and fictionality, its eerie thickets of self-contradiction. Some of what he notices has been noticed in print before (and since), of course — analysts of the novel's incoherent racial politics have had occasion to point out the narrator's painstaking separation (even alienation) of Oroonoko/Caesar from other, more "slavish" African slaves or from other, more "African"-looking Africans, for instance, or the strangeness of Behn's black male protagonist/white female narrator equation, given the presence of a black female protagonist whose name in fact is the novel's last word.[17] Marxist literary historians Michael McKeon and Lennard Davis, with their different but related interests in the connections between the early novel, on the one hand, and historicity, "news," and the realm of the epistemological generally, on the other, have important things to say about Behn's truth claim in her preface to the Catholic convert Lord Maitland, and its various supporting echoes in the text of the novel.[18] But the *tortured* quality of the novel's understanding of representational truth, or of any ascertainable truth, period, is captured best in Chibka's article, which invokes fear as well as fictionality in its title (although it does not actually set out to account for the fear-effect of the narrative it so carefully describes): "'Oh! Do Not Fear a Woman's Invention!'"

Since forms of prevarication, misrepresentation, and confusion permeate the action and the narration of the novel it is difficult to summarize briefly. Besides the label mentioned above (and customary in all colonial writing of the time) of "naked" as applied to native Americans whose clothing is later described, and leaving aside the various deceptions practiced in the romance plot by Oroonoko, Aboan, and Imoinda, a list of examples concentrating on deception and contradiction in the representation of slavery and colonial

---

17. On the crossed axes of race and gender, see Laura Brown, "The Romance of Empire" (an essay revised into a chapter in her recent book *Ends of Empire*); Ros Ballaster, "New Hystericism: Aphra Behn's *Oroonoko*"; Catherine Gallagher, "The Author-Monarch and the Royal Slave" (chap. 2 in *Nobody's Story*), and Margaret Ferguson, "Juggling the Categories of Race, Class, and Gender."

18. See McKeon, *Origins of the English Novel*, 111–13, and Davis, *Factual Fictions*, 106–10.

domination will perhaps be helpful. Consider the following: (1) the slaveship captain's kidnapping of Oroonoko, a prince from whom he regularly buys slaves and whom he has invited on board his ship for a feast; (2) the captain's and later, in Surinam, Trefry's and even the narrator's assurances to Oroonoko that he will be set free very soon (and Oroonoko's amazing persistence in belief); (3) the rhetorical condemnations of slavery by both Oroonoko and the narrator, in tandem with their mutual use of such derogatory phrases as "like a common slave" (frequent) or "by Nature Slaves" (109), and Oroonoko's even more contradictory practice, back home, of selling Africans to European slavers; (4) the narrator's opening account of the "perfect Amity" with which the English and the Indians coexisted, which contradicts the frightening level of violence invoked at the beginning of the passage describing her visit to the Indian village ("we were in many mortal Fears, about some Disputes the *English* had with the *Indians*, so that we could scarcely trust ourselves . . . to go to any Indian Towns . . . for fear they should fall upon us, as they did immediately after my coming away; . . . [they] cut a Footman, I left behind me, all in Joints, and nail'd him to Trees" [100]).[19]

The opening account of Amity explains the situation persuasively enough: the Indians "being on all occasions very useful to us, we find it absolutely necessary to caress 'em as Friends, and not to treat 'em as Slaves, nor dare we do other, their numbers so far surpassing ours in that Continent" (60). Yet on the way back from the visit to the village they meet up with some people of a different tribe and "our *Indian Slaves*, that row'd us, ask'd 'em some Questions" (103). And though Africans in Surinam are treated throughout as unambiguously slaves, it turns out that, like the Indians, they too "exceed the Whites in vast numbers" (46)—which lends some credibility to the fear later expressed by some members of the colonial council that "no Man was safe from his own Slaves" (112). Indeed, Oroonoko's mutilations, before and after death, are considered necessary displays of power in precisely this context. As Chibka points out, one important effect of the narrative's corrosive murkiness is to blur or erase distinctions: especially the distinction between Negro slave and native American in the English colony (a blur perfectly adumbrated in Oroonoko's African name—which is the Carib word for "coiling snake" and the name of the great river to the west in

---

19. This prefiguration of Oroonoko's death and mutilation at the hands of the English is especially interesting for the discussion below of collapsing distinctions, as the perpetrators here are native Americans rather than colonial intruders. It is one of several instances in which European fears (and desires) are displaced onto the lives or bodies of their African servants—in this case, the fear of indigenous revenge.

Venezuela, Ralegh's "Large, Rich and Bewtiful" Guiana).[20] In the matter of self-mutilation especially, Oroonoko slicing open his abdomen and pulling out his intestine is pointedly like the Americans who "cut through each Cheek, with long Slashes, through which their Teeth appear'd." And the native American attention to the face as painful bearer of text is matched by Oroonoko's strange mutilation of Imoinda's body after he kills her: he is described as cutting her throat "and then severing her yet smiling Face from that delicate Body" (114).

These points, slashes, and masks are all echoed in features of English fashion—which as we have seen in Chapter 7 is dependent in some ways on the customs of the colonies' native inhabitants for its novelties. And of course the words and actions, sentiments and behaviors of Oroonoko and Imoinda are precisely those of characters in a heroic tragedy (which Thomas Southerne soon turned it into).[21] However much it may be simply true that Africans from the areas around "Coramantien" had European features, it is surely not "true" that they (or for that matter any European lovers or kings) spoke in a crisis in the precise accents of English heroic drama.[22] The blurring of cultural distinctions sometimes complained of as racist and artificial, other times argued away in defenses of Behn's ethnological accuracy, is more interesting to consider as yet another feature in the murky text's systematic and progressive abandonment of functioning distinctions—another dynamic it shares with *Heart of Darkness*, though the blurring here is not represented moralistically as degradation and horror, as it is in the case of Mr. Kurtz.

This telling abandonment is one aspect of the novel's freedom from the demands to which anthropology or even its forerunners, the colonial reports and plantation pamphlets, would have to answer—would perhaps need to answer. To quote Taussig one more time, speaking of the early twentieth-century explorer and amateur anthropologist Captain Thomas Whiffen (whose photographs belong to the Cambridge University Museum of Archaeology and Anthropology): "Where does the heart of darkness lie, in the fleshy body-tearing rites of the cannibals or in the photographing eye of the beholder exposing them naked and deformed piece by piece to the world? It

---

20. Elaine Campbell, "Aphra Behn's Surinam Interlude," 28.
21. Amazingly, he represented Imoinda as European (French!) because of the "impropriety" of white actresses working in blackface—thus forcing the representation of miscegenation on the theater-going public as the lesser of two (closely related, as we have seen in the antipatch literature) evils. See Ballaster, 289, and, for discussion, Ferguson, esp. 218–24.
22. For "Coramantien" see Campbell, "Aphra Behn's Surinam Interlude."

is a clinical eye and one never so lewd as in the closeness of the distance it maintains while dissecting the body of the Indian. . . . In his fear of the Indians, alone and lost in the jungle, the calming thought came to Captain Whiffen of their proper place: in glass cases in an Anthropological museum" (*Shamanism*, 117). There is nothing so efficient at making distinctions as that piece of glass between the specimen and the anthropologist. It was not easy to make it, though, that transparency of representation. In Behn's thick texture, knotted with contradiction and saturated with narratorial personality, there is no glass.

And yet there is a link to be made between the nailing of various Black men to trees in this narrative and the fixing of specimens in a case: here as in the Cambridge Museum of Archaeology and Anthropology we see persons as objects "expos[ed] piece by piece to the world"—the phrase makes a chilling translation of the decontextualized object of wonder into the dismembered body of terror. And Behn protests in several places that her narrative ("writ . . . in a few Hours," [56]), "comes simply into the World . . . there being enough of Reality to support it . . . without the Addition of Invention" (57). Indeed, in another and perhaps the weirdest of all blurrings of distinction, she proclaims the text and its male protagonist simply equivalent to each other and to herself: "'Tis purely the Merit of my Slave that must render [the book] worthy of the Honour it begs" (56), and "I hope, the Reputation of my Pen is considerable enough to make his Glorious Name to survive to all Ages" (119).

This is not what Mary Louise Pratt means by "autoethnography"—her genre is dependent on the non-European status of the writer. But perhaps we should expand the category to include its opposite—the "Story" of a cultural Other given by a narrator so projective and appropriating (like the cosmographer Thevet) that the work becomes a portrait of the narrator's encounter. What we saw in the *Blazing-World*—characters all avatars of the narrating Duchess, fusing beyond the boundaries of the body—is hidden here beneath the lineaments of a conventional characterology and "true history." But it is not hidden deeply—it lies, like Imoinda, under a "coverlid of Nature," of leaves and flowers that do not hide the "noisom [Stink]" of death, with the "Face . . . left bare to look upon" (114–15).

This grounding of the Other's story in the narrative matter or perhaps lattice of the self—more comically blatant in Thevet's cosmographies—accounts for the identification of European female narrator with African male protagonist that has so fruitfully puzzled critics like Gallagher and Ferguson, and it may tell us something about the colonial imaginary at this relatively early stage. It is for one thing more pleasant for Behn to consider herself an enslaved king (the Emperor/Empress of a blazing African or American world) than an enslaved chattel like Imoinda, whose story is always the story of someone fidgeting in her chains, trying to find a comfortable position, as one man after another—good and bad, African and English, at home and at the court, and in Surinam—assumes possession of her and the right to dispose of her. And if we ask why Behn did not find an enslaved Carib king with whom to identify, it is easy enough to answer that as a transplanted product of the Old World, she wanted an equally transplanted personage to express her dislocation, her sense of having lost a measure of power by leaving home (even if she has gained a measure of social freedom). "And though there was none above me in that Country," (56) still, she is not able to alter by a jot or tittle the awful fate that awaits her friend the fallen prince. "Because my Father dy'd at Sea, and never arriv'd to possess the Honour was design'd him (which was Lieutenant-General of Six and thirty Islands . . .)" (95); because "his late Majesty . . . [had] parted so easily with [that vast and charming World] to the Dutch" (95); because "the Governour [was] drown'd in a Hurricane" (104) before the project of gaining privileged access to Amazonian gold could be put in place. Because the Dutch (in 1667) "kill'd, banish'd and dispers'd all those that were capable of giving the World this great Man's Life," it was consequently Oroonoko's "Mis-fortune . . . to fall in an obscure World, that afforded only a Female Pen" (88). Not only are all the sources of power in this narrative fallen, they are castrated—this bunch of "mangl'd Kings" has quite a bit in common with our Female Pen as she often seems to conceive herself.[23]

Such boundary problems as the narrator manifests are not what distinguish ethnography from novel (the articulations of modern science from those of fiction), and Behn's novel has been tapped for historical ethnography as well as read for (that most suggestive of readerly purposes) "Diversion."[24] Whatever distinctions of genre or agenda can be made must be largely made at the writer's end; it is easy, as James Clifford et al. have shown, to read an ethnography with the same techniques as critics use to produce readings of novels. It is also easy to read a novel like *Oroonoko* for the same kind of information we can perhaps most reliably obtain from an ethnography: information about the ethnographer and his/her cultural as-

---

23. Gallagher's argument in *Nobody's Story*, that the woman writer is just an extreme case of the writer per se, suggests in this context the intriguing image of the writer as always a "mangl'd King," or at any rate the colonial writer as always such.

24. See, for instance, James Williamson, *English Colonies in Guiana*: "*Oroonoko* . . . is a genuine authority for the social life of the colony" (151). For Behn on "diversion" as a readerly aim, see opening paragraphs of *Oroonoko*.

sumptions and categories, information about his/her culture's precise state of anguish and confusion at the historical moment in which the ethnography was produced.

EXPOSED, EXCISED, EXPERIENCED

If Behn's narratorial appropriation of her characters as avatars of the narrating self is reminiscent of Cavendish's hall of mirrors, nothing could be more unlike the Duchess's cozy, disembodied cohabitation with her imaginary friend the Empress inside the body of her husband the Duke than the isolated, public, and excruciatingly physical disembodiment of Behn's enslaved protagonist. The passages of torture and dismemberment that finally exterminate Oroonoko and Imoinda are acts of a different imagination, one that seems itself tortured by a desire to get, like the microscopist, "inside" the alien Other and yet (like the microscopist again) finds there always only surfaces, the figments of projection, the same old Same, the reflection of its laboratory windows on the fly's dead eye.[25] *Oroonoko* is a colonialist narrative through and through, brilliantly so, and "exposes, piece by piece," the anatomy of the colonial voyeurism and fear that found a repressed perfection in nineteenth-century ethnography. The sensationalism of the dismemberment passages is something we do not expect to see in serious modern and contemporary ethnography—and would not have been offered in even an early promotional text like Hariot's, which needs to hide the terror of colonial "conquest." (The Algonquin village burned to the ground by the English is simply not mentioned, even the wrecks of the expedition's departing ships are hinted at only in a detail of the frontispiece illustration.) What we can find in it here is news that ethnography in its classic form tended to hide about the experiences that underwrote it, or to diffuse through formal means, as the new rhetoric of natural history diffused and hid the related experience of wonder.[26]

---

25. For the material climax of this trajectory, see Ludmilla Jordanova (*Sexual Visions*, chap. 2) on the wax "Venuses," later eighteenth-century anatomical models "adorned with flowing hair, pearl necklaces, removable parts and small foetuses," (45) whose removable torsos revealed detailed innards. Jordanova uses the image of clothing to discuss the predissected wax bodies: "these models were already naked, but they gave an added, anatomical dimension to the erotic charge of unclothing by containing removable layers that permit ever deeper looking into the chest and abdomen" (55).

26. Although even this novel often hides the narrator's conversations with her informant, where they would have had to be scenes exposing the complexity and aggressiveness of her/our curiosity, as well as attestations of her privileged status of witness and confidante.

They were no sooner arriv'd at the Place, where all the Slaves receive their Punishments of Whipping, but they laid Hands on *Caesar* and *Tuscan*, faint with heat and toyl; and, surprising them, Bound them to two several Stakes, and Whipt them in a most deplorable and inhumane Manner, rending the very Flesh from their Bones; especially Caesar, who was not perceiv'd to make any Mone, or to alter his Face. . . .

When they thought they were sufficiently Reveng'd on him, they unty'd him, almost Fainting, with loss of Blood, from a thousand Wounds all over his Body; from which they had rent his Cloaths, and led him Bleeding and Naked as he was; and loaded him all over with Irons; and then rubbed his Wounds, to compleat their Cruelty, with *Indian Pepper*, which had like to have made him raving Mad; and in this Condition, made him so fast to the Ground that he cou'd not stir, if his Pains and Wounds had given him leave. They spar'd Imoinda, and did not let her see this Barbarity. (110)

The event narrated here is not the one that motivates the events in the next three passages I will quote, though together these four passages could be said to provide the real or anyway effective plot and climax of the last part of the novel. Oroonoko is moved to his revenge (which he thinks necessitates his first killing his pregnant wife) not by pain, but by the betrayal of his enemies/owners, who had promised him a truce and his freedom after outnumbering him and his weak-willed army of escaping slaves. But where he is moved to new action by the anguish of being lied to in a solemn contract (by having been told "the thing which is not"), Behn's readers are moved by a detailed representation of physical invasion and agony. (In New World writing, Las Casas had preceded Behn's exposure of European sadism, but the force of his rhetoric had depended on numbers rather than sensational details.) The intensity of tactile response to this passage, even for readers in an age of filmic "hyperviolence," separates us from the actual and very stoical sufferer, who makes no "Mone," and reveals us as far more susceptible to physical than to moral incitements. This has actually, if more quietly, been a feature of the narrative from almost the first page—the readerly pleasures and wonders of Surinam have consistently been sensational, even where the most fundamental mythic figures are being invoked (Eden, or the hero's battle with the serpent).[27] It would be an exaggeration to say that the work reads like a pornographic novel, its structure and plot supporting, unnoticed, a filigree of arousing physical detail. But it would not be untrue.

---

27. Interestingly represented here as a "numb-Eel," a serpent whose grasp robs victims—sometimes fatally—of *sensation*. Oroonoko's last experience before his slave rebellion and the excruciating consequences are narrated is of being robbed of sensation and consciousness by the touch of this numb-Eel.

Here is the most intimate scene we are offered of interaction between the enslaved lovers:

> All that Love cou'd say in such cases, being ended; and all the intermitting Irresolutions being adjusted, the Lovely, Young, and Ador'd Victim lays her self down, before the Sacrificer; while he, with a Hand resolv'd, and a Heart breaking within, gave the Fatal Stroke; first, cutting her Throat, and then severing her, yet Smiling, Face from that Delicate Body, pregnant as it was with Fruits of tend'rest Love. As soon as he had done, he laid the Body decently on Leaves and Flowers; of which he made a Bed, and conceal'd it under the same coverlid of Nature; only her Face he left yet bare to look on. (114)

Although violence has always gotten more page-space with less trouble than sexual experience, Behn is a libertine writer, frequently reprimanded for her lewdness (in a notoriously lewd time and place, Restoration London in the Europe that invented modern pornography). We can assume, then, that this is the scene the writer understands her readers most to desire,[28] or least that it is the scene that best manages the forces set in motion by the love-plot of the Africans. It is both an *exposé* of the horrors of slavery and an indulgence of a self-protective readerly sadism towards the helpless. It is also an interesting gesture on the part of the female authorial narrator who so repeatedly identifies herself with the loving male murderer. But any close-up detail of this more fatal "carving" of Imoinda's skin would perform these functions: the most densely significant feature of the representation is the particular act of "severing her Face . . . from that Delicate Body." Rather than a penetration or dissection, microscopist style, such as we get when the Europeans wield the knives, this murder performs operations closer in kind to release or liberation (Behn informs us, as will naturalist Maria Sibylla Merian later in the century, that the Africans in Surinam believe death will return them home):[29] Imoinda's throat is cut, releasing her arterial blood. Her face becomes mask-like under Oroonoko's scalpel, and it is that mask, that exterior surface, that rivets him for the next eight days while the body decomposes separately beneath its "coverlid." Despite the novel's structural reliance on the notion of female sensibility as the proper vehicle for narrative of experience, Imoinda's death (like her face) is represented as external to her—not

---

28. In the epistle "To the Reader" of *Sir Patient Fancy* (1678), responding to attacks on the play for its "Bawdry," Behn writes about feeding her audience's tastes: offended ladies should have "attributed all its faults to the Authors unhappiness, who is forced to write for Bread and not ashamed to owne it, and consequently ought to write to please (if she can) an Age which has given severall proofs it was by this way of writing to be obliged" (*Works* 6:5).

29. For another intrepid seventeenth-century woman's experience and representations of Surinam, see Natalie Zemon Davis's account of Merian in *Women on the Margins*.

as her own tactile and psychological experience, but as Oroonoko's largely visual one.

This passage brings up most forcefully the question of Behn's source, or rather the circumstances of the narrator's access to the details represented. As I noted in chapter 5, we assume that sensation can only be reported by the owner of the sensorium. The so-called omniscient narrator, by now an ignorable convention in both prose fiction and film, took a long time in learning how to know the bodily details of characters' experience; medieval romance and history either objectified physical experience, rendering it as it would appear to a distant and disengaged observer, or passed over it in explicit silence: "the two of them felt a joy and wonder, the equal of which has never been known. But I shall let it remain a secret forever, since it should not be heard or written of" (Chrétien, *Lancelot*, 264–65 [ll. 4684–89]). In the first passage quoted above we are given an opportunity to feel torture vicariously, by virtue of the level of detail, but all such detail is still based on observation of public and visible events. In the passage on Oroonoko's private and hidden killing of Imoinda, the narrator (represented as the historical character Behn, the author) has no source for the details, including the "breaking Heart," other than the killer himself, and what is more, the killer after he has been tortured almost to death:

> *Look ye, ye faithless Crew*, said he, *'tis not Life I seek, nor am I afraid of Dying*; and at that Word, cut a piece of Flesh from his own Throat, and threw it at 'em, *yet still would I live if I cou'd, till I had perfected my Revenge. But oh! it cannot be; I feel Life gliding from my Eyes and Heart; and if I make not haste, I shall yet fall a Victim to the shameful Whip.* At that, he ripp'd up his own Belly; and took his Bowels and pull'd 'em out, with what Strength he cou'd. (116)

It is an extraordinary situation to imagine, and one never dramatized in the novel. Suffering from such a wound, which a doctor tells him and his "friends" he will not survive, Oroonoko presumably lies on a couch at the plantation house and tells the narrator/"friend," who has run away every time he needs her promised protection, the physical and emotional details of murdering his beloved wife. This certainly makes Behn's narrator what ethnographers call a "participant observer," but at the same stroke suggests the profound unseemliness of that function in the social web that produces ethnography. It is precisely in the narrator's role as "observer" that she most participates in the horror of plantation slavery; in her effacement of the contact experiences in which she gains access to the information the novel communicates, she makes a crucial contribution to "epistemic murk." We are always technically able to tell ourselves that there is some rational way for her to know the things she narrates about Oroonoko's experiences—"what I cou'd not be Witness of, I receiv'd from the Mouth of the chief Actor in this History" (57); but this blanket explanation feels forced and thin with respect to passages like the one just quoted. One receives more strongly an impression of absolutely successful surveillance: the very walls of Parham must have ears. Oroonoko can do nothing without our knowledge. This constitutes a suggestive moment in the gradual production of the novelistic "omniscient narrator."

That Behn herself became a spy for the government of Charles II shortly after her return from Surinam (or was spying, Todd speculates, *while* in Surinam) and before she became a published writer is an interesting bit of trivia; certainly *Oroonoko* (and not only *Oroonoko* among her fictions) shares with her espionage letters the writerly stance of observer in a world that does not realize how open it is to her observation. There is something like sadism in observation: no image of it seems thoroughly wholesome or at any rate benign. Espionage is secret by definition, whereas the ethnographer as exotic foreigner is usually public ("their Numbers so far surpassing" his or hers). But in both cases, as in the fictional or fictionalized case of Behn's narrator and Oroonoko, the observer is more humanly disengaged than in "true" social relations, so to speak, and is learning more from the encounter than the other realizes. What Behn's passages of physically intimate experience do is to expose (by literally exposing flesh) the actual degree of disengagement, the sadistic potential in the observational posture.

> He had learn'd to take Tobacco; and when he was assur'd he should Dye, he desir'd they would give him a Pipe in his Mouth, ready Lighted, which they did; and the Executioner came, and first cut off his Members, and threw them into the Fire; after that, with an ill-favoured Knife, they cut his Ears, and his Nose, and burned them; he still Smoak'd on, as if nothing had touch'd him; then they hacked off one of his Arms, and still he bore up, and held his Pipe; but at the cutting of this other Arm, his Head Sunk, and his Pipe drop'd; and he gave up the Ghost, without a Groan, or a Reproach. (118)

This scene the narrator has neither witnessed nor "receiv'd from the mouth of the chief Actor." As so often when the going gets rough, the narrator is off visiting Colonel Martin ("whom I have celebrated in a character of my New *Comedy*, by his own Name" [111]). As a result she has to posit a mother and sister, of whom we have not until this penultimate paragraph heard anything, to stand by "Caesar's" execution and report the details (the noble Trefry, too, has been removed from the scene by a fiction of the Lieutenant-Governor's). Behn's reluctance to offer us the sight of her narrator/alter ego standing thus by "but not suffer'd to save him" (118) lets us know that the passive detachment required by such witnessing strikes her as unseemly. And yet there is no seemly alternative—escaping at such a moment for a social visit is the action of a Pontius Pilate, and the nameless, suddenly-appearing (and then disappearing) "mother and Sister" (118) read like split-off fragments of the narrator.

The literal fragmentation of the narratee is offered in the novel's starkest prose—a strangely concrete fulfillment of Sprat's wish for "so many things, almost in an equal number of words" (*History*, 113). Only one value-term, indeed only one adjective ("ill-favor'd," strangely aesthetic for describing the executioner's inadequate knife) blurs the anatomical clarity of the list of cuts and parts. The passage sounds like a dissection report intended for the pages of the *Philosophical Transactions*, and this I think is no accident in a novel that begins with a claim to Sprat's rhetorical vision—there will be no "adornments," "Invention" or amplifications—and goes on immediately to invoke an affiliation with the Royal Society's "*Antiquaries*" (their cabinet). That the subject of this dissection of an exotic creature by Europeans is still alive at the time of the procedure does not in fact weaken the point: recall the experiment on a living dog Hooke could only report in the passive voice. For Behn, the phantasmal "mother and Sister" perform a similarly self-exculpating (and passive) function. They look, in tableau, like two of the Marys standing by the cross at the crucifixion of the King of the Jews, but apparently they report like virtuosi. And Behn, like Oldenburg at the *Transactions*, gathers reports together and publishes them, "tak[ing] care [they] should be the Truth" (56). Sensation has been evacuated in this passage both from the character who experiences it and the experience of those who observe. The facts arrive in monosyllables, in simple declarative sentences. No one but the reader feels pain, no one feels pity or terror; indeed, Colonel Martin explicitly refuses to countenance the use of Oroonoko's dismembered body to induce these feelings in his slaves.[30]

And once again, the narrator escapes just when, you could say, she is needed. Her readers are presumably reeling from the horror of the facts so briskly outlined in this penultimate paragraph. But as soon as the body parts

---

30. See Richard Andrews on French Enlightenment reform of public punishment ("The Cunning of Imagery"); and for the English social dynamics, see Peter Linebaugh, *The London Hanged*. It is notable that the textual scene of Oroonoko's mutilation and death is framed neither as Foucault's *punir* nor as *surveiller*, but as an observation in which we become, murkily, the "virtual witnesses" Steven Shapin and Simon Schaffer (*Leviathan and the Air Pump*) term the readers of experiment reports.

have been removed from the scene, the narrator removes herself as well, ending the novel in a short, rhetorical paragraph of womanly/writerly (im)modesty that once again blurs the distinction between herself and her narrative objects, whose immortality her own must ensure: "I hope, the Reputation of my Pen is considerable enough to make his Glorious Name to survive to all Ages" (119). The novel puts very little distance between climax and closure—so little that we feel it almost on our pulses that the narrator cannot persist separated from her protagonists. *Oroonoko* and Oroonoko were close to being the same. The novel, like a colony, exploits the raw material provided by warm and distant places and the labor of slaves. Without its "royal" slave, it ends.

FICTION, SCIENCE, AND AGENCY

If a fictional work is a machine for providing experience, what kind of experience does *Oroonoko* provide? From the familiar structure and intermittently the familiar rhetoric of heroic (and exotic) romance we see hanging here a number of kinds of representation and allusion or invocation that consort strangely with the pleasure-gadget of romance. What seems most at odds with the exotic romance, to a modern reader and I suspect to a seventeenth-century reader as well, is the situational and sensational detail as yet largely confined to scientized travel writing (such as, however humbly, George Warren's *Surinam*, from which so many earlier critics thought Behn had "stolen" her Surinam) and the journals of the various scientific academies. That such writing should appear whenever the narrative most closely approaches its mythic materials has an overt emotional suggestiveness in such an early text that it has lost by the time writers such as Joyce and Faulkner appear. What would a royalist writer of the English Restoration, with its dependence on tradition and symbolic panoply, mean by naturalizing, scientizing the ancient materials?

As seen in Chapter 2, the scientizing observational style (of Hariot, later of Hooke) is antinarrative. Students regularly note it: description retards (even replaces) action. The senses, particularly the eye, are rapt, as opposed to the ideational mind that asks for sequence, argument (as Milton and Spenser call the plot summaries in their epics), *narratio*. Myths simply *are* actions, the raw material of plots, and romance is a plot-based, even plot-obsessed, genre (or group of genres). So a conflict is apparent in the warp and woof of this novel—a conflict, perhaps, which is transported into the new genre generally through the formal characteristics it generates.

Hariot's antinarrative impulse functioned partly to excise the colonial action constitutive of his "Virginia"—burned towns, manipulated and betrayed Algonquin allies, and violence that could discourage prospective planters disappear there into a still pool of natural history, ethnographic portraiture, and the listing of commodities.[31] Behn is writing an "Entertainment," which must have a plot and does. But many potential or half-realized aspects of the plot or web of action are disturbing rather than, to use her word, diverting. The disturbance seems to have been deep and general, given the concern of so many candidates for "the first novel in English" with "new worlds" more and less equated with the exploitable and exploited New World of the Americas. What *happens* in the New World is hard to say, it's *murky*. The *tableau vivant* does not erase the action, impossible in an extended fiction, but diverts us from it. Similarly, Behn's narrator diverts Oroonoko and Imoinda from the humiliation of their real story with "the Lives of the Romans" and "Stories of Nuns" (93) and glorified nature-walks, which provide opportunities for transmitting the natural history, ethnography, and lists of commodities we might expect in plantation propaganda like Hariot's or George Warren's.[32]

The bad faith of persons who see and decry the brutality of slavery but neither free their slaves nor refuse to participate in the slave-based economy creates a narrative complication Behn apparently cannot resolve, one sign of which is, I think, the resort to information, description, the specular—the digressive pause. (This strategy first manifests itself at the novel's initial mention of the slave trade [57]). For a worker in the new genre I am taking to be partially constituted by the situations of colonial exploration and plantation, an even more fundamental or formal challenge is the problem of enslaved characters. The ideal form of the slave is something like Wren's robot-scientist, or Descartes's vision of what may lie under the clothes of the passersby on the street—a machine. Logically, a slave exists only to function in the interests of her/his owner and has no "inner being," no sensibility, private life, or personal motives. A slave, *qua slave*, is not a subject; a slave does not *have* a "tremulous private body." Her owner has it. Equally logically, to be the protagonists of a heroic romance, Oroonoko and Imoinda must be all

---

31. See Chapter 2. Alternate sources for the situation of Roanoke and the events attending Hariot's trip there can be found in David Quinn, *Roanoke Voyages*.

32. "The Lives of the Romans" is a particularly poignant digression for "Caesar," who is a member of that fictional and parodic world of upside-down power relations created by the European colonial custom of renaming slaves after the "Romans" of the Europeans' imperial model.

action and subjectivity. If they are, though, then we must identify with, feel with, persons who have been brutalized and paralyzed by slavery. No wonder people have sometimes thought of this novel as an important abolitionist work. But it is not — the hero himself unapologetically sells people into slavery and makes social engagements with European traffickers.

What are the implications of a tragic plot enacted by enslaved characters? The picaresque, following the lead of the medieval *conte* and *fabliau* and extending their length and range, has already provided a successful genre for the representation of nonaristocratic persons as significant agents — but the tone of the genre is comic, as had been the *conte* and *fabliau*, and as are the scenes between laborers or peasants in staged tragedy. Behn is obviously not striving for comic effects here, and she manages to sidestep that association with characters of low social status by providing her protagonists with another world, their proper world, in which (as in Freud's family romance) they are figures of very high status indeed.

A bigger problem with slave characters is the issue of agency — on the plot's moral register, how actualized can these characters be, who are entirely overpowered by the force of their masters? Theirs is the opposite of the romance situation, in which the intentions of the hero and often even the major female characters have more influence and power in their environments than is the case for the powerful in reality (a fact partially expressed in the relentlessly *eventful* narrative texture of romance). To whatever degree a character's range of action is limited, the salience of his or her sensibility or sensation must be heightened: passive experience is more "internal" than the events of romance. But at a certain point of passivity we begin to experience the characters in the actions to which they are subjected as quasi-pornographic objects.[33] It is worth thinking about seriously, the fact of the very restricted range of action of the protagonists of every English or French literary work that has been advanced as a candidate for the honor of being the "first" modern novel. All are either young, unmarried women or men who have been one way or another incarcerated or enslaved. Even *Don Quixote*, traditionally seen as the John the Baptist of the modern novel, makes its mark as a romance about someone of less power and competence than the genre requires.

---

33. The case par excellence for comparative anatomy and physical anthropology is the nineteenth century's "Hottentot Venus," although "Venus" was not the protagonist on those anatomy theater stages — the scientist was. This is a different subtype of pornography, which invites identification only with the aggressor or sadist. For an interiorized version of this scene, see the poet Elizabeth Alexander's *Venus Hottentot*.

In a way, then, the emerging genre of novelistic fiction participates with the emerging procedures of experimental science to restrain the independent motion and freedom of its objects so that they may be probed, penetrated by insight. But fictional insight is accomplished by a form of empathy with the object (however wrongheaded the results) while the scientist maintains a strict separation between subject and object. This scientific relation is strained to its limit where the object is, like the scientist-subject, human. Anthropology and, later, psychology are limit cases for the assumptions behind the methodical investigation of the phenomenal world; it is no surprise that ethnography and the psychological case study share so many features with the modern prose fiction that takes human personal and social relations as its material.

But fiction, too, has its empathic limits. In *Oroonoko* we see several places where fiction temporarily hides itself behind or, rather, subsumes the rhetoric and even the content of a scientific objectivity, most overtly in the passages of natural history and "manners and customs" once thought to be stolen from George Warren, but also in the stark, evacuated passage that describes the hero's dismemberment and death. Obviously these passages still belong to the fictional text and so they "are" fictional, they serve the fiction, as do the famous cetological chapters of *Moby Dick*. What they permit the narrator to avoid, at least temporarily, is the awkwardness of the moral position held by an observer in this scene. Authors of fiction always have more power than their characters, who might be envisaged as their slaves (in spite of — or perhaps in line with — the weary cliché writers produce when interviewed, about how their characters "took on a life of their own" and acted in quite unexpected ways, "running away with the story"). In Behn's case this power difference is closer to a terrible actuality; her dedication to Lord Maitland refers to Oroonoko as "my slave" ("'Tis purely the Merit of my Slave that must render [the book] worthy of the Honour it begs"). The best "my Slave" can do in the line of running away with the story is to try running away from the narrator's plantation — and when he fails, to take on a death of his own.

The powerful relation I mentioned at the start of this chapter, between formal features (vehicles) and the provocation of what we now call the "material," is clearly instanced here. What one sees in looking carefully at the construction and textures of *Oroonoko* is a scene of formal excruciation. What can be done to communicate such a story? Neither tragedy (which requires free agents), nor romance (which requires powerful ones), nor "true history" (which would require unseemly exposure of the narrator's complicity), nor plantation report (which would necessarily exclude narrative and individualized character) can accommodate the epistemic murk of Surinam's plantation culture or the struggles, within it, of two enslaved lovers.[34] To observe the classical "unities" of time and place would obliterate the central meaning of the narrative situation (lovers moved from one world in which they are powerful to another in which they are powerless and endangered). The Greek novel and its Renaissance prose romance descendants, with their disregard for these dramatic economies, are comic forms for achieving vicarious *triumphs* over spatial dispersal and loss of proper status. The subtitle of *Oroonoko, or The Royal Slave* elaborates an oxymoron that will underwrite European wealth and empire for centuries to come. Like a good scientist, Behn seems to aim at an accurate account of the system and its costs (it requires criminal deception and aggression and the death of the slave, or the death of the slave as social/human being). Also like a good scientist, she attempts to provide this account by means of "occular testimony" and objective detachment. Unlike the attention of the Restoration virtuosi, however, hers and her readers' attentions are focused (with some pained interruption) on narrative phenomena, the discrete particulars of individual history, rather than the pattern or generality or law that focuses scientific investigation. And unlike the scientists of her own period (though in something of the spirit of a Kepler), her sense of "the other World" seems to be a sense of the alternative world: for herself, liberating (imagine a respectable Englishwoman at home having the adventures of Behn's narrator in Surinam), and for the protagonists, devastating.

The year before *Oroonoko* was published, Behn's wonderful farce, *The Emperor of the Moon*, had been played at Drury Lane, with a dramatic "Prologue" of disgruntled meditation on genre, levels of style, and recently uninspiring audiences. Productions have descended the chain of literary being from tragedy to comedy to farce to the display of a mechanical wonder, a "Speaking Head," which sings and laughs irrelevantly while the actor playing its voice forgets to maintain his part of the illusion. "Well," says Harlequin philosophically, "this will be but a nine-days wonder too; / There's nothing lasting but the puppets' show" (whose mechanical plots and stock characters he then begins to parody). It's a funny prologue, to a funny play, but going back

---

34. Behn consistently employs an ironic metaphoric language of transport and conquest in narrating the details of Oroonoko and Imoinda's love affair that reiterates the sexualized power relations of colonialism, as invoked in earlier European representation of exploration, discovery, and conquest. See my "Carnal Knowledge" for an extended treatment of the topic.

to it from *Oroonoko* one detects a hint of fear and fatalism in it, too, a crudely anthropological sense that in some way all those people down on the street are indeed the robots Descartes feared might be "concealed under the hats and clothing"—robots of a regulating culture, robots of slavery. The discernible patterns of human "custom" come to seem, as the Enlightenment peeps over the horizon, a truth too close to the slavery that also constrains the will, or to the everlasting puppet show. In the final chapter I will dissect a text that describes a similarly frightening colonial world, through the phantasmatically "lasting" (and universalizing) patterns of a scientific study, a text which must, like but also fundamentally unlike *Oroonoko*, make room among the puppets for the sublime of pain.

Fig. 26. American fashions (Caribs, Acephales, Brazilians, Floridians, Virginians). Plate 3 from Joseph Lafitau's *Moeurs des sauvages amériquains*, vol. 1 (Paris, 1724). By permission of the Houghton Library, Harvard University.

IX • *E PLURIBUS UNUM*
Lafitau's *Moeurs des sauvages amériquains* and Enlightenment Ethnology

> And when we preach to them of one God, Creator of Heaven and earth, and of all things, and even when we talk to them of Hell and Paradise and our other mysteries, the headstrong savages reply that this is good for our Country but not for theirs; that every Country has its own fashions. But having pointed out to them, by means of a little globe that we had brought, that there is only one world, they remain without reply.
>
> —Jean de Brébeuf (1635)

THE NEW POPULARITY of the "one world" paradigm signaled the decline of the trope of the island, or at least of the island as "little world" (and vice versa). Increasingly, islands must be identified as part of a larger whole, of a system. In this chapter Lafitau's Iroquois of southern Canada belong to alliances of tribes, their language to a *family* of eastern North American languages; even hoaxer George Psalmanazar's Formosa (Taiwan) is an island in an archipelago, tributary to the emperor of Japan. French Jesuits have already settled in both far-flung places, which for the reader therefore preexist the European encounters in these texts and already belong to "the world."

But the differences between the works of Lafitau and "Psalmanazar" the imposter are more significant than their inevitably Same participation in the *episteme* of early eighteenth-century western Europe. Lafitau takes the world as a single place, despite his long immersion in specific Iroquois villages of New France, and the enmities he records in gruesome detail between neighboring villages or tribes. And despite the magnificent ethnographic detail of his two-tome opus, it is the book-length theoretical chapter on the likeness between New World and Old World beliefs and symbols that he seems most proud of and that got the most attention in contemporary reviews.[1]

---

1. See *Journal des Sçavans* 75 (November 1724): 593–98; *Mémoires de Trévoux* 22 (December 1722): 2189–92, 24 (September and November 1724): 1565–1609 and 2001–29, and 25 (February 1725): 197–207. Even life-forms mirrored each other everywhere: Lafitau's first

Psalmanazar, a character without, in several senses, a "name," relies on the distant and unassimilable otherness of his imaginary island of Formosa. His particulars are the badge of his credibility and his ticket to fame: they increase from edition to edition, as does their sensationalism.² The more quickly the actual Formosa is assimilated into the new mercantile world system, the sooner Psalmanazar will have to run and hide, or recreate himself again.³ And of course, in what may look initially like the biggest difference of all, Lafitau's fieldwork-based account of New France is "true," Psalmanazar's invented *Description of Formosa* is "false." Put in terms of a different polarity, Lafitau's work is "science," and Psalmanazar's hoax is "fiction"—a term of high opprobrium in Lafitau's survey of his sources and a term which, over the course of the century, will come to mean simply a made-up story, often told in prose.⁴

ON THE BRINK OF ANTHROPOLOGY

> Le Père Lafitau, que le zéle avoit conduit parmi ces prétendus Sauvages, avec les préjugez ordinaires, s'est heureusement trouvé y avoir porté aussi un certain oeil philosophique, tout propre à rétablir les objets dans leurs situations et leurs dimensions naturelles, & à reduire beaucoup d'idées vagues en un corps de systême; c'est-à-dire, de Science.
>
> —*Mémoires de Trévoux* (1724)

American writing was about Panax Ginseng, a plant he discovered in Quebec after having decided that the popular Chinese herbal remedy logically *must* grow in the New World as well as the Old.

2. Among contemporary reviews, see the trusting remarks on the first edition in *History of the Works of the Learned* 6 (April 1704): 244–52; and, on the French translation of the second edition, the *Journal des Sçavans*, 15 February 1706 ("il est á propos de suspendre toujours son jugement en lisant cet ouvrage") and the infuriated Jesuit journal *Mémoires de Trévoux* (April 1705): 587–95. The gamut of trust ranges from Protestant English to French Catholic to Jesuit reviewer, suggesting it is the work's promotion of Anglican piety as much as its sensationalism or fictionality that moved its reviewers, although it is relevant that the second edition is richer with sensational detail.

3. For information on Psalmanazar's self-refashionings, see his own posthumous *Memoirs* and two biographies: Richard Swiderski's *False Formosan* and Frederic J. Foley's *Great Formosan Imposter*. Good critical accounts of his practice include Susan Stewart's "Antipodal Expectations: Notes on the Formosan 'Ethnography' of George Psalmanazar" and an unpublished paper by Christopher Wells, "Dwindling into a Subject: George Psalmanazar, the Exotic Prodigal."

4. On categorical distinctions between false ethnography and fiction, see Justin Stagl's chapter on Psalmanazar, "The Man Who Called Himself George Psalmanazar" in Stagl, *A History of Curiosity*. In English, the word "fiction" acquired its relation to art towards the end of the eighteenth century, though it had been in use earlier in a related if still somewhat judgmental sense regarding the "fictions of the poets."

What is human, and how various? It is clear that from the beginning of European expansion in the mid-fifteenth century a set of new and intense historical pressures were brought to bear on the question, and resulted eventually in a socially functional set of regulated and standardized practices, including periodical publication of data belonging to a particular regimen of knowledge. A question once central to the task of poetry became the property of a science opposed to the indeterminate and the inspired. Lafitau's enormous work, *Moeurs des sauvages amériquaines comparées des moeurs des premier temps*, was a preinstitutional but recognizably "scientific" case of ethnological research and theory.

As we have seen in preceding chapters, many kinds of representation-bearing information about customs other than "one's own" were available over the centuries between the sudden expansion of travel and the production of Lafitau's book. Their aims were certainly not those of a theorist; in some cases their aims were no more than entertainment ("diversion") of a purchasing audience. Lafitau made use of most such kinds of representation and refers to them explicitly: obviously, "the old Relations" (*Moeurs* 1:337) of explorers, colonial administrators, and missionaries who had gathered relatively unsystematic information before him in the newly contacted lands, East and West, along with the more comprehensive cosmographies (or "natural and moral histories") of geographers like Thevet and Peter Heylen or administrators such as Father Joseph Acosta; somewhat less obviously, commentaries on classical literature, especially Homer and the books of the Hebrew Bible, where considerable interest in the folkways and material conditions of those important cultural cousins had sprung up among the humanists.[5]

Perhaps most interesting of contributors to an eventually theorized and methodized social science was the rapidly developing art of prose fiction (which overlaps with the utopia). Lafitau cites Fénelon's quasi-utopian *Télémaque* and (with some censure) Baron de Lahontan's sometimes fictionalized account of his journey to Canada; in England Behn's ethnographic *Oroonoko* and, in the same year as Lafitau's *Moeurs*, Swift's *Gulliver's Travels* are famous examples of overt fiction handling the materials of which Lafitau makes use (and from which he explicitly separates himself). A special and particularly

---

5. See, for instance, Fontenelle's reworking of Antonius Van Dale's 1683 Latin treatise (*De oraculis ethnicorum*) on the cessation of oracles, the *Histoire des oracles* (1686), which makes the claim, counter to Lafitau's arguments of universal religious feeling and spiritual presences, that there never had been real oracles in the first place, only "The Cheats of the Pagan Priests" (subtitle of Behn's 1688 translation, *History of Oracles*).

notable subcategory of the new fiction (or of the emerging anthropology?) are the travel hoaxes studied as a corpus by Percy Adams; the most fabulous character among the authors was Psalmanazar. But even Madame de Lafayette's novel of court manners, *La Princesse de Clèves* (1678), bound though it is to the royal court in the major metropolitan center of western Europe, undertakes work instructive to an as yet unmethodized ethnology. Its analysis of manners and their circumscription of narrative outcome adumbrates the antinarrative fascination of anthropology with pattern, though it nonetheless manages to render as narrative the lived and sometimes surprising experience of cultural rules.

None of the texts described above aimed to engender an anthropology, of course. They belonged to a culturally self-conscious moment, stimulated economically and intellectually by contact with unknown kinds of personhood and community—with people, moreover, who had been theoretically and, in particular, theologically impossible. It was a moment stimulated as well by the violent fragmentation of the culture of Christendom during the religious wars of the Reformation. The most theoretical antecedents to the work of Lafitau (other than utopias) were those produced in the debate between jurist Hugo Grotius and Johann de Laet, director of the Dutch West India Company, over the "origins" of the American people.[6] The Americans must have come from the Old World (otherwise there would have been more than one Adam and Eve!), but an intellectually respectable narrative was hard to come by: studying traces and survivals in Amerindian customs might provide evidence of an Old World genealogy. Grotius's brief study linked Americans to ancient Aethiopian and Chinese settlements, de Laet's to Siberia/Scythia. Lafitau seems to favor Greek origins, perhaps because he has more material on ancient Greek language, *"fables,"* and customs, perhaps because he is anxious to raise the reputation of his "sauvages."[7] But this particular choice of origin is not the destination of his argument. It provides the (shaky) ground for a systematic treatment of human homogeneity as expressed in cultural variation.

---

6. See Grotius, *De Origine gentium americanarum dissertatio*, and Laet, *Notae ad dissertationem Hugonis Grotii*, both published in Paris, 1643.

7. Raising the American reputation has a serious function for this missionary and Christian apologist in a libertine era: "To give, then, to religion all the advantages that it can draw from a proof as strong as is that of the unanimous consent of all peoples and to take from atheists other means of attacking it in this way, it is necessary to destroy the false idea given by authors of the Indian since this alone is the basis of such a disadvantageous prejudice" (1:29 [6]). All subsequent quotations from Lafitau will be cited parenthetically by volume and the page numbers of both the William Fenton and Elizabeth Moore edition and, in brackets, the French first edition. Both are two volumes, splitting at the same place in the text.

FATHER LAFITAU AND THE
MOEURS DES SAUVAGES AMÉRIQUAINS

The science of the manners and customs of different peoples has some quality so useful and interesting that Homer thought he ought to make it the subject of an entire poem.

—Joseph Lafitau (1724)

Joseph François Lafitau (1670–1746) spent six years among the Iroquois in a Canadian mission at Sault Saint Louis (outside of Montreal) in the early eighteenth century and who knows how many more years reading "the old Relations" for data about the earlier, contact-period lifeways of the Iroquois and other American peoples. His big illustrated book is considered by many to constitute the first work of ethnology proper (especially in its articulation of a classificatory system to describe Iroquois kinship). Though little read in the century and more intervening between its appearance and Lewis Henry Morgan's *League of the Iroquois* (1851), it has earned the respect of many anthropologists and is still in use as a reliable source for the folkways it set out, in part, to represent and interpret.[8] Even more than interpretation, Lafitau's book offered a theory of global culture, and a methodology, called in more recent times "upstreaming" or, in the introduction to ethnologist William Fenton's translation (with Elizabeth Moore), "reciprocal illumination."[9] "I have not limited myself to learning the characteristics of the Indian and informing myself about their customs and practices.... I have read carefully [the works] of the earliest writers who treated the customs, laws and usages of the peoples of whom they had some knowledge. I have made a comparison of these customs with each other" (1:27 [3]). Homer, Herodotus, Pliny, and Tacitus allowed the anthropological Jesuit a look at Iroquois customs "upstream" in Francis Bacon's River of Time, closer to their origins and better recorded than the past integrity of the Iroquois' oral culture.

---

8. On Lafitau's place in the history of anthropology see Sol Tax, "From Lafitau to Radcliffe-Browne," and Margaret Hodgen, "Early Anthropology" (346–49, 446, 490–91). He remains a source for ethnologists and ethnohistorians today: see, for example, on his identification of matrifocal and matrilocal structures in Iroquois government and genealogy, Martha Harroun Foster, "Lost Women of the Matriarchy," 122–23.

9. The introduction to Fenton and Moore's edition/translation of the *Moeurs* (*Customs of the American Indians*) is a definitive account of the life and work; it is cited by French as well as American historians. A French-language annotated edition is in preparation for the University of Montréal Press. For other views of transfers between the historically and spatially distant (the "exotic," in Bohrer's terms), see Frederick Bohrer, "The Times and Spaces of History" and Michel de Certeau, "Writing vs. Time."

The point of Lafitau's relatively novel method, not unprecedented but unique in its exhaustive and continuous application, was a polemically theological one: to show that the fundamental principles of religion are common to all cultures, past and present. "To give, then, to religion all the advantages that it can draw from a proof as strong as is that of the unanimous consent of all peoples and to take from atheists other means of attacking it in this way" (1:29). Secondarily, the work functions to some extent as a "mirror for princes" (and everyone else): Lafitau admires much in the training and conduct of Iroquois warriors and chiefs, whom he has elevated in prestige for his readers through the comparisons with ancient Greek culture, and points out in his introduction that "people should study customs only in order to form customs" (1:28).[10] It is not easy to imagine what use he conceives for polite French society in his discussions of cliterodectomy and cannibalism, but, as we will see later, we *can* learn something quite important to Lafitau from his representations of pain and mutilation.

One feature of *Moeurs* that in twentieth-century hindsight seems a step toward the modern social sciences is its combination of near-total independence from first-person narrative, along with its open acknowledgment of the fieldwork and theoretical speculation of others.[11] One is confronted in Lafitau's book with what amounts to the work of a (retroactively scientific) community, aimed at the understanding of first principles rather than the representation of historical particulars or subjective experience. Lafitau's Renaissance predecessor Thevet had also traveled and produced big illustrated books that attempted to cover the world for (at least phantasmatically) a community of practical or serious readers, but his debts to the voyages and texts of others had been scrupulously hidden, their authors sometimes accused by Thevet of plagiarizing *him*.[12] In the intervening century and a half, the socialization of scientific communities had taken off, and the physical sciences were already recognizably modern in their institutional structures: re-

---

10. Elizabeth Moore is careful to point out this function in her notes. Here is an example of the text working as Lafitau intended in this regard: "Although [the chiefs] have a real authority which some know how to use, they still affect to give so much respect to liberty that one would say, to see them, that they are all equal. While the petty chiefs of the monarchical states have themselves borne on their subjects' shoulders . . . , they have neither distinctive mark, nor crown, nor sceptre, nor guards, nor consular axes to differentiate them from the common people. Their power does not appear to have any trace of absolutism" (1:292–93 [473]).

11. Certeau sees the work as infused with its narrator's testifying presence, in "Writing vs. Time," 50–52. This essay seems unaware of the widespread iconographic tradition to which Lafitau's frontispiece (its initial topic) belongs, and fails to take into account Lafitau's considerable historical research and the framing of his work and its object in historical time.

12. See Chapter 2, 30–31 and 34–35.

search centers, journals, relatively open availability of data, standardization (and thus depersonalization) of techniques and report formats were all more or less normal features of "natural philosophy."[13] Lafitau writes in this spirit, but on social realities ("manners and customs") the study of which was not yet institutionalized—most historians date the birth of anthropology to the mid-nineteenth century.[14]

Lafitau's relation to these sources out of which so much of the texture of his book is composed is characteristically critical. Of travel relations and voyages, "Travellers would have spoken differently [about the Indians' lack of religion] if they had been in less haste to relate these things to the public and to impart discoveries by which they intended to gain honour for themselves.... One must never take it upon oneself to describe the manners and customs of a country on which there are no systematic studies unless one knows the language: a knowledge which demands long study" (1:94 [112]). On the quality of attention of classical sources, Plutarch, for instance, "sought to explain [the meanings of these mysteries], saying that the turtle which was carrying its house signifies that women should remain at home ... but it is clear that this is an inference that Plutarch has made in accordance with the code of conduct of his times" (1:84 [98]). Similarly, "authors who have written on the customs of the Americans have not paid any attention to the gynocracy established among these peoples" (1:344 [569]). On (literate) native informants, "the testimony of the Inca, Garcilasso, impresses me more than all the rest.... For though, he did not come into the world until sometime after the ... fall of that great empire ... he is still an author born in the very country of which he writes" (1:274 [440–41]).

Lafitau compares reports on the same topics, notices and registers the significance of unanimities and discords, and makes allowance (usually) for the passage of historical time between one source and another. He uses oral sources when he needs to, discriminatingly and plurally: "I have asked the

---

13. See for instance the recent work of historians of early modern science Steven Shapin, *Social History of Truth*, and Lorraine Daston, "Baconian Facts, Academic Civility, and the Prehistory of Objectivity." For pre-seventeenth-century norms of secrecy in scientific investigation and dissemination, see William Eamon, *Science and the Secrets of Nature*.

14. Official birthdays of anthropology usually relate to the dates of origin of ethnological societies: the Ethnological Society was founded in England in 1843, the Société Ethnologique de Paris in 1838, and the Gesellschaft für Anthropologie, Ethnologie, und Urgeschichte in Germany in 1869. Historians of early modern Europe sometimes see anthropology, as I do, emerging earlier as a discourse or "language" (see, e.g., Joan-Pau Rubiés, "New Worlds and Renaissance Ethnology," and Anthony Pagden, *Fall of Natural Man*). Hodgen, though disapproving of earlier phases in its development, does allow anthropology a pre-nineteenth century history, as does John Rowe in his classic article "Renaissance Foundations of Anthropology."

best informed Canadians and those who have communicated with the Indians, to find out what could be the significance of that white stick covered with swan feathers" (1:151 [266]). His sources are not sources of heroic text, appropriable signs of heroically witnessed or collected knowledge, but sources of what will be transformed into putatively disinterested information. Information has had the knower sieved out of it; it is certified not by authority but by rigorous method. Lafitau seems to be battling the mythic element in earlier "encounter" narratives, first and foremost by removing the hero—removing the rhetorical presence of point of view (as far as that term is understood to involve a personal subject). Given his respect for immediate eyewitness testimony, he naturally uses his own experience as respectable data, where it has been extensive enough. But the "I," at least as *witness*, appears rarely. The collecting and classifying I, though certainly a presiding genius and sometimes literally present in the first person pronoun, is not a protagonist, neither an agent of narrative nor a personal intermediary for our vicarious experience of the culture systematized in Lafitau's book.

The layout of *Moeurs* is carefully presented in chapter 1, "The Design and Plan of the Work." Lafitau describes the rational ordering of his apparently heterogeneous topics ("difficult to assemble . . . under a single point of view" (1:36 [18])) as achieved by "connecting things in a natural order and making such connection between them that they *appear to* follow each other naturally" (1:37 [18]; emphasis mine). Within the topics focusing the fifteen chapters, "in the descriptions of the customs of the Americans the parallel with those of the ancients is always sustained, because there is not a single detail of the former which has not its roots in antiquity" (1:37 [18]). The account of the Indians he wishes to give is one very carefully extrapolated from secondary sources to represent "this culture ["ces Moeurs"] and these customs as they were before their alteration" through "trade with the European nations" (1:41 [25–26]). Though he is a diffusionist, believing similarities between heroic and mythic Greek culture and the warrior culture of the Iroquois are historically related rather than "evolutionary" parallels, he wants to describe a pure form of Iroquois folkways, unrecorded in any primary sources he respects or has access to, rather than what he sees before him. "I shall be able to speak of those changes made among them in another work," he says (1:41 [26]), but that work never appeared.[15]

15. The fate of the manuscript and Lafitau's publication plans are unclear. He himself describes that "other work" as one "in which I propose to relate the establishment of the Christian religion among [the Indians] and the efforts made by the evangelical workers to change these Indian mores to bring them into conformity with the laws of Jesus Christ" (1:41 [26]). Fenton and Moore's footnote says only, "The book on culture change and the establishment of the

However actually interested this globally theoretical account of *moeurs* by an author in the embattled service of the "holy, *catholic and apostolic* Church" may be, its integrated diversity of sources and voices—European and Iroquois; male and female; trader, explorer, and missionary; oral and written; past and present; personal and official—provides a surface of such collective testimony as to sound like objectivity itself (the fools' gold of colonial science). In *Moeurs des sauvages amériquains* we are faced with the compost of hundreds of encounters, including but in no way privileging the author's—not the case we found with Thevet, who subsumed all sources into his personal narrative voice and made of himself the "sole Self" of colonial France. What happens to our fabled object of obsession, the Other, when no marked or identifiable self-interested Self is shown to be looking? Lafitau widens the stage of this psychodynamic to represent a culture looking at a culture, prefiguring the long-dominant un-self-consciousness of an anthropology that probes "other cultures" as if it is itself from Noplace. A Self disguised as a culture's representative and instrument of inquiry may find an Other disguised, not as subhuman but as a subcategory: Linnaeus's *homo americanus*, or the "Savage Mind."

"LA RELIGION INFLUOIT EN TOUT"

Lafitau's deepest interest, it seems clear from his longest chapter, "Religion," is in proving the universal and persistent presence of the religious impulse in all cultures. Chapter 4, with its relentless demonstrations that everything is everything, can be exquisitely boring reading for a (post)modern critic, and it is not what modern ethnologists and ethnohistorians have gone to Lafitau to find, being almost entirely about Old World pantheistic religions and their symbolism, of which Lafitau finds strong traces in many features of the Amerindian cults. But it will behoove us to cut through the boredom, for it was this focus, with its rational, erudite embrace of the mystical, the ecstatic

---

Christian religion was never published" (1:41). Elsewhere, in their introduction, they speak of a never-published work they describe as "a repackaged version of the long and tortuous chapter on religion in the *Moeurs*," prepared "within two years after the printing" of the *Moeurs* (1:xli). "Armed with all the required permissions from within the Jesuit order, the work, nevertheless, failed to clear the Royal censor and it was still in his hands in 1740." Lafitau claims he was scooped by another Jesuit "who appropriated my system"; the editors wonder "whether these difficulties with the civil authorities ... stem from conflicts arising from his duties as procurator in a time when tension was rising between the Jesuits and the Crown" (1:xli–xlii). Whether the editors are talking about the same work in two different places is not easy to tell; I have found no other references that do not refer to this note.

and the sublime, that mesmerized and motivated the earliest modern scientist of culture.[16]

Culture in Lafitau's work is clearly (though not etymologically) derived from "cult" [*culte*], and religion casts a wide net over what Lafitau in fact describes, ingenuously, as "everything" (1:36; his metaphor is better than mine—"la Religion *influoit en* tout" [1:17]).[17] One senses in this work, harbinger of Enlightenment values and projects, a resistance to the encroaching *secularism* of the Enlightenment (and not just to the atheism Lafitau explicitly opposes with his theory), a desire (for all colonial texts are formulated around desire) for a less "disenchanted" world.[18] Father Lafitau's loving attention to "our Indians," combined with his exhaustive detailing of parallels from many locations of antiquity as well as early modern China and Oceania, gives first and foremost an impression, at least in the giant chapter on religion, of wallowing—in a flood of religious and ecstatic materials. The science of culture is an oxymoron, for the system of investigation and classification that quantifies and regulates information is here applied to phenomena whose character and function lie in the production of (multiple) *significance*, of that which exceeds quantification and ignores regulation.

Capitalism, and the consumer culture that must float it, require the subtle, unconscious despair of atheism (or its contemporary cognates). As long as there exists a structure of desire and partial fulfillment other than that organized around the commodity (such as that organized around salvation), a viable form of nonparticipation in the economy blocks the absolute extension of the market. But, as we noted in Chapter 1, the acquisitive impulse remains constant across this important gap of interest: "the harvest of souls is more ripe here than in any other place," says one of Lafitau's chief sources, Jean de Brébeuf, of his village in the Huron country (*Jesuit Relations*, 8:101). Gabriel Sagard, another of Lafitau's sources, chooses a different metaphor—and in-

---

16. Contemporary reviews, especially those in the Jesuit *Mémoires de Trévoux*, tend to cite this as the book's most crucial feature.

17. "The matter of customs is a vast one comprising everything in its scope" (1:36 [17]). "We shall see several similar and curious traits of it [religion] in the other chapters, those on [the Americans'] government, marriages, warfare, medicine, death, mourning and burial so that it seems that, formerly and in the first times, religion played a part in everything" (1:36 [17]).

18. Here for instance is Thomas Blackwell, from his 1735 commentary on Homer (in which he frequently considers Homeric Greece in an ethnological light, as a "primitive" culture he occasionally compares to those of the Arabs and Indians (there are one or two references to native American cultures as well): "It may look odd to say, that even the *Ignorance* of these Ages contributed not a little to the Excellency of [Homer's] Poems: But it was certainly so. The Gods were not called in doubt in those days; *Philosophers*, and speculative incredulous People had not sprung up, and decry'd Miracles and supernatural Stories" (*Enquiry into the Life and Writings of Homer*, 149).

scribes it on the land: "I scratched crosses and the name of Jesus with the point of a knife in the bark of the largest trees, to signify to Satan and his imps that we were taking possession of that land for the Kingdom of Jesus Christ" (41).[19] The dominant language, even in the missionary letters, remains the language of gain and conquest, of possession.

The bureaucratization of knowledge favored by the conditions of early modern French government and commerce is part of the cultural equipment of capitalist expansion. Lafitau gathers and organizes a great deal of ethnographic information and renders it coherent and intelligible — available and usable. The book is for sale, not for the Jesuit archives. But the flooding of religious significance into all parts of the territory displayed in its anthropological mirror of early modern European culture undermines the essential atheism of the growing state. Lafitau's glass, in which we see the world as numinous, was not wholly *à la mode*.

It did however share another important feature with the totalizing impulses of imperial organization. If Lafitau wanted to restore the world's diminishing divinity (a feature not apparently on the wane in the world of the Iroquois), then it was incumbent upon him to link (*enchaîner*) those areas in which he found it, including areas of the past. Christianity's historicized account of the incarnate presence of divinity on this earth (and its corollary sense that we are always leaving it farther and farther behind) would lead such a theorist inevitably to ancient scriptural and classical instantiations and documentation of human management of the divine. The comparative method not only multiplies divine presence, it also ties the world together in the mind. Michel de Certeau's essay on Lafitau's *Moeurs* ("Writing vs. Time") discusses the basic structure of the text as that of the *copula*: "the [text] will not be sustained by the antiquity nor the social identity of the documents it treats, but by 'the relation alone' that it establishes among them. . . . It constitutes a distinct 'language' or *system of relations*" (48). Since this copulative text aims to represent the world, its world becomes not only universally worshipful but *One*. As in the aim of empire.

VENTRILOQUISM AND POSSESSION:
EMPTY BODIES, MOVING SOULS

In fact that mystical death in the initiations, those expiations, lustrations, the Evasme of the Bacchantes which were true wailings, as well as the tears shed in the mysteries of Atys, Adonis and Osiris, also the enigmatic myths of Adonis and

---

19. See Brébeuf's "Relation de ce qui s'est passé aux Hurons, en l'année 1635" (embedded in "Le Jeune's Relation"), and Sagard, *Long Journey to the Country of the Hurons*.

Osiris, dead and reborn, the regeneration, the new life of the initiations, the tests of rigour and penitence, the state of perfection taught in the great mysteries; all these, I say, united, could not have for object this perishable life for all would have been useless and foolish if all was to perish with it.

—Joseph Lafitau (1724)

Citing and marveling over sixteenth-century Huguenot colonist Jean de Léry's aesthetic excitement at a choral song of the Tupinamba in Brazil has become a trope of contemporary analysis of the "contact period."[20] Léry, despite his religion, is one of Lafitau's most respected sources for South American customs, and it may be instructive to compare their passages on related field experiences. What was for Léry a communicable and relatively unproblematic personal experience is, amidst Lafitau's conflicted allegiances (to rationality and Revelation, theory and testimony), a rather different matter. Musical performance and transport are important phenomena in his account of the world as divinely and demonically infused, as are dance and shamanistic trance (his word for "shaman" is "Jongleur"). But it is a Cartesian world ("what do I see other than these hats and clothing? Could not robots be concealed under these things?" [Descartes, *Meditations*, 66]): a world of ventriloquism, marionettes, irruption, and possession.[21] In Lafitau's discussions of music, dance and medicine, spirits and voices penetrate and take possession of alien bodies (as do European states—and successful missionaries).[22] Rationality and even personality are always under threat, where the spirit is so "independent" (Lafitau 1:230 [362]) of the body, and the body so capable of performance and deceit.

Here is Léry, who has sneaked into a lodge where for two hours "five or six hundred men danc[ed] and [sang] incessantly" (Léry, 143): "At the beginning of this witches' sabbath, when I was in the women's house, I had been somewhat afraid; now I received in recompense such joy, hearing the measured harmonies of such a multitude, and especially in the cadence and refrain of the song, when at every verse all of them would let their voices trail, saying *Heu, heuaure, heuaure, heura, heura, oueh*—I stood there transported with delight. Whenever I remember it, my heart trembles, and it seems their voices are still in my ears" (144). This passage follows a close description of the dance, and is followed in turn by some discussion of the songs' meanings, as revealed by "the interpreter."

20. See Certeau, "Ethno-Graphy"; Janet Whatley's introduction to Léry, *History of a Voyage*, xxx–xxxi; Stephen Greenblatt, *Marvelous Possessions*, 14–20.
21. I have occasionally adjusted toward the literal the Donald Cress translation of the Latin text of Descartes's *Meditationes* in Charles Adam and Paul Tannery's edition of the *Oeuvres*.
22. One thinks again (and with increasing wonder) of Guyana's Dutchman Ghosts.

Lafitau's chapter on "Political Government" ends by recapitulating Léry's experience "of some Brazilian dances that seem no different from those of the Iroquois" and demurring:

> I have not felt at all such keen pleasure as Mr. de Léry did at our Indians' festivals. It is difficult for me to believe that everyone was as much impressed as he at those of the Brazilians. The music and dancing of the Americans have a very barbarous quality which is, at first, revolting, and of which one can scarcely form an idea without witnessing them. Nevertheless, little by little, one grows accustomed to them and, finally, one witnesses them willingly ["dans la suite on y assiste voluntiers"]. . . . I have never been able to discern either finesse or delicacy in the violence of these impetuous dances nor could I distinguish one from the other but the natives ["les Naturels"] of the country know how to differentiate them and their young people become impassioned about them as we do about our plays. (1:326 [534])

(Note that even here where his own sensations are recounted Lafitau compares, and generalizes: "on y assiste.")

This account of the ecstatic art of music is one of the most confused and ambivalent sections of the text. Even his strongly felt statement of disagreement ("I have not felt at all such keen pleasure as Mr. de Léry") modulates immediately into a grudging admission of his own pleasure, announced in the impersonal case and habitual present: "finally, one witnesses them willingly." These two accounts by religious Frenchmen of opposed sects were separated by a century and a half in the history of musical style and taste — never mind the gap in American physical and cultural geography — though Lafitau is aware that a great deal has changed in that same time for the Amerindians he has lived among. The wobbling value judgments in this chapter point together to the fact that Lafitau, committed in many ways to Montaigne's notion of the resonance between Tupinamba songs and the Anacreontic lyrics of classical Greece, cannot figure out where to come down on the issue of the music's pleasurability, even in his own experience of it.[23] It is more palpably alien than a kinship system, but must be reduced to Greek origins by the neoclassicist, which gesture, in turn, estranges by association his own cultural origins, orphaning him: "if this turtle[-shell lyre] of the Indians is the same thing as Apollo's lyre . . . the poets have wasted their time in praising so highly its music. . . . They are no less wrong in invoking it with its muses, if their songs of *hié, évohé,* etc. were none other than the *hé, hé, éoué,* drawn by

---

23. "Now I am familiar enough with poetry to be a judge of this: not only is there nothing barbaric in this fancy, but it is altogether Anacreontic. Their language, moreover, is a soft language, with an agreeable sound, somewhat like Greek in its endings" (Montaigne, "Of Cannibals," 115).

our Indians from the depths of their throats, for certainly I do not know of a more detestable music anywhere in the world" (1:154–55 [218]). The performances must finally be good, since they are traces of Greek antiquity, and this passage goes on to describe and praise their flute music. But then there is the problem of seduction—music that makes the listener "tremble" with "delight" as Léry had is not recommended as the regular fare of a missionary who is there to repossess the culture for the God of the Christians.

Lafitau must have been surrounded by transports, many of them more demonic and therefore less problematic than the transports of music. He must defend the reality of demons and immaterial spirits against the sophisticated skepticism of such as Fontenelle, but then they are his spiritual enemies, inculcating through the shamans superstitious (un-Christian) beliefs and illicit magical observances among "our Indians."[24] He must censure both the superstitious and the atheist cynic at once, but he shows less antagonism to the "simple": "It is true that in all times, there were unbelievers as well as simple and credulous people. But the display of the incredulity of the former and the foolishness of the credulity of the latter should not prejudice the truth. . . . We should make the world appear too foolish if we wished to suppose it, for some centuries, the dupe of miserable thimble-riggers ["joüeurs de gobelets"]" (1:239 [375]). Lafitau at any rate believes: "The foreign spirit appears to take possession of them [the shamans] in a palpable and corporeal manner and so to master their organs as to act in them immediately. It casts them into frenzies of enthusiasm and all the convulsive movements of the sybil. It speaks from the depths of their chests thus causing fortune tellers to be considered ventriloquists" (1:243 [383–84]). Like Descartes's imaginary robots, these animated bodies are uncanny. Their potentially psychological "depths" are for rent: there is no stable, unitary Other for this polyglossic ethnologist/text (or the reader) to encounter.

Lafitau's judgments are logically unpredictable though. He may believe demonic spirits animate the body of the "Jongleur," but he smiles at the Iroquois idea that soul may animate the animal body: "If they express themselves very much as Descartes does on the subject of man's soul, they are very far from thinking as he does about that of beasts. Far from making of them automatons, and pure machines, [they think that] they, judging by their performance, possess a great deal of reason and intelligence" (1:230 [361]). There is a spectrum of Iroquois body–soul, or body–spirit relationships to consider, and Lafitau seems tremulous with ambivalence about

---

24. See Pierette Désy, "A Secret Sentiment (Devils and Gods in Seventeenth Century New France)."

where he should draw his own lines. He uses a self-erasing indirect discourse to describe the Iroquois sense of the soul's independence: "The soul of the Indians is much more independent of their bodies than ours is and takes much more liberty. . . . It transports itself into the air, passes over the seas and penetrates into the most inaccessible and tightly closed places [again, like European missionaries!]. Nothing stops it because it is a spirit" (1:230 [362]). On the other hand, the next page provides this more dubious account: "As they lack sufficient knowledge of physical science ["de Physique"] to explain the meaning of dreams, they persuade themselves that, in fact, their soul, sensing their body plunged deep in sleep, takes advantage of those moments to take a walk. . . . On their awakening they really believe that the soul has really seen what they thought in their dreams and they act accordingly" (1:232 [363]).

If one "went native" in country like this, one would lose more than a cultural identity. One might lose a sense of self-possession. One might begin to understand the spirit as mobile, inquisitive, perceptive, liberty-loving, yet subject to the invocation of religious officials (not always Christian). One might begin to look at the body as a mere vehicle, not an identity but a location, not racial and genealogical but simply inhabitable, like a colony, or edible, like the Host. Perhaps this sense of the American body had something to do with the astonishing capacity for enduring torture that Lafitau represents as the fruit of the Iroquois warrior's training, and represents graphically in the two longest and most important chapters of the *Moeurs*, on "Religion" and "Warfare."

TORTURE

> I realize at this point that there is a great difference between a mind and a body, because the body, by its very nature, is something divisible, whereas the mind is plainly indivisible.
>
> —Descartes, *Meditations* (1641)

The final section of the big chapter on religion is called "The Cross in America" and mostly debunks earlier missionaries' unpersuasive sightings of crosses and their eager identification of the symbol with a preexisting or (to use Bartolomé de Las Casas's term) "wild Christianity" of the Americas.[25] Lafitau is a good mythographer, and knows that Christian symbolism does not exhaust the uses and significances of this elemental form. But in another

---

25. The phrase is from the 1583 translation of his *Brevísima relaçión* (1552), *The Spanishe Colonie*.

and subtler way the crucifixion is made "native" in the richly detailed and multiple accounts of initiation rituals for princes and warriors and in the descriptions of the kinds of torture that await prisoners before their deaths and, often, their consumption as sacramental food.[26] The vividness of these long descriptions provides the most intimate, because most painfully *sensational*, encounter in the *Moeurs* with alien bodies and their experience.

The descriptions are mostly too long to quote. The following shorter account of an ordinary Carib warrior's initiation is only prolegomenon to the princely sequences of pain, which often last for years (Lafitau is quoting Rochefort's *Histoire morale des Antilles*):

> Before the young men are admitted to the rank of those who can go to war, they [296] must be declared soldiers ["soldats"] in the presence of all their relatives and friends.... The father who has convoked the assembly ... has his son sit on a little seat placed in the middle of the hut or shed; and, after pointing out to him all the duties of a Carib soldier ..., he seizes, by the feet, a certain bird of prey, called *mansfenis* in their language, which was readied long before to be used for this purpose and strikes his son repeatedly with it until the bird is dead and its head entirely crushed. After this rough treatment which stupefies the young man he scarifies his whole body with an acouti tooth; and to cure the wounds made, he dips the bird into an infusion of pimento grain and rubs all the wounds rudely, causing a sharp and cutting pain to the poor patient.... He is then made to eat this bird's heart and [297], to conclude this ceremony, he is put into a hammock where he has to stay, fasting, stretched out to his full length, until his strength is almost exhausted. (1:196 [295–97])

The gist of these depictions of ingeniously augmented pain is clearly that one can learn—and American *soldats* have learned—to dissociate from the body; the initiation accounts (which run for almost seventy quarto pages in the original edition) are followed immediately by discussions of other practices involving dissociation, or the mobility of the soul: magic, divination, dreams and dream contests, familiars, shamanism, divination by frenzy, witches, fetishes, charms, "the state of the soul after death" (1:251), dancing in heaven, metempsychosis. Lafitau has already said that his method involves "making such connection between [things] that they appear to follow each other naturally" (1:37 [18]). It is apparently a "natural" segue for him, from bodily torture ("the body is something divisible") to spirit possession and invocation and the experience of the disembodied soul. Two kinds of *unheim-*

---

26. For the controversy over the reality of Iroquois or any kind of cannibalism, see William Arens's book *The Man-Eating Myth* and two (among many) responses, Thomas S. Abler, "Iroquois Cannibalism: Fact or Fiction," and Donald W. Forsyth, "The Beginnings of Brazilian Anthropology: Jesuits and Tupinamba Cannibalism." Frank Lestringant's *Cannibals* takes no stand on the issue.

lich dissociation share the stage in this chapter (indeed in the work as a whole): the absenting of the body by the excruciated sensorium or mind and the disembodiment of various kinds of spiritual entities. This may sound like Frantz Fanon's analysis of the consciousness of the colonized and oppressed (in which the mind is *not* "plainly indivisible"), although in this work we are trying to understand it first as a perception of an ethnologically minded French missionary, applied to a "culture and customs" being described "as they were *before their alteration.*"[27]

The initiation processes are shown to bear fruit in the endurance of war captives. The most sensational stretch of *Moeurs* comes in its other big chapter, on "Warfare," which includes about fifty-five quarto pages of almost unremitting description of torture and mutilation. These grisly, pornographically vivid descriptions surpass such more familiar passages as the torture and execution of Behn's Oroonoko. Indeed, one is reminded at least as quickly of Sade (born a year before Lafitau's death).[28] As this section of *Moeurs* offers the most extended and intense representation of Cartesian (or Fanonian) dissociation, we would do well to look closely at it. This will involve attending to a lengthy and unpleasant quotation from a kind of *blazon* of graphic violence, which may represent some even more unpleasant experiences of authorial witness.

The long closing cadence of the chapter moves from sections on "Sacking" and "Scalps" through the guarding of prisoners on the march back to the conquerors' village, their ritualized entry, the council on choosing "destina-

---

27. "Any study of the colonial world should take into consideration the phenomenon of the dance and of possession. The native's relaxation takes precisely the form of a muscular orgy in which the most acute aggressivity and the most impelling violence are canalized, transformed, and conjured away. The circle of the dance is a permissive circle; it protects and permits. At certain times on certain days, men and women come together at a given place, and there, under the solemn eye of the tribe, fling themselves into a seemingly unorganized pantomime, which is in reality extremely systematic, in which by various means—shakes of the head, bending of the spinal column, throwing of the whole body backward—may be deciphered as in an open book the huge effort of a community to exorcise itself, to liberate itself, to explain itself.... There are no limits, for in reality your purpose in coming together is to allow the accumulated libido, the hampered aggressivity, to dissolve as in a volcanic eruption.... One step further and you are completely possessed. In fact, these are organized séances of possession and exorcism; they include vampirism, possession by djinns, by zombies, and by Legba, the famous god of the voodoo. This disintegrating of the personality, this splitting and dissolution, all; this fulfills a primordial function in the organism of the colonial world" (Frantz Fanon, *Wretched of the Earth*, 57–58).

28. On the relations between cannibalism, Sade, and the history of the European novel see Claude Rawson, "Cannibalism and Fiction." Although the adulation of Flaubert, with the consequent assumption of his art as the telos of novelistic aesthetics, is a distraction, the essay is insightful on the different ontologies of novelistic kinds, and the importance of the extreme and sensational to the history of fiction.

tions" for the prisoners (generally understood as replacements for lost members of the village), their formal torture (separate sections on different techniques in North and in South America) and their death songs (a section including depictions of ritual cannibalism). Eerily, the chapter ends with a section on "Adoption," the alternative method of disposing of prisoners of war: adopted captives face "a gentler [condition] in proportion as that of those thrown into the fire is more cruel" (2:171 [308]). Adoptees are usually treated as regular members of the families to which they are assigned, intermarrying with other village families and even becoming "important . . . in the village" (2:171–72 [308]). The rhetorical effect of this calm, painless alternative to the descriptive frenzies which precede it in the chapter is disquieting. The narrator himself seems dissociated, able to slip back into contemplation of alternatives as though the contrast were anything but hyperbolic, as though the reading imagination needed no buffer between the dismemberment of the still-living captive for the village feast and the ritual bathing and reclothing of the future adoptee.

> If the captive walks in the lodge, or in the square, someone stops him or, if he is already tied to the post, goes to him to torment him. But, so that this cruel pleasure may last longer, they touch him only at long intervals without emotion or haste. They begin at the extremities of the feet and hands, going up little by little to his body; one tears off a nail, another the flesh from the finger with his teeth or a bad knife, a third takes this fleshless finger, puts it in the bowl of his well lighted pipe, smokes it in the guise of tobacco or has the captive himself smoke it; thus, successively, [until] they do not leave him any nails. They break the bones of his fingers between two stones, cut all his joints. Several times in the same place, they pass and repass over him burning irons or glowing torches until they are extinguished in the blood or discharge running from his wounds. They cut off, bit by bit, the roasted flesh. Some of these infuriated people ["furieux"] devour it while others paint their faces with his blood. When the nerves are uncovered, they insert irons into them to twist and [278] break them, or tie his arms and legs with cords which they draw by the two ends with extreme violence.
>
> This is, however, only a prelude. (2:156 [277–78]) [29]

29. I have tried to extrapolate some information about early eighteenth-century readers' horizon of possible response to such a passage: although contemporary reviews in the journals cited above (see n. 1) mainly addressed themselves to the comparison of Indian and classical religions ("le Parallele" as the *Trévoux* reviewer keeps saying), it is suggestive that almost the only marginal highlighting in the copy owned by Lieutenant-Governor William Dummer of Massachusetts (now in Harvard's Houghton Library) is next to this and other passages describing torture. There is also marginal marking next to two surprising and comic anecdotes; in the final chapter, on language, the reader has written equivalent tribal names next to those in the text labeling particular language groups or surviving clans. The work has been read through carefully, but from the specific point of view of a colonial administrator in country not far from Iroquois territory.

The increasing capacity of early modern scientific prose to construct detail seems harnessed here to functions beyond transmission of data or support of nearby articulations of principle. This is more the detail of pornography than of the experiment report. Certainly Lafitau has a point to make about American courage, but the number of pages of sensational agony with which he supports that point exceeds the demands of his argument. His rhetoric shares the mode of sublimity with the prose of religious ecstasy but, where that prose is forced into metaphor and the *topoi* of inexpressibility, Lafitau's is the simple grammar and denotative language urged on natural philosophy by the manifesto-writers of the seventeenth century: here again are "so many things" expressed, as Bishop Sprat would have it, in "an almost equal number of words."

That denotative plainness is only part of what goes on here: because of the particular nature of the "things," the words for them have the *effect* on a reader of hyperbole, "outdoing," and other features of sublime rhetoric. The mimesis is *physiologically* powerful (my legs and arms hurt when I read or transcribe these passages). What does Lafitau accomplish by imposing upon his readers, in the midst of a text mainly erudite and exhaustive, such an intense invitation to engagement: such shared pain or, for some readers, such sadistic gratification? That we are really being *invited* is clear from the composition of this passage (and others like it). There are delays, advances, augmentations of sensation, adverbial modifiers, and interpretive statements that slow the pace of transmission "little by little," "successively," "bit by bit," "so that this cruel pleasure may last longer." There is the almost literally stunning effect of the transitional sentence at the end of the passage: "This is, however, only a prelude."

What is the engagement for? Many functions of such a passage present themselves at once—here, for instance, is the return of the repressed pain of conquest and deracination, absent more generally from a work that aims to dehistoricize its account of the Americans "before their alteration." Here as well is "the horror, the horror," the imaginative collapse of the displaced European faced with what Taussig calls the "epistemic murk" of the colonization he (or she) facilitates. Here is the inevitable Manichaean counterpart to Lafitau's restoration of the world's divinity—the world's *demonicity*. For the spirit of cruelty inhabiting the scenes recounted is always distinct from the "marionettes" and "ventriloquists," as it were, who perform the physical motions: the "furieux" who, so often figured by Lafitau as an ideal of civility, are here seen as possessed by a sadistic frenzy.

But the most salient function of this invitation to engagement is the most immediately linked to the reading experience. Represented sensations of pain, however we place ourselves in relation to them, make us intimate with the persons represented as receiving them.[30] They force upon us the impression of the victim's subjectivity. We may repudiate that victimization, but if we do then we share the weaker sensations of the torturers and flesh-eaters, who are also Americans, and at their most "sauvage." Either way we are sharing a virtual experience with American persons, rather than coolly observing a systematic pattern of lineage or following the speculative etymology of a religious term. The torture scenes will have a future (indeed, as we have seen, already have a past) in novels, especially exotic novels. The interior and immediate experience of indubitable self is not the anthropologists' task to provide in the future of the science Lafitau's book helps to initiate. Here it is, nevertheless, waving both hello and goodbye in the middle of the ways, where colony and scientific anthropology first meet. The image is a familiar one to Christian readers — the excruciated, resolute body bound to a stake and tortured to death, its flesh consumed by the community in a ritual feast; the uncomprehending, demonic actors in a catastrophe that tests the powerful spirit "inside" but somehow not *of* the bound body. Lafitau's Iroquois are brought up to be Jesus, though he never says so in "so many words." They are also brought up to be the centurions who mock the incarnate god on the cross and cruelly augment his pain. In addition to having displayed through erudition the double truth of universal religious instinct and the monogenic origins of religious cult, Lafitau seems to have demonstrated through sensationalism the global brotherhood of people who suffer and, more oddly, of people who torture.[31]

Lafitau's sources (for example, Le Jeune, Brébeuf, Sagard, Thevet, and Léry) also describe the tortures visited upon prisoners of war in Iroquois and

30. Elaine Scarry's magnificent *Body in Pain*, with which I wouldn't disagree, concentrates productively on the absence of empathy in the torturer and his team. I am talking about a readerly relation to *depictions* of torture, a different matter. However much a reader may in fantasy identify with a torturer, she is not complicit in the torture, which is over, and her motives are unrelated to the motives (analyzed by Scarry in her first chapter) of the actual torturer of an actual suffering human. It is important to bear in mind the urgent actuality of such a scene, the irreducible difference between torture and reading about it. I have learned much from this book, but it does not deal with the possibility of another structure than that of modern state-controlled torture: those who have read *The Body in Pain* will recognize that the situation of mutuality in the small-scale and continuous warfare of a nomadic tribal people like the early modern Iroquois alters the picture in some ways. Note, however, that while interrogation is *not* part of that picture, the torturers still require speech of the victim: among the Iroquois it is ritualized as song.

31. Of course, the depiction of savages as *in*humanly sadistic has led to the modern connotations of the adjective "savage." But I am trying to show how Lafitau's writing harnesses the powers of identification to make even sadistic action and feeling shareable for the predisposed. And anyone who has read the first few pages of Michel Foucault's *Discipline and Punish* knows that eighteenth-century French readers were aware such tortures could be conceived and executed in their own polity.

Huron villages, or among the Tupinamba of Brazil, but the passages are briefer and usually less detailed. Sagard provides a description (161–62) almost as carefully gruesome as Lafitau's, and which Lafitau certainly read. Brébeuf's and Le Jeune's accounts are shorter and less vivid and cadenced: they halt themselves, mid-paragraph, with "Bref." Lafitau is unrelenting, saying, well past where they would have said *bref*, "But this is only a prelude." Much of the most refined cruelty is attributed by other writers to the women and girls, and the writers produce distancing statements of pity, so that readerly (presumably male) involvement with the torturers is clearly marked as off limits.[32] The sheer length and density of Lafitau's series of subchapters in "Warfare" transforms the effect of any one passage within them. And Lafitau is more elaborate in his warning to (or seduction of) the reader: he points explicitly to the sensational aspect of his data and the potential of writing to reproduce it. "Among the tribes of North America whose customs we know, the [form of] torture of captives is to burn them over a slow fire, but this scene takes place under such enormously barbarous conditions that the very thought makes us tremble. It is so disagreeable as to make it difficult to give an exact description of it. As it is necessary to speak of it, however, here is an adequate description and one which will convey some knowledge of it" (2:155 [274]).[33]

The instinct to involve the register of sensation in the project of the *Moeurs* is consistently observable and separates Lafitau's representation of Amerindian life from those of many of his sources, even where those sources share his purpose of humanizing the once debatably human American Other.[34] Whatever his own complex theology of the body might have been, the re-

---

32. See Thwaites 10:226–29 for the passage from Brébeuf; for Le Jeune's passages from *Relations* of August 1632, see documents 107 and 109 in Lucien Campeau, *Monumentae Novae Franciae*).

33. "Comme il faut en parler, voici à peu près ce qu'on peut dire, and cela suffira pour en avoir quelque connaissance."

34. Lafitau was not alone in wishing to connect the Indians and their cultures to the larger human family and the reader's own lifeways and experience. Of Lafitau's two main sources for South American torture of prisoners, Léry put considerable stress on the family resemblances among human societies, pointing out at some length the cannibalism of the French during the religious wars: like Montaigne, he preferred Tupinamba to French cannibalism (for the French "plunged into the blood of their kinsmen" [Léry, 133]). But Léry, who was an observer, and Lafitau's other main source, Hans Staden, who had himself been captured and prepared for eventual sacrifice by the Tupinamba, both concentrate especially on the rhetorical battle between prisoner and captors, and on the details of the *post*sacrificial barbecue: "He who has to kill the captive again takes the club, and then says, 'Yes, here I am! I will kill thee, for thine have also killed and eaten many of my friends.' Answers he, 'When I am dead, I shall yet have many friends, who will revenge me well'" (Staden, 158). "Now after all the pieces of the body, including the guts, have been thoroughly cleaned, they are immediately put on the *boucans*" (Léry, 126).

sulting "preludes" and "scenes" of his "spectacles" and "tragedies" play as a picture of a people martyred as well as martyring (Lafitau doesn't mention any Iroquois torture of Europeans, including that of one of his predecessors and chief sources, Father Brébeuf, whose notorious death by torture entered the epic Matter of Canada).[35] Perhaps paradoxically, these "scenes" are presented as scenes of a people acting—in costume and makeup, performing. The object-ivity of Lafitau's account of a culture he had lived amidst and of kinds of experience he must actually have had, or witnessed, tends to empty both Self and Other of individual and interior dimensions. "Lafitau" the narrator is a patchwork of paraphrase, extended quotation, and generalized experience; his characters have become the actors of rule-bound social paradigms (often manifested in such theatrical forms as dances, pantomimes, and pageants), possessed bodies represented as almost hyperbolically transient homes for a variety of entities, spirits, and states. The ceremonial lifeways rendered prominent by such an impersonal and dehistoricized account are easy to confuse with the fictions of theater, and Lafitau's dependable metaphors are mostly theatrical. "Scenes" of torture, however, challenge the work's ruling imperative of impersonality, the fiction of automatons in motion observed by science. "What can be more intimate than pain?" asks Descartes (so admired by Lafitau) in his final meditation (92). And Lafitau rehearses pain for us in a way that at once demonstrates the vacancy or dissociation of the Iroquois interior and also makes present to us through our own sympathetic or mimetic pain that the *sauvage* does, after all, *have* an interior.

The *sauvage* has an interior to the extent that we readers do, and our own is made self-conscious in the act of reading graphic representations of torment. This kind of mutual dependence reaffirmed through pain has a long history in the analysis of colonial or other oppressive social relations, from Hegel to Homi Bhabha. Claude Reichler has pointed out, in an essay on the figure of the circle in Lafitau's illustrations, that in Lafitau's book the descriptions of torture function to complete his representation of all-encompassing savage civility: "his analysis . . . includes the enemy, and the aggressiveness with regard to him in the wider social exchange, which war and vengeance constitute. . . . It is the circulation of violence reciprocally due and exercised which holds Lafitau's attention. . . . Across the care constantly

---

35. For a similar torture passage and metaphors of theatrical spectacle, see Baron de Lahontan, *Nouveaux voyages . . . en l'Amérique septentrionale* (1703), vol. 1, letter 23, or the same letter in the expanded English edition (*New Voyages* [1703]).

Fig. 27. "Tortures." Plate 14 from Lafitau's *Moeurs des sauvages amériquains*, vol. 1. By permission of the Houghton Library, Harvard University.

manifested by the Savages to integrate the dead with life, to preserve the memory of the deceased, he exposes yet one more time the question of social unity" (Reichler, "La cité sauvage," 70). See Lafitau's plate "Tortures" (figure 27) for a good example.

Reichler's sense of Lafitau as imposing a perfectly ordered image of savage civility on "our Indians" ("before the alteration," in a time before time) is close to mine of his production of "one world." An intelligent analyst does not make one world out of a multiplicity by *fiat* or even by rhetoric: there must be dynamics, a sort of cultural ecosystem at work. Reichler sees Lafitau's encapsulation of Iroquois culture and social life in these pictorial and logical circular forms as rendering their culture "timeless," perfect—and therefore, dynamics or no, both static and dead.

But as we have seen above, passages of torture can operate as a channel for empathy (benign or not)—literally as *passages*, in this case between the European reader's world of books and historical time, on the one hand, and the imagined American world of circular self-sufficiency and the utopian prehistorical, on the other. Such a passage was nowhere in sight in either Thevet's or Hariot's accounts of American life, nor could it be included in the forms of natural philosophy and history struggling to be born in the interim, which may be one reason why it is so structurally and narratively dramatized in Margaret Cavendish's oppositional *Blazing-World*. Of course the sine qua non of planetary voyage narratives is the effecting of the passage (which affords the climax of the film *2001*), but so far only Kepler and Cavendish have represented it as an interior or subjective motion: in Kepler's case a kind of reading or listening, in Cavendish's one might say an effect of writing, in both at any rate a transaction between two minds.

Of course Behn too uses the readerly experience of vicarious pain and also of terror (an emotion unreported and even denied in Lafitau's prose) to prove the reality of the Other.[36] But Behn's entire Surinam is one of terror and pain—even the nature walks that serve her tormented characters as a kind of scientific relief produce alarming encounters with tigers, "Numb-eels" and reputedly dangerous Indians (who *do* in the end turn out to be dismemberers, as do the Europeans and the Africans). Surinam is only visually and only in some places Edenic. The "spectacle of a mangl'd King" with which the novel ends renders its "other world" a dreamlike reconstruction of the royalist author's own Europe, though that is not to deny that the text with all its dreamy involutions offers a representation of "America." Behn's America is of the near present, located in historical time rather than "before the alteration." The passage to it is indeed, for Europeans and Africans both in different ways, a torturous one, and the torture never ends: when the Americans appear out of the jungle they are lacy with self-mutilation, and their next and only other act is to dismember a footman of Behn's household and

---

36. It may convey, as well, the *difficulty* of the relation as experienced over the long term—Surinam and Quebec are established *settler* colonies.

nail the parts to a tree. This is an uncontainable other world, tied to England by a cord of blood, whose governor never arrives—the passage is too difficult and he dies on the way.

The colonialist missionary's America is and must be reachable, governable, and transformable, so Lafitau tries to give it to us as it was, self-governed and as yet untraumatized by invasion. But readers live in history and cannot have fellow-feeling for those who, like Hariot's stop-time Algonquin, paradisally do not. Pain must enter the picture, and images of domination. In Lafitau's accounts, ceremony dominates the natural spontaneity of pain, as the organized state and his systematic science require—even to the point of a scalded child's brave insistence on singing his "death song."[37] But the *copula* of pain guarantees that others are possible, *copulae* of faith and piety—both matters ultimately that require an interior from which to go forth. As a reading experience, Lafitau's work of both theory and data functions first to construct a set of *copulae* (its comparisons of past with present, Old World with New) within which ethnic difference can register without necessarily provoking opposition or fracture.

Lafitau's book is not only a "mirror for princes," then. It makes two facing mirrors of ancient Greece and contemporary Canada; it also makes of evangelical Catholicism a mirror for capitalist acquisition and colonial expansion, and of the Iroquois body a mirror for the colony—a location, a habitation, a place for converting raw materials into civilized products.[38] The most intense and ambiguous of the book's effects is finally the linking of bodies—of Iroquois to Huron, Iroquois (via music and dance) to French Christian, Iroquois and Huron to reader, witness to reader (or "virtual witness"). For Catholic Christians generally as well as for Descartes, spirits animate bodies and are dichotomous with them. Torture, such as that of the incarnate Christ at the pillar and on the cross, attempts to drive the "ghost" away so the wit-

---

37. "I have myself seen a child of five to six years whose body had been burned by a sad accident when boiling water was poured over him, who, every time that they dressed his wounds, sang his death song with unbelievable courage" (Lafitau 2:158 [280]).

38. Or most grimly, into an edible carcass: "Nevertheless, so that those who read these horrible things . . . may also think more carefully about the things that go on every day among us, over here . . . consider in all candor what our big usurers do, sucking blood and marrow, and eating everyone alive—widows, orphans, and other poor people" (Léry, 132–33). As the colony makes people into slaves, the metropole makes them metaphorical and even, Léry points out, literal food: "During the bloody tragedy that began in Paris on the twenty-fourth of August 1572 . . . among other acts too horrible to recount . . . the fat of human bodies (which, in ways more barbarous than those of the savages, were butchered at Lyon after being pulled out of the Saône)—was it not publically sold to the highest bidder?" (132).

nesses can eat the body. The representation that makes us most intimate with the Iroquois here also most deeply figures the relationship of the dominant colonizing culture to the one it has come to "alter." It figures as well the predicament of the anthropologist, who witnesses the death of the culture and then takes up the task of displaying the corpse. Lahontan, the urbane traveler (Keats's "sole self"), confesses his queasiness at what he has seen and told us: "nothing is more grating to a civil man, than that he is obliged to be a Witness of the Torments which this kind of Martyrs suffer" (*New Voyages* 1:270). But Lafitau the composite does not apologize for witnessing. It is his professional obligation, and he is only one of many. Here in the *Moeurs* we have as in a massive piece of amber the primal scene of power, helplessness, and intimacy, to be shrouded by the ever more systematic and objective anthropology of the colonizing nations but not quite forgotten.

The message of Lafitau's work, which tends to replace history with comparative ethnology, is not only that all men are brothers in the worship of God, and therefore that all cultures are affiliated. The work bears as well a more unsettled and unsettling message: that the knowledge of other persons, other villages, other worlds is gained, like knowledge of God, through agonies of subjection and even death, at the cost of disintegration or digestion. No wonder the anthropologist of One World separates out and then loses the historical manuscript on the Iroquois' "alteration," and no wonder the historical message of colonial erasure is nonetheless reinscribed in his choreography of the world's customs. Perhaps only a Catholic apostle could have borne the implications of the science, back when it was being carved from a wilderness of flesh.

PSALMANAZAR, FORMOSA, AND THE OTHER SELF

Lafitau's magisterial tomes have had an uneven *Nachleben*, but he is remembered as a "father" of anthropology, an Enlightenment intellectual, a man of worth, and, despite his unfashionable religious concerns, a man of his time. "George Psalmanazar" (1679–1763), an imposter as Behn was during her espionage in Holland and elsewhere (and like her, of unknown parentage), was the perfect alternative. Also an intellectual, who worked on many large projects during what became a long writing life as a hack, and educated as Lafitau probably was by Jesuits in France, this wandering blonde Protestant (or so says Frank Lestringant) made his name, literally, as the pseudonymous author of an almost completely false account of the supposedly Japanese island of Formosa (Taiwan), supposedly his native home, from which he

Fig. 28. Formosan fashions. From George Psalmanazar's *Historical and Geographical Description of Formosa* (London, 1704). By permission of the Houghton Library, Harvard University.

had been seduced by a Jesuit tutor who wanted to bring him to France and convert him.[39]

This "false" *Historical and Geographical Description of Formosa* (1704) offers a picture of a single island's culture—and its recent historical relation to European cultural invasion—unlike Lafitau's system, which unifies all the world's known cultures, past and present, and occludes the historical present of European colonial relations in which its author participates so actively. It claims to be eyewitness and indeed native testimony (where it is not simply plagiarized from the seventeenth-century account of George Candidius, and even where it is plagiarized),[40] and it presents the point of view of a "pagan" who is abducted, physically and culturally, by European visitors engaged in attempts at cultural invasion. The first edition presents a long autobiographical account of the "passage" first and the description of Formosa second, in the pattern of, say, Sagard's *Long Journey*; the next edition (1705) reverses this order as well as hyperbolizing the most exotic and sensational parts of the "Description" (it is from Psalmanazar's description in the second edition of fattening crucified Formosan virgins and selling their flesh as meat that Swift may first have gotten the germ of his *Modest Proposal*, which alludes to Psalmanazar on this topic).

The work is full of violence, even in the first edition: the "Description" proper ends with an account and explanation of the Formosan massacre of Jesuits and Christians in the mid-seventeenth century as the by-product of a manipulative Jesuit political strategy. But it is violence to and violence caused by Jesuit missionaries, including the kidnapping of the narrator himself. The aggressive desire to alter, even if only by force, the cultural world of the non-European is held up to a harsh light and shown to backfire in the ways it in fact *does* backfire. Psalmanazar's pièce de résistance is his representation of the one truly revolting aspect of the mostly admirable Formosan culture as a liturgical cannibalism managed by the Priests or, as Lestringant's reading of the work makes clear, its Eucharistic feasts. This particular estranging device has been seen before in European literature, but never to my knowledge put to the use of mocking the Eucharist (it usually serves to *deflect*

---

39. See Lestringant, "Travels in Eucharista." Among Psalmanazar's later writings were the entry on Formosa in Emanuel Bowen's *Complete System of Geography* (1747) and the section on ancient history, particularly the history of the Jews, in Archibald Bower et al., *Universal History* (1736–1744).

40. Psalmanazar's sponsor in his deceptions, Alexander Innes, provided him with "A Short Account of the Island of Formosa" by Candidius, a mid-seventeenth century Dutch missionary. The account had just appeared in English in Awnsham and John Churchill's *Voyages and Travels* (1704) and in the longer Latin compendium of Bernhardus Varenius, *Descriptio regni Japoniae et Siam* (Amsterdam, 1673).

xenophobic reactions to foreign rites of burial and mourning, as in Sir John Mandeville's Far Eastern pagans, who honor their dead fathers by ritually drinking from cups made of their skulls).

Like those of More's *Utopia* or Lafitau's *Moeurs*, the (little) world of Psalmanazar's book offers both alternative and fun-house mirror to its culturally insular readers (French and Dutch as well as English). In that, it belongs simply to a genre, one more brilliantly exploited and historicized twenty years later by Jonathan Swift (whose map of Gulliver's third voyage overlaps along the eastern border with Psalmanazar's illustrative map of Formosa). But as a partisan and oppositional spokesman for the story the present book has been suggesting, the story of the conceptual and aesthetic containment of wonder by a civilization ever more rational, utilitarian, and knowing (and ever more hungry for "diversion"), George Psalmanazar can be seen as up to much more than satirical utopia, orientalism, or "the world upside down."

Despite Lestringant's expert arguments, I doubt that the man was a Protestant whose false *Description* begins with a false memoir mostly consisting in an outline of Christian (Anglican) doctrine he alleges himself to be newly espousing (after a redundant baptism in Scotland, from an army chaplain who believed his claim to be a Formosan). I doubt he was any kind of believing Christian at all. But whether he was or wasn't, he was certainly conscious of what kind of grand impostures and lies he was attaching to his exposition of Anglican doctrine, and he must have had at least some knowing readers in mind: as commentators have noticed, the work is filled with clues and hidden jokes.[41] Not only is the book blasphemous, but it exposes for that knowing reader the spuriousness of the religious and scientific legitimations of colonial appropriation. Missionary work, which could enculturate prospective native workers and often even train them, and the gathering of scientific knowledge (eventually institutionalized as a separate mission in such organizations as the Royal Geographical Society, but usually carried out by missionaries, military officers, and attachés) were both describable to the culture of enlightenment as modes of enlightening, for Europeans and non-Europeans alike, rather than as facilitators and covers for colonialist takeover. Psalmanazar's knowledge was false, his conver-

---

41. Beginning with the second sentence of the preface: "But when I had met with so many Romantic Stories of all those remote Eastern Countries, especially of my own, which had been impos'd upon you as undoubted Truths, and universally believed, then I was much discourag'd from proceeding in my Description of it; yet since Truth ought to dispell these Clouds of fabulous Reports . . . I thought myself indispensably oblig'd to give you a more faithful History of the Isle of *Formosa*, than as yet you have met with."

sion was false, and his doctrinal text was breathtakingly inauthentic and insincere.

On the other hand, his faith in the power of language to actualize alternative realities—such as his identity, or his "Formosa"—is correspondingly powerful.⁴² As well it might be. Even after his fall from credibility, he was hired to write the entry on Formosa in Emanuel Bowen's *Complete System of Geography* (1747), and as late as 1887, after he had become the archetype of Fraud, a lengthy account in *The Journal of the Royal Asiatic Society* of some newly found Formosan manuscripts suggested that Psalmanazar's Formosan alphabet had been accurate.⁴³ That faith in and love of language as a vehicle of transport produces the most charming of all phenomena proferred in the *Description*, the imaginary alphabet (figure 29) and language of Formosa, as well as the pictographic language of the "Formosan" coins and even the island's fashions (figure 28). What does it mean, in 1704, to translate the Lord's Prayer into a self-invented imaginary language (figure 30)?

I cannot do justice here to the complexities of Psalmanazar's literary (and social) performance. But we can at least briefly consider some answers to this question. The Lord's Prayer was one of the primary translation texts (and still is, in teaching medieval languages) for learners of new languages and for missionary teachers. It made sense in Psalmanazar's context, as "himself" a pagan and non-European who had been subject to missionary tutelage, in the French language as well as in the Christian faith. But the writer behind "Psalmanazar," raised in Catholic France, whatever his religious identity by the time of writing, must have felt attached to these words an aura of sacred authority: the prayer was first uttered, after all, by Jesus Christ and had since become so talismanic that saying it backwards was understood to be the actionable practice of witches and satanists. To translate it into nonsense syllables is an act perilously close to saying it backwards. On the other

---

42. In Susan Stewart's study of his fictional ethnography as a clarifying special case of ethnography in general, she points out that "the function of a description in the confessional mode is not the replication of an 'outside,' an objective world unaffected by authorial consciousness, but rather the invention of the speaking subject as the location of veracity" ("Antipodal Expectations," 67).

43. See the remarkable article of Terrien de Lacouperie, "Professor of Indo-Chinese Philology, University College, London" in the *Journal of the Royal Asiatic Society*: an article written "at the suggestion of my excellent friend, the much-venerated scholar Col. Henry Yule . . . after he had received from our common friend, E. Colborne Baber, H. B. M. Consular Service in China, then at Se-ul in Corea, nine sheets of manuscript secured from Formosa" (413). My reading of Psalmanazar suggests that he laughed long in his grave with satisfaction at the news of this publication.

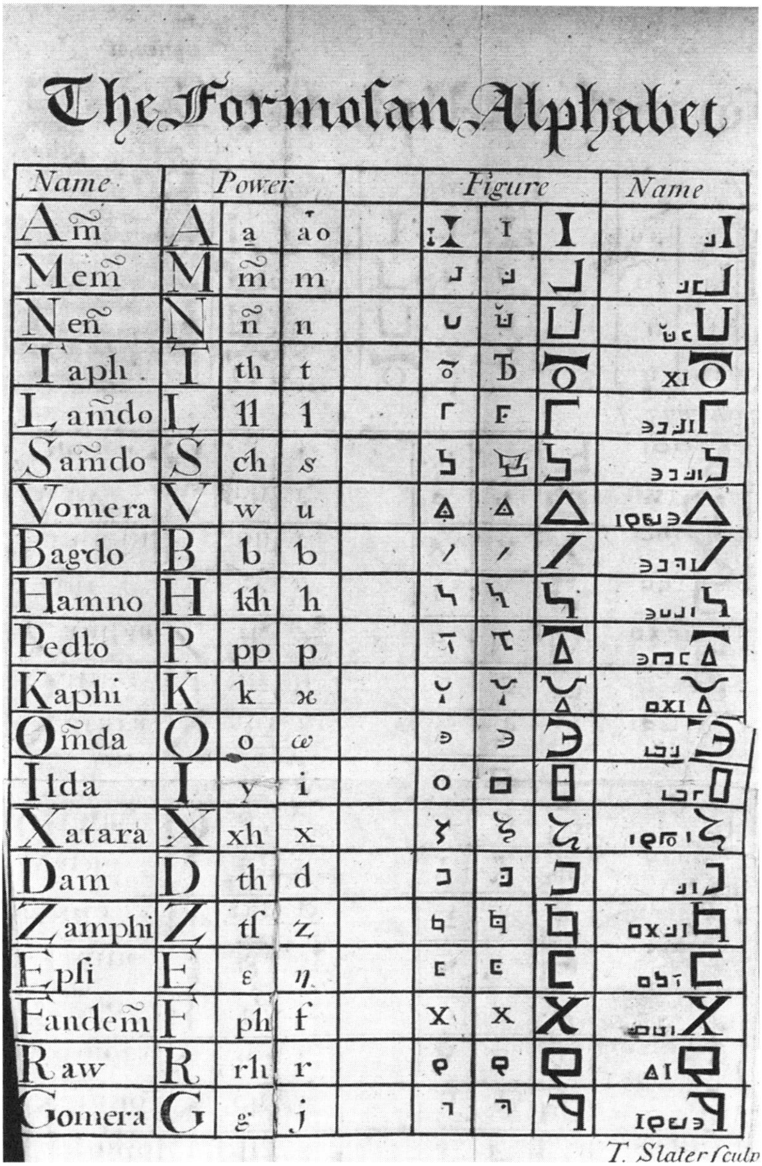

The Formosan Alphabet

Name		Power			Figure		Name
Am	A	a	ao	ɪX	I	I	ᴊI
Mem	M	m̃	m	˩	ᴊ	ᴸ	ᴊᴄᴸ
Nen	N	ñ	n	ᴜ	ŭ	ᴜ	ŭᴄᴜ
Taph	T	th	t	ᴛ̃	ᴛ̃	O	xɪO
Lamdo	L	ll	l	ᴦ	ꜰ	ᴦ	ɘᴊɪᴦ
Samdo	S	ch	s	Ⴢ	Ⴢ	Ⴢ	ɘᴊɪᴢ
Vomera	V	w	u	△	△	△	ɪᴏᴜɘ△
Bagdo	B	b	b	/	/	/	ɘᴊɪ/
Hamno	H	kh	h	ꝗ	ꝗ	ꝗ	ɘᴜɪꝗ
Pedlo	P	pp	p	π	π	ᴀ	ɘᴨᴄᴀ
Kaphi	K	k	x	⋎	⋎	⋎	ᴅxɪ⋎
Omda	O	o	ω	Ɔ	Ɔ	Ɔ	ᴌᴊɪƆ
Ilda	I	y	i	ᴏ	▫	ᴏ	ᴌᴊɪᴏ
Xatara	X	xh	x	Ⴣ	Ⴣ	Ⴣ	ɪɘᴏɪႣ
Dam	D	th	d	ᴝ	ᴝ	ᴝ	ᴊɪᴝ
Zamphi	Z	tf	z	ᴃ	ᴃ	ᴃ	ᴅxᴊɪᴃ
Epsi	E	ε	η	ᴇ	ᴇ	ᴇ	ᴏʙɪᴇ
Fandem	F	ph	f	X	X	X	ɘᴜɪX
Raw	R	rh	r	ꝗ	ꝗ	ꝗ	ᴀɪꝗ
Gomera	G	g	j	ᴎ	ᴎ	ᴎ	ɪɘᴜɘᴎ

T. Slater sculp.

Fig. 29 The Formosan alphabet. From Psalmanazar's *Description of Formosa*. By permission of the Houghton Library, Harvard University.

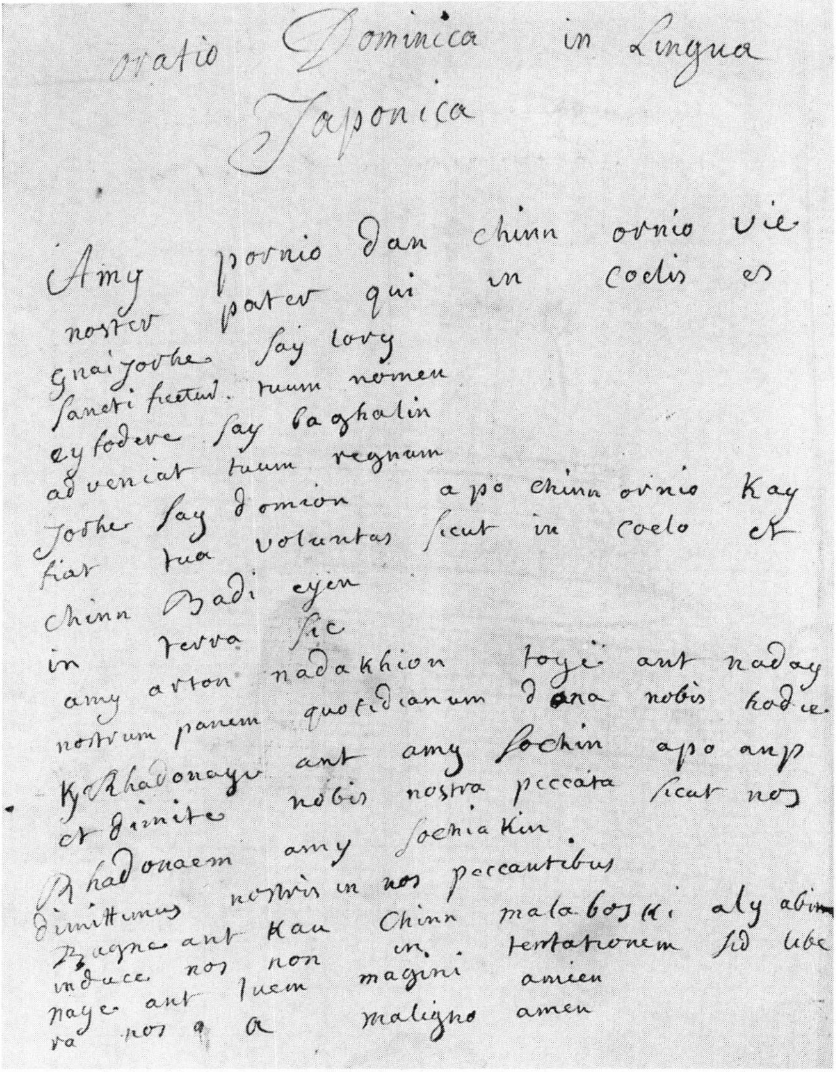
oratio Dominica in Lingua Japonica

Amy pornio dan chinn ornio vie
noster pater qui in coelis es
Gnaijorhe sai lory
sanctificetur tuum nomen
eyfodere sai baghalin
adveniat tuum regnum
Jorhe sai domion apo chinn ornio kay
fiat tua voluntas sicut in coelo et
chinn badi eyen
in terra sic
amy arton nadakchion toye ant naday
nostrum panem quotidianum da nobis hodie.
kykhadonaye ant amy sochin apo anp
et dimitte nobis nostra peccata sicut nos
khadonaem amy sochiakin
dimittimus nostris in nos peccantibus
bagnae ant kau chinn malabosky aly abim
induce nos non in tentationem sed libe
naye ant luem magini amien
ra nos a maligno amen

Fig. 30 Psalmanazar's manuscript copy of the Lord's Prayer in Formosan. By permission of The Archbishop of Canterbury and the Trustees of Lambeth Palace Library. MS 954, item 97, *oratio Dominica in Lingua Japonica*.

hand, to create an internally consistent language of one's own, and subsume this sacred utterance within the published texts of that language, is to make of one's imaginary language a vehicle capable of powers beyond denotation, a *parallel* and not simply mimic reality.

Finally, if crazily, the act makes of the pseudonymous writer (he kept the name Psalmanazar until his death, and beyond) one who has invented a full, and fully oppositional, world out of words (which he attempts to justify at real life dinner parties and as a teacher of "Formosan" at Oxford). Like Dickens or Tolstoy, in a way, he has imagined a world thickly, in all its constituent details. But like Swift, he has imagined a world outside of his own country, outside of Europe, and like Defoe, a world with an actual life of its own, unwritten and unimagined by the opportunist fictional exploiter. Psalmanazar has lied spectacularly and consciously about Taiwan; Defoe, incidentally and perhaps unknowingly, about Trinidad and Tobago (and China). And Psalmanazar has gone a step beyond any of these writers in sealing himself and his fortunes into that character and that imaginary culture: to admit to his imposture would be also to admit to blasphemy or impiety of a shocking elaborateness. So his alphabet and language and imaginary country go publicly unbetrayed until his death: what does it mean, then, that he *does* betray them posthumously, in a memoir concealed until his will revealed it and requested its publication?

Such powers of language to be fictional and still constituent of shared realities were not George Psalmanazar's invention from whole cloth, and he was not a literary genius. But it is part of our story that in the extremity and devotion of his demonstration of these powers he chose for its medium the fully imaginary representation of a distant and alien culture, and the fully imaginary empathy with the (always fully imaginary) Other instanced in his performance of Formosan identity, in life as well as letters. And people believed him.

Readers in England and France and Holland, one must conclude, were at least capable of confusing earnest and mendaciously imaginary accounts of distant places and cultures at this point in the still mutual developments of fiction and anthropology. Or rather, the efforts of imagination involved in the production as well as the reception of these texts were differentiated along some other axis than that to which a later period clings—in spite of docudrama, TV news, and virtual reality. Perhaps that axis ran between One World and Little World. We will end with a quick look at some smaller and smaller worlds, less and less believable, more and more wonderful: the microcosmic anthropologies of Robinson Crusoe, the Wild Boy of Aveyron, and the Princess Caraboo.

CODA
The Wild Child

It happened one day, about noon, going towards my boat, I was exceedingly surprised with the print of a man's naked foot on the shore, which was very plain to be seen in the sand. I stood like one thunderstruck, or as if I had seen an apparition. I listened, I looked around me, I could hear nothing, nor see anything. . . . How it came thither I knew not, nor could in the least imagine. But after innumerable fluttering thoughts, like a man perfectly confused and out of myself, I came home to my fortifications, not feeling as we say, the ground I went on, but terrified to the last degree . . . mistaking every bush and tree, and fancying every stump at a distance to be a man.

— Daniel Defoe, *Robinson Crusoe* (1719)

The boy was about four and a half feet tall and his knee ligaments were normal. He murmured while he ate, was subject to sudden fits of anger, liked fires, slept according to the sun, tried constantly to escape and could not understand his reflection in the mirror — he always looked for the person he was sure was hiding behind the glass.

. . . He loved playing in the snow and enjoyed gazing Narcissus-like into the still waters of the pond. At night he would stare long and admiringly at the moon. He was incapable of imitation.

— Lucien Malson, *Wolf Children and the Problem of Human Nature* (1972)

> Did some philosophic analysis draw
> Her component degrees from *some hot-water spa*?
> Did some chemical process occasion her birth?
> Did *galvanic* experiments bring her on earth?
> . . .
> Astronomers sage may exhibit her soon
> As daughter-in-law to the man in the moon.

— "Caraboo," *The Bath Herald*, June 10, 1817

DEFOE'S ROBINSON CRUSOE, solitary on his distant American island, dressed in animal skins like the iconographic Wild Man of European folklore, terrified of other human beings and restlessly escaping all nets of social or cultural existence (over three volumes and at least fifty years), is nonetheless Defoe's thought experiment, designed to show that Man is Man, a producer of "manners and customs" and technology, that the minimum unit of culture is the solitary castaway. Alexander Selkirk, whose frightening story served to inspire Defoe's extended fiction, lost the power of speech after a single year alone on such an island. At the end of his sojourn of "twenty-eight years, two months, and nineteen days" (311), by contrast, Crusoe discovers he has "kept a true reckoning of years" and, though he had lost exact track of days during an illness, still "kept the anniversary of my landing here."[1] His successful conversion of Friday to Protestant Christianity registers the degree to which he was still a creature of discourse, and in the increasingly populous and multilingual final years of his "captivity" he is amused but also satisfied to consider "how like a king I looked. First of all, the whole country was my own mere property, so that I had an undoubted right of dominion. Secondly, my people were perfectly subjected. I was absolute lord and lawgiver; they all owed their lives to me, and were ready to lay down their lives . . . for me. It was remarkable, too, we had but three subjects, and they were of three different religions. My man Friday was a Protestant, his father was a Pagan and a cannibal, and the Spaniard was a Papist. However, I allowed liberty of conscience throughout my dominions" (269).

Crusoe's narrative reveals itself to be not only the science fiction of its long passages of technical invention but utopia—a strange utopia, with no inhabitants, painfully increased toward the end to an ecumenical community of four. Without the others, Crusoe manages to piece together a culture for himself, complete with regulations and observances—and strong fortifications (as Thevet was wont to recommend). With them, he instantly constructs a strong monarchy. Throughout the text, his income and his capital increase, both back in Brazil where he co-owns a plantation and on his "own" island. The obsession with culture and the aversion toward other people (especially people of other cultures—"pagans" and "cannibals") makes for a chaotic ambivalence, echoed in the peculiar fit of his "spectre-like" Wild Man looks (283) with his capacity to invent porcelain on a desert island and keep a calendar for twenty-eight years.[2]

In the year of *Robinson Crusoe*'s publication, two feral children were discovered in the forests of the Pyrénées; Linnaeus recorded them ("Pueri 2

---

1. I quote from George A. Aitkin's edition of Defoe's works, *Romances and Narratives*, vol. 1.
2. In "Robinson Crusoe's Earthenware Pot," Lydia Liu has uncovered, through a Chinese translation of *Robinson Crusoe*, the science fiction of his largely accidental invention of pottery as a technically unknown (to the English at the time) but exact representation of one of the most sought after industrial secrets of the century. Her essay is fascinating on Defoe in relation to China, realism, and "the poetics of colonial disavowal."

Pyrenaici. 1719") in creating his category of *Homo sapiens ferus* under the genus *Primates* (*Systema Naturae*, B2ᵛ). Lucien Malson's "List of Recorded Cases" includes only two feral children discovered before 1672 and no "detailed reports" dated earlier; it seems to become a newly salient phenomenon in the early stirrings of the Enlightenment. Defoe's choice to bring his fictional Wild Man home through the Pyrénées and to stage his final adventure as a face-off between Crusoe and three wolf packs acquires resonance from these reports of wolf-children in the same area; Defoe's book has to dramatize its protagonist's difference from the wild. Despite the clear difference already communicated through Crusoe's calendars, account books, farms, and fortifications, the solitary man in skins is a signal of the potentially acultural and profoundly antisocial longings of *Homo acquirens*, or of the peculiar psychological contradictions between the isolated, antisocial miser and the cash *nexus* in which he acquires his wealth, between the homebody values of the bourgeois and the reckless nomadism of the (ad)venture capitalist.

Defoe's island is very like Godwin's St. Helena, and his character a *picaró* in (initially at least) middle-class clothing. But it does not serve as a station on the way to a larger inhabited world like the Moon (though in the second volume Crusoe does spend time, as Domingo had, in a despised China). Its wonders, emphatically so labeled ("my story as a whole is a collection of wonders" [288]), are always the wonders of Crusoe's own exertions and individual experiences. The compact power of this novel comes in part from its simplicity of pattern: all dimensions reduce to a single point. Crusoe is the Self as world, the Self as culture, the Self as source of all wonder; as subject but also object of his own gaze; the Self as Self and as Other. One might see in Crusoe's *Life and Surprizing Adventures* the assimilation into the interior, the individual, and the personal of all the other worlds and all the outward gazing encompassed in this book.

This does not mean, clearly, that the issues engaged or the disciplines emergent have faded away. But such objects of wonder and interest as the nature of human culture and the "human race," Linnaeus's *Homo sapiens*, came to be examined by means of the anomalous individual at home as much as the exotic collective far away; and the Romantic exotic is one of individual Self or Other, privately and even intimately experienced. George Psalmanazar *was* Formosa for a while. Crusoe was the culture and society of Tobago, and his scientific observations and investigations were entirely self-centered and utilitarian. When "culture" comes, as in Lafitau's hands, to be a single phenomenon postulated of a single race or species, it can then be investigated in the negative form of close analysis of a feral child—ethnography in the singular.

Jean Marc Itard's case study (1801) of the "wild boy" of Aveyron is an important text in the development of the modern sciences of culture.³ Like *Robinson Crusoe*, it focuses on the daily and bodily experiences of a restless, insatiable, alienated solitary with no apparent sexual instinct. It, too, glorifies ambivalently its liberty-loving protagonist (a being as averse to Western-style clothing as Crusoe became after his island life). Itard's close observations of Victor, though not *self*-observations like Crusoe's, offer similar joys of detail and fine articulation of daily or unconscious activity—in a link with Hooke's depiction of the invisible domestic. And Itard, like Defoe, is interested in the concepts of culture and of human nature. Both writers make their savage but European protagonists—little worlds themselves—objects of wonder, and objects of science.

With regard to arguments, Itard's work comes to a significantly different conclusion from Defoe's. His "wild boy" is like Selkirk, the feral castaway of real life. Despite Victor's love for the moon, and for gazing into still water ("I saw new worlds beneath the water lie, / New people, and another sky / And sun"),⁴ he is made to represent the pathological and desolate anomaly of the human without culture, who cannot speak except to utter the syllable of wonder: "Oh!"⁵ Victor has none of the energy, invention, or (imaginary) self-containment of Crusoe. But then, he is *truly* captive, in the power of an entire interlinked and mechanically sophisticated fortification, called Paris, or "medicine," or "anthropology," or the Imperial Institute of Deaf-Mutes, of which Itard was the chief physician. He tries to escape but he is outnumbered. Civilization is his desert island, and the moon and still water the other worlds he needs to survive.

Itard's social constructivism was an idea with a noble and liberatory future; from the beginning he saw its power to dispel cultural narcissism and contempt for difference. But one feels for the little boy with the big scar on

---

3. It features crucially in Christopher Herbert's intelligently but severely disenchanted overview of the cultural roots of modern anthropology, chap. 1 ("Original Sin and Anomie") of *Culture and Anomie: Ethnographic Imagination in the Nineteenth Century*. I will quote both Itard and Malson from Malson, *Wolf Children and the Problem of Human Nature*, which includes the complete text of the 1802 translation of Itard's report, "Of the First Developments of the Young Savage of Aveyron" (1801).
4. Thomas Traherne, "Leaping over the Moon," a mid-seventeenth-century poem first found and published in 1910.
5. Itard cannot move Victor from that exclamation into articulate speech, which he tries ingeniously to do though the coincidental sound of the French word for water: "In vain, even at those moments when his thirst was most intolerable, did I frequently exclaim *eau, eau*, bringing before him a glass of water" (121).

his neck, who ran to the window in a country house "with a view, if it were open, of escaping . . . or, if it were not, to contemplate . . . all those objects towards which he was irresistibly attracted, by recent habits, and, perhaps, also by the remembrance of a life independent, happy, and regretted" (Itard, 115). Itard could not cultivate Victor, but he could not set him free. For Itard, there is no other world.

But what the sciences began to reject was still open for the alienated solitary soul with an imagination. Our final castaway is the wondrous young criminal of the laboring class, born in Devonshire three or four years after Victor, the Princess Caraboo, alias Mary Baker, Mary Willcocks, etc. This out-of-luck vagrant, born a poor farmer's daughter, taught to read as a domestic in London, "ruined" by a sailor from the South Seas (who got her pregnant and then disappeared), manages to fuse in one illustrious and admirable person Defoe's sense of the person as miniature culture, and Victor's experience as captive stranger and scientific object. She is a nearly literal fulfillment of John Donne's metaphoric self, a "little world, made cunningly / Of elements, and an angelic sprite" (*Holy Sonnets*, no. 16). Taking a leaf, perhaps, from Psalmanazar's book, she appeared in a village near Bristol and Bath in exotic dress and not speaking a word of English, and she was taken in as a kidnapped and abandoned foreigner by the softhearted wife of a wealthy Bristol banker, Mrs. Worralls of Knole Park. For about two months the Oriental "Princess Caraboo" of "Javasu" held court at the Worralls's house; she was entertaining and (chastely) enticing to linguists and Orientalists who tried to trace the etymologies of her self-created language and to place the island of Javasu on the map of the known world.

Like Crusoe, Caraboo had her religious observances and customs, her homemade fashions, her alphabet, and her language no one could share (one of the few words the linguists decided on a meaning for was "Buis or Bugos, any Wild People").[6] It was an internally consistent cultural world. She was watched closely by Mrs. Worralls and the townspeople of Bath, as well as the curious professors and virtuosi. But they wanted to believe her, and she made no major slip in her exuberant performance until she was caught while writing something in her make-believe script: she instantly corrected herself in response to a visiting philologist's overheard remark that Oriental writing

---

6. A number of materials — alphabets, word lists, newspaper articles, and letters about Caraboo and her strange language — are brought together along with a narrative of her life and of the events in Bristol ("compiled in great part from Conversation" and "in many instances . . . printed *verbatim*, as it was spoken or dictated by the party") by the printer J. M. Gutch in *Caraboo: A Narrative of a Singular Imposition* (1817).

generally moves from the right to the left of the page.[7] Things soon began to unravel as philological inquiries drew the notice of one of Mary Wilcocks Baker's former employers, and the True History was out soon enough. Not a princess, kidnapped and cast away on England's strange shore, but a *bricoleur*, like Lévi-Strauss. Time spent with the gypsies, her time with the South Seas sailor, her reading of illustrated travels in the houses where she'd worked had all contributed matter to her fancy. And when she was forced by her trespasses against propriety and property, and her resulting criminal vagrancy, to create a world in which she was safe and loved and neither a criminal nor a beggar, she made that world, as Nelson Goodman says all worlds are made, "from other worlds." Her punishment was deportation to America.

Caraboo's inner world, with all its alluring qualities of outerness, distance, otherness, fullness, was richer than the undiscovered world of Victor. If the Wild Boy's resistance held out an irreducible darkness to those who were beginning to tire of light, Caraboo held out more: an alphabet, in which some reader may begin to form, as Margaret Cavendish instructed us, "a World of her own."

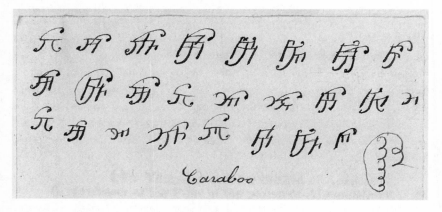

Fig. 31. Manuscript copy of the alphabet of "Javasu," probably by Caraboo (c.1817). By permission of the Houghton Library, Harvard University.

---

7. For an account of the triumphant career of Mary Wilcocks Baker, with fascinating attention to the ways in which what might have been dangerous slips were smoothed away by the silent collusion of Mary's fellow servants and "foreigners," see Anne Janowitz, "Caraboo: A Singular Imposition."

WORKS CITED

Unless otherwise indicated, early modern printed works in English are published in London; works in French are published in Paris.

Abbot, George. *A Brief Description of the Whole World*. 1599.
———. *The Reasons Which Doctor Hill Brought, for the Upholding of the Papistry*. 1604.
Abler, Thomas S. "Iroquois Cannibalism: Fact or Fiction." *Ethnohistory* 27 (4) (Fall 1980): 309–16.
Académie des Sciences. *Journal des Sçavans*. Paris: Iean Cusson, 1665–.
Ackerman, James. "Artists in Renaissance Science." In *Science and the Arts in the Renaissance*, ed. John W. Shirley and F. David Hoeniger. Washington, D.C.: Folger Shakespeare Library, 1985.
Acosta, Jose d'. *Historia Natural y Moral de las Indias*. Seville, 1590.
Adams, Percy G. *Travel Literature and the Evolution of the Novel*. Lexington: University Press of Kentucky, 1983.
Agricola, Georg. *De re metallica*. Basel, 1556.
Albanese, Denise. "The *New Atlantis* and the Uses of Utopia." *English Literary History* 57 (1990): 503–28.
———. *New Science, New World*. Durham, Duke University Press, 1997.
Alcover, Madeleine. *L'Autre monde ou les estats et empires de la lune*. Paris: Librairie Honoré Champion, 1977.
Alden, John, ed. *European Americana: A Chronological Guide to Works Printed in Europe Relating to the Americas, 1493–1776*. 2 vols. New York: Readex Books, 1980.
Alexander, Elizabeth. *The Venus Hottentot*. Charlottesville: University Press of Virginia, 1990.
Allen, Don Cameron. *The Legend of Noah: Renaissance Rationalism in Art, Science, and Letters*. Illinois Studies in Language and Literature, vol. 33, nos. 3–4. Urbana: University of Illinois Press, 1949.
Allen, Phyllis. "Scientific Studies in the English Universities of the Seventeenth Century." *Journal of the History of Ideas* 10 (2) (April 1949): 219–53.

Alpers, Svetlana. *The Art of Describing: Dutch Art in the Seventeenth Century.* Chicago: University of Chicago Press, 1983.

Ambassades du Roy de Siam envoyé à la Haye le 10. Septemb. 1608. The Hague, 1608.

American Heritage Dictionary. Ed. William Morris. Boston: American Heritage and Houghton Mifflin, 1969.

Anderson, Benedict. *Imagined Communities: Reflections on the Origin and Spread of Nationalism.* Rev. ed. London: Verso, 1991.

Andrews, Richard. "The Cunning of Imagery: Rhetoric and Ideology in Cesare Beccaria's Treatise *On Crimes and Punishments*." In *Begetting Images: Studies in the Art and Science of Symbol Production,* ed. Mary B[aine] Campbell and Mark Rollins. New Connections: Studies in Interdisciplinarity, vol. 2. New York: Peter Lang, 1989.

Anghiera, Pietro Martire d'. *The Decades of the newe worlde of west India* (see entry under Richard Eden).

Anonymous. *England's Vanity: or the Voice of God Against the Monstrous Sin of Pride in Dress and Apparel.* 1683.

Appadurai, Arjun. "Introduction: Place and Voice in Anthropological Theory." *Cultural Anthropology* 3 (1) (February 1988): 16–20.

———, ed. *The Social Life of Things: Commodities in Cultural Perspective.* Cambridge: Cambridge University Press, 1986.

Apter, Andrew. "Que Faire? Reconsidering Inventions of Africa." *Critical Inquiry* 19 (1) (Fall 1992): 87–104.

Arbeau, Thoinot [Jehan Tabourot]. *Orchésography.* 1588. Trans. Cyril W. Beaumont. 1925. Reprint, New York: Dance Horizons, n.d.

Arens, William. *The Man-Eating Myth: Anthropology and Anthropophagy.* New York: Oxford University Press, 1979.

Aretino, Pietro. *Aretino's Dialogues.* Trans. Raymond Rosenthal, with an epilogue by Margaret Rosenthal. Rev. ed. New York: Marsilio, 1994.

Aristotle. *Generation of Animals.* Trans. A. L. Peck. Cambridge: Harvard University Press, 1963.

Armstrong, Elizabeth. *Before Copyright: The French Book-Privilege System, 1498–1526.* Cambridge: Cambridge University Press, 1990.

Asad, Talal, ed. *Anthropology and the Colonial Encounter.* London: Ithaca Press, 1973.

Atkinson, Geoffroy. *The Extraordinary Voyage in French Literature.* 2 vols. 1920. Reprint, New York: Burt Franklin, n.d.

Aubrey, John. *Brief Lives, chiefly of Contemporaries, set down by John Aubrey between the years 1669 & 1696.* Ed. Andrew Clark. 2 vols. Oxford, 1898.

Axtell, James. "The Invasion Within: The Contest of Cultures in Colonial North America." In *The European and the Indian: Essays in the Ethnohistory of Colonial North America.* New York: Oxford University Press, 1981.

———. "The English Colonial Impact on Indian Culture." In *The European and the Indian.*

Bachelard, Gaston. *The Poetics of Space.* Trans. Maria Jolas. Boston: Beacon Press, 1969.

Bacon, Sir Francis. *The Advancement of Learning.* 1605. In *Works,* ed. Spedding et al. (see entry), vol. 3.

———. *The New Atlantis*. 1627. In *Works*, ed. Spedding et al. (see entry), vol. 3.
———. *New Organon and Related Writings*. Ed. Fulton H. Anderson. Indianapolis: Bobbs-Merrill Educational Publishing (Library of the Liberal Arts), 1960.
———. *Novum Organum*. 1620.
Ballaster, Ros. "New Hystericism: Aphra Behn's *Oroonoko*: The Body, the Text, and the Feminist Critic." In *New Feminist Discourses: Critical Essays on Theories and Texts*, ed. Isobel Armstrong. London: Routledge, 1992.
Barbeau, Marius. "How the Huron-Wyandot Language Was Saved from Oblivion." *Proceedings of the American Philosophical Society* 93 (3) (1949): 226–32.
Barthes, Roland. *The Fashion System*. Trans. Richard Howard and Matthew Weld. New York: Hill and Wang, 1983.
Battigelli, Anna. "Between the Glass and the Hand: The Eye in Margaret Cavendish's *Blazing World*." *1650–1850: Ideas, Aesthetics, and Inquiries in the Early Modern Era* 2 (1996): 25–38.
———. *Margaret Cavendish and the Exiles of the Mind*. Lexington: University of Kentucky Press, 1998.
———. "Political Thought/Political Action: Margaret Cavendish's Hobbesian Dilemma." In *Women Writers and the Early Modern Political Tradition*, ed. Hilda Smith. Cambridge: Cambridge University Press, 1998.
Behn, Aphra. *The Emperor of the Moon*. 1687.
———. *Oroonoko or, The Royal Slave. A True History*. 1688.
———. *The Works of Aphra Behn*. Ed. Janet Todd. 7 vols. Columbus: Ohio State University Press, 1995.
Behn, Aphra, trans. *A Discovery of New Worlds* [*Entretiens*], by Bernard le Bovier de Fontenelle. 1688.
———, trans. *The History of Oracles, and the Cheats of the pagan Priests* [*Histoire des Oracles*], by Bernard le Bovier de Fontenelle. 1688.
Bender, John. "Enlightenment Fiction and the Scientific Hypothesis." *Representations* 61 (Winter, 1998), 6–28.
Benjamin, Walter. "The Work of Art in the Age of Mechanical Reproduction." In *Illuminations*. Ed. Hannah Arendt. Trans. Harry Zohn. New York: Harcourt, Brace, and World, 1968.
Benzoni, Girolamo. *Historia del mondo nuevo*. Venice, 1565.
Berk, Philip R. "De la Mode: La Bruyère and the Myth of Order." *Actes de Davis: Papers on French Seventeenth-Century Literature*, ed. Claude Abraham, 17 (1988): 131–39.
Bernal, Martin. *Black Athena: The Afroasiatic Roots of Classical Civilization*. New Brunswick, N.J.: Rutgers University Press, 1987.
Bernheimer, Richard. *Wild Men in the Middle Ages*. Cambridge: Harvard University Press, 1952.
Biagioli, Mario. *Galileo, Courtier*. Chicago: University of Chicago Press, 1993.
———. "Galileo the Emblem Maker." *Isis* 81 (307) (1990): 230–58.
Biddick, Kathleen. "Humanist History and the Problem of Virtual Worlds." In *The Shock of Medievalism*. Durham: Duke University Press, 1998.

Bie, Cornelis de. *Het Gulden Cabinet van de Edel vry Schilderconst*. Antwerp, 1661.

Bierce, Ambrose. "An Occurrence at Owl Creek Bridge." In *The Complete Short Stories of Ambrose Bierce*, ed. Ernest Jerome Hopkins. 1970. Reprint, Lincoln: University of Nebraska Press, 1984.

Birrell, T. A. *The Library of John Morris: The Reconstruction of a Seventeenth-Century Collection*. London: British Museum Publications, for the British Library, 1976.

———. "Reading as Pastime: The Place of Light Literature in Some Gentlemen's Libraries of the 17th Century." In *Property of a Gentleman: The Formation, Organisation, and Dispersal of the Private Library, 1620–1920*, ed. Robin Myers and Michael Harris. Winchester, U.K.: St. Paul's Bibliographies, 1991.

Bishop, Elizabeth. "Crusoe in England." *Geography III*. New York: The Noonday Press, Farrar, Strauss, and Giroux, 1988.

Black, Dory. "Working Women's Writing in Early Modern England." B.A. Thesis. Brandeis University, 1997.

Blackwell, Thomas. *An Enquiry into the Life and Writings of Homer*. 1735.

Blaeu, Johannes [Joan]. *Atlas Major*. 1662–63. Reprint, New York: Rizzoli, 1990.

Bloom, Terrie. "Borrowed Perceptions: Harriot's Maps of the Moon." *Journal for the History of Astronomy* 9, pt. 2 (25) (1979): 117–22.

Blount, Thomas. *Glossographia*. 1656.

Boesky, Amy. *Founding Fictions: Utopias in Early Modern England*. Athens, Ga.: University of Georgia Press, 1996.

Bohrer, Frederick N. "The Times and Spaces of History: Representation, Assyria, and the British Museum." In *Museum Culture: Histories, Discourses, Spectacles*, ed. Daniel Sherman and Irit Rogoff. Minneapolis: University of Minnesota Press, 1994.

Boissard, Jean-Jacques. *Habitus variarum orbis gentium*. Mechlinburg, 1581.

Bonan, Gordon B. "Effects of Land Use on the Climate of the United States." *Climatic Change* 37 (1997): 449–84.

Boon, James. "Cosmopolitan Moments: Echoey Confessions of an Ethnographer-Tourist." In *Crossing Cultures: Essays in the Displacement of Western Civilization*, ed. Daniel Segal. Tucson: University of Arizona Press, 1992.

Boorsch, Suzanne. "America in Festival Presentations." In *First Images of America*, ed. Chiappelli (see entry), vol. 1.

Bora, Ranu. "Outing Texture." In *Novel Gazing*, ed. Sedgwick (see entry).

Bordo, Susan. *The Flight to Objectivity: Essays in Cartesianism and Culture*. Binghamton: SUNY Press, 1987.

Borel, Pierre. *Bibliotheca chimica. Seu Catalogus librorum philosophicorum hermeticorum*. Paris, 1654.

Bossy, John. *Giordano Bruno and the Embassy Affair*. New Haven: Yale University Press, 1991.

Bowen, Emanuel. *A Complete System of Geography*. 2 vols. 1747.

Bower, Archibald, et. al. *A Universal History from the Earliest Accounts of Time to the Present*. 7 vols. 1736–1744.

Bowerbank, Sylvia. "The Spider's Delight: Margaret Cavendish and the 'Female' Imagination." *English Literary Renaissance* 14 (3) (Autumn 1984): 392–408.

Bowerbank, Sylvia and Sara Mendelson. *Paper Bodies: A Margaret Cavendish Reader.* Peterborough, Canada: Broadview Press Literary Series, 1999.

Boyle, Robert. "General Heads for a Natural History of a Country." *Philosophical Transactions* 2 (1666): 186–89.

Brathwaite, Edward. "The English Gentleman" and "English Gentlewoman." In *Times Treasury, or, Academy for Gentry.* 1652.

Brébeuf, Jean de, Saint. "Relation de ce qui s'est passé aux Hurons, en l'année 1635." Embedded in the text "Le Jeune's Relation." In *Relations*, ed. Thwaites (see entry), vol. 8, pp 68–155.

Breiner, Laurence A. "Italic Calvino: The Place of the Emperor in *Invisible Cities*." *Modern Fiction Studies* 34 (4) (1988): 559–573.

Brown, Laura. *Ends of Empire: Women and Ideology in Early Eighteenth-Century English Literature.* Ithaca: Cornell University Press, 1993.

Browne, Janet. *The Secular Ark: Studies in the History of Biogeography.* New Haven: Yale University Press, 1983.

Browne, Thomas. *Hydrotaphia [Urn Burial]* and *The Garden of Cyrus.* 1658. In *Works*, ed. Keynes (see entry), vol. 1.

———. *Musaeum Clausum, or Bibliotheca Abscondita.* 1684. In *Works*, ed. Keynes (see entry), vol. 3.

———. *Pseudoxia Epidemica.* 1672. In *Works*, ed. Keynes (see entry), vol. 2.

Bruno, Giordano. *De l'infinito universo et mondi.* "Venetia" [London], 1584.

———. *De innumerabilis immenso et infigurabili; sue de universo et mundis libri octo*, with *De monade.* Frankfurt, 1591.

———. *Opere Italiane.* Ed. Giovanni Gentile. 3d ed. *Dialoghi Italiani*, rev. Giovanni Aquilecchia. Florence: Sansoni, 1958.

Brunt, Samuel. *A voyage to Cacklogallinia.* 1727. In *The Virgin-seducer*, etc., ed. Malcolm J. Bosse. Foundations of the Novel. New York: Garland, 1972.

Bry, Theodor de. *America, Part I.* Frankfurt, 1590.

Bry, Theodor de, and sons, eds. *Les grands voyages.* Frankfurt, 1590–1634.

Bucher, Bernadette. *Icon and Conquest: A Structural Analysis of de Bry's "Great Voyages."* Trans. Basia Miller. Chicago: University of Chicago Press, 1981.

Bulwer, John. *Anthropometamorphosis. Historically Presented, In the mad and cruel Gallantry, Foolish Bravery, ridiculous Beauty, Filthy Finenesse, and loathsome Lovelinesse of most Nations, Fashioning and altering their Bodies from the Mould intended by Nature.* 1650.

———. *Chirologia and Chironomia.* 1644. Ed. James W. Cleary. Carbondale: Southern Illinois University Press, 1974.

———. *Pathomyotomia or A dissection of the significative muscles of the affections of the minde.* 1649.

———. *Philocophus: or, The deaf and dumbe mans friend.* 1648.

———. *A View of the People of the Whole World, or, A Short Survey of their Policies, Dispositions, Naturall Deportments, Complexions, Ancient and Moderne Customes, Manners, Habits and Fashions.* Expanded ed. of *Anthropometamorphosis.* 1653 and 1654.

Burton, Robert. *The Anatomy of Melancholy: What It Is, with All Its Kinds, Causes, Symptomes, Prognostickes and Severall Cures of It.* 1620.

Bynum, Caroline Walker. "Presidential Address: Wonder." *American Historical Review* 102 (1) (February 1997): 1–26.

Campanella, Tommaso. *De civitate Solis*. Frankfurt, 1623.

Campbell, Elaine. "Aphra Behn's Surinam Interlude." *Kunapipi* 7 (2–3) (1985): 25–35.

Campbell, Mary B[aine]. *"Anthropometamorphosis*: John Bulwer's Monsters of Cosmetology and the Science of Culture." In *Monster Theory: Reading Culture*, ed. Jeffrey Jerome Cohen. Minneapolis: University of Minnesota Press, 1996.

——. "Carnal Knowledge: Fracastoro's *De syphilis* and the Discovery of the New World." In *Crossing Cultures: Essays in the Displacement of Western Civilization*, ed. Daniel Segal. Tucson: University of Arizona Press, 1991.

——. *The Witness and the Other World: European Travel Writing, 400–1200*. Ithaca: Cornell University Press, 1988.

Campeau, Lucien, ed. *Monumentae Novae Franciae*. Vol. 2. Quebec: Laval University Presses, 1979.

Cappetti, Carla. *Writing Chicago*. New York: Columbia University Press, 1992.

Carroll, Margaret D. "The Erotics of Absolutism: Rubens and the Mystification of Sexual Violence." *Representations* 25 (Spring 1989): 3–30.

Carter, Angela. *The Sadeian Woman and the Ideology of Pornography*. New York: Pantheon Books, 1978.

Caspar, Max. *Kepler*. Trans. C. Doris Hellman. 1959. Rev. Owen Gingerich and Alain Segonds. New York: Dover, 1993.

Cavendish, Margaret. *CCXI Sociable Letters*. 1664.

——. *The Description of a New World Called the Blazing-World*. 1666.

——. *Observations on Experimental Philosophy*. 1666 (bound with *Blazing-World*).

——. *Poems and Fancies*. 1653.

Certeau, Michel. "Ethno-Graphy, or The Speech of the Other." In *The Writing of History*. 1975. New York: Columbia University Press, 1988.

——. "Writing vs. Time: History and Anthropology in the Works of Lafitau." Trans. James Hovde. *Yale French Studies* 59 (1980): 37–64.

Chappuys, Gabriel. *L'estat, description et gouvernement des royaumes et républiques du monde, tant anciennes que modernes*. 1585.

Chartier, Roger. *The Cultural Uses of Print in Early Modern France*. Trans. Lydia G. Cochrane. Princeton: Princeton University Press, 1987.

Chiappelli, Fredi, ed. *First Images of America: The Impact of the New World on the Old*. 2 vols. Berkeley: University of California Press, 1976.

Chibka, Robert. "'Oh! Do Not Fear a Woman's Invention.'" *Texas Studies in Literature and Language* 30 (4) (1988): 510–37.

Chrétien de Troyes. *Lancelot, or the Knight of the Cart*. In *Arthurian Romances*, trans. William W. Kibler. Harmondsworth: Penguin Books, 1991.

Chrisman, Miriam Usher. *Lay Culture, Learned Culture, 1480–1599*. New Haven: Yale University Press, 1982.

[Churchill, Awnsham and John Churchill]. *A Collection of Voyages and Travels*. 4 vols. 1704.

Clifford, James. "On Ethnographic Allegory." In *Writing Culture*, ed. Clifford and Marcus (see entry).

Clifford, James, and George E. Marcus, eds. *Writing Culture: The Poetics and Politics of Ethnography*. School of American Research Advanced Seminar. Berkeley: University of California Press, 1986.

Cohen, Selma Jean. *Dance as Theatre Art*. Denver: A. Swallow, 1964.

Colas, René. *Bibliographie genénérale du costume et de la mode*. Paris: R. Colas, 1933.

Columbus, Christopher. *De insulis nuper repertis* ("Letter to Sanchez"). Basel, 1493.

———. *Select Documents Illustrating the Life and Voyages*. (See entry under Jane, ed., *Documents*.)

Compagnie de Jesus. *Mémoires pour l'histoire des Sciences et des beaux Arts* [*Mémoires de Trévoux*]. Trévoux, 1701–31; Paris, 1732–67.

Copeland, Thomas. "Francis Godwin's *The Man in the Moon*: A Picaresque Satire." *Extrapolation* 16 (2) (May 1975): 156–63.

Copernicus, Nicolaus. *De revolutionibus orbium coelestium*. Leipzig, 1543.

Cornelius, Paul. *Languages in Seventeenth- and Early Eighteenth-Century Imaginary Voyages*. Geneva: Librairie Droz, 1965.

Cortés, Hernan. *Letters from Mexico*. Ed. and trans. Anthony Pagden. New Haven: Yale University Press, 1986.

Cotton, Charles. *Erotopolis: The Present State of Betty-land*. 1684.

Couliano, Ioan P. *Eros and Magic in the Renaissance*. Trans. Margaret Cook. 1984. Chicago: University of Chicago Press, 1987.

C[oward], W. *A Just Scrutiny: Or, a Serious Enquiry into the Modern Notions of the Soul*. [1705?].

Crisciani, Chiara. "History, Novelty, and Progress in Scholastic Medicine." *Osiris*, 2d ser., vol. 6 (1990): 118–39.

Crombie, Alistair C. "The Study of the Senses in Renaissance Science." In *Actes du dixième congrés international d'histoire des sciences*. Paris: Hermann, 1964.

Crowe, Michael J. *The Extraterrestrial Life Debate: The Idea of Plurality of Worlds from Kant to Lowell*. Cambridge: Cambridge University Press, 1986.

Cunningham, J. V. *Woe or Wonder: The Emotional Effect of Shakespearian Tragedy*. Denver: University of Denver Press, 1951.

Curtin, Tyler. "The 'Sinister Fruitiness' of Machines: *Neuromancer*, Internet Sexuality, and the Turing Test." In *Novel Gazing*, ed. Sedgwick, (see entry).

Cyrano de Bergerac, Savinien. *L'autre monde ou les estats et empires de la lune*. 1657. Ed. Madeleine Alcover. Paris: Librairie Honoré Champion, 1977.

———. *Les estats et empires du soleil*. 1662.

———. *Other Worlds: The Comical History of the States and Empires of the Moon and the Sun*. Trans. Geoffrey Strachan (see entry). London: Oxford University Press, 1965.

Dale, Antonius van. *Antonii van Dale M. D. De oraculis ethnicorum dissertationes duae*. Amsterdam, 1683.

Dalgarno, George. *Ars signorum, vulgo character universalis et lingua philosophica*. Edinburgh, 1661.

Dampier, William. *A new voyage round the world.* 1697.
Daniel, Gabriel. *Voiage du Monde de Descartes.* 1690.
Darnton, Robert. *The Literary Underground of the Old Regime.* Cambridge: Harvard University Press, 1982.
Darwin, Charles. *The Expression of the Emotions in Man and Animals.* London: John Murray, 1872.
Daston, Lorraine. "Baconian Facts, Academic Civility, and the Prehistory of Objectivity." *Annals of Scholarship* 8 (3–4) (1991): 337–63.
———. "History of Science in an Elegiac Mode." *Isis* 82 (313) (1991): 522–31.
Daston, Lorraine, and Katharine Park. *Wonders and the Order of Nature, 1150–1750.* N.Y.: Zone Books, 1998.
Davis, Lennard. *Factual Fictions: The Origins of the English Novel.* New York: Columbia University Press, 1983.
Davis, Natalie Zemon. *Fiction in the Archives: Pardon Tales and Their Tellers in Sixteenth-Century France.* Stanford: Stanford University Press, 1987.
———. "Printing and the People." In *Society and Culture in Early Modern France.* Stanford: Stanford University Press, 1975.
———. *Women on the Margins: Three Seventeenth-Century Lives.* Cambridge: Harvard University Press, 1995.
Dear, Peter. "*Totius in verba*: Rhetoric and Authority in the Early Royal Society." *Isis* 76 (282) (1985): 144–61.
Debus, Allen G. *Man and Nature in the Renaissance.* Cambridge: Cambridge University Press, 1978.
Dee, John. "Mathematical Preface." In *The elements of geometrie of the most auncient philosopher Euclide of Megara*, trans. H. Billingsly. 1570.
Defoe, Daniel. *The Life and Surprizing Adventures of Robinson Crusoe.* . . . 1719. Vol. 1 of *Romances and Narratives*, ed. George Aitkins. 1895. Reprint, New York: AMS Press, 1974.
DeJean, Joan. *Libertine Strategies: Freedom and the Novel in Seventeenth-Century France.* Columbus: Ohio State University Press, 1981.
———. "Method and Madness in the *Voyage dans la Lune*." *French Studies* 2 (3) (1977): 224–37.
———. "The Politics of Pornography: *L'Ecole des Filles*." In *Invention of Pornography*, ed. Hunt (see entry).
Delft, Louis van. "Caracterologie et cartographie à l'âge classique." In Anna Balakian et al., eds., *Proceedings of the 10e congrés de l'Association internationale de littérature comparée*, vol. 2: *Comparative Poetics / Poetiques comparées.* New York, 1982.
———. "Moralistique et topographie: *Caractères* et *lieux* dans l'anthropologie classique." In Fritz Nies and Karlheinz Stierle, eds., *Französische Klassik: Theorie, Literatur, Malerei.* Munich: Fink, 1985.
Descartes, René. *Meditationes in prima philosophia.* In *Ouevres de Descartes*, ed. Charles Adam and Paul Tannery, vol. 7, pp. 1–90. Paris: Librairie Philosophique J. Vrin, 1983.
———. *Meditations on First Philosophy.* In *Discourse on Method and Meditations on First Philosophy*, trans. Donald A. Cress. Indianapolis: Hackett, 1980.
Desprez, François. *Recueil de la diversité des habits, qui sont de present en usage, tant as pays d'Eu-*

rope, Asie, Affrique & isles sauuage. 1562. Published also as *Omnium fere gentium*. Antwerp, 1572.

Désy, Pierette. "A Secret Sentiment (Devils and Gods in Seventeenth-Century New France)." *History of Anthropology* 3 (1987): 83–121.

Dick, Steven J. *Plurality of Worlds: The Origins of the Extraterrestrial Life Debate from Democritus to Kant*. Cambridge: Cambridge University Press, 1982.

Dickason, Olive. *The Myth of the Savage and the Beginnings of French Colonialism in the Americas*. Edmonton, Alberta: University of Alberta Press, 1984.

Dickason, Olive, and L. C. Green. *The Law of Nations and the New World*. Edmonton, Alberta: University of Alberta Press, 1989.

Dickinson, Emily. *Complete Poems*. Ed. Thomas Johnson. Boston: Little, Brown, 1960.

Digges, Leonard. *A prognostication everlasting of right good effect*. Rev. Thomas Digges. 1576.

Donne, John. *Devotions*. 1624.

——. *Holy Sonnets*. 1635.

——. *Ignatius his Conclave*. 1634.

Douthwaite, Julia V. *Exotic Women: Literary Heroines and Cultural Strategies in Ancien Régime France*. Philadelphia: University of Pennsylvania Press, 1992.

Drake, J. *Anthropologia Nova, or, A New System of Anatomy*. 1706.

Drake, Stillman. "Copernicus: Philosophy and Science: Bruno-Kepler-Galileo." No. 28. Norwalk, Conn.: Burndy Library, 1973.

——. *Galileo at Work: His Scientific Biography*. Chicago: University of Chicago Press, 1978.

Dryden, John, and Robert Howard. *The Indian Queen*. 1664.

Duffy, Maureen. *The Passionate Shepherdess: Aphra Behn, 1640–1689*. London: Jonathan Cape, 1977.

Dumée, Jeanne. *Entretiens sur l'opinion de Copernic touchant la mobilité de la terre*. 1680.

Eamon, William. *Science and the Secrets of Nature: Books of Secrets in Medieval and Early Modern Culture*. Princeton: Princeton University Press, 1994.

Earle, John. *Microcosmographie, or, A piece of the world discover'd*. 1628.

Easlea, Brian. *Witch Hunting, Magic, and the New Philosophy*. Brighton: Harvester Press, 1980.

Eden, Richard, trans. *Decades of the newe worlde of west India*, by Pietro Martire d'Anghiera (see entry). 1555.

Edgerton, Samuel Y. *The Heritage of Giotto's Geometry: Art and Science on the Eve of the Scientific Revolution*. Ithaca: Cornell University Press, 1991.

Egmond, Florike, and Peter Mason. *The Mammoth and the Mouse*. Baltimore: Johns Hopkins University Press, 1997.

Elliott, J. H. *The Old World and the New*. Cambridge: Cambridge University Press, 1970.

England's Vanity: or the Voice of God Against the Monstrous Sin of Pride in Dress and Apparel. 1683.

'Espinasse, Margaret. *Robert Hooke*. Berkeley: University of California Press, 1956.

Ethnohistory 33 (2) (1986). Special issue on travel literature, ethnography, and ethnohistory.

Evans, R. J. W. *Rudolph II and His World*. Oxford: Oxford University Press, 1873.

Evelyn, John. *Diary of John Evelyn, Esq., F.R.S., to which are added a selection from his familiar letters.* . . . Ed. William Bray. 4 vols. London: Bickers and Son, 1879.

Fabian, Johannes. *Of Time and the Other: How Anthropology Makes Its Object.* New York: Columbia University Press, 1983.

Fanon, Frantz. *The Wretched of the Earth.* Trans. Constance Farrington. 1961. New York: Grove Press, 1963.

Favaro, Antonio. *Le Opere di Galileo Galilei.* 20 vols. Florence: Tipografia di G. Barbèra, 1890–1909.

Fénelon, François de Salignac de la Mothe. *Avantures de Télémaque.* 1699.

Ferguson, Margaret. "Juggling the Categories of Race, Class, and Gender: Aphra Behn's *Oroonoko.*" In *Women, "Race," and Writing in the Early Modern Period,* ed. Margo Hendricks and Patricia Parker. London: Routledge, 1994.

Fernández de Oviedo y Valdés, Gonzálo. *Historia general de las Indias.* Seville, 1535.

Findlen, Paula. "Humanism, Politics, and Pornography in Renaissance Italy." In *Invention of Pornography,* ed. Hunt (see entry).

———. "Jokes of Nature and Jokes of Knowledge." *Renaissance Quarterly* 43 (2) (1991): 292–331.

———. *Possessing Nature: Museums, Collecting, and Scientific Culture in Early Modern Italy.* Berkeley: University of California Press, 1994.

Fisher, Philip. *Wonder, the Rainbow, and the Aesthetics of Rare Experiences.* Cambridge: Harvard University Press, 1998.

Fitelieu, M. de. *La contre-mode.* 1642.

Flecknoe, Richard. *Relation of Ten Years Travels in Europe, Asia, Affrique and America.* . . . 1656.

Fletcher, Angus. *Allegory: The Theory of a Symbolic Mode.* Ithaca: Cornell University Press, 1964.

Fletcher, Phineas. *The Purple Island, or, The Isle of Man.* Cambridge, 1633.

Flint, Valerie. *The Imaginative Landscape of Christopher Columbus.* Princeton: Princeton University Press, 1992.

Floyd-Wilson, Mary. "Temperature, Temperance, and Racial Difference in Ben Jonson's 'The Masque of Blackness.'" *English Literary Renaissance* 28 (1) (1998): 183–209.

Fogel, Aaron. *Coercion to Speak: Conrad's Poetics of Dialogue.* Cambridge: Harvard University Press, 1988.

Foley, Frederick J., S.J. *The Great Formosan Imposter.* Sources and Studies for the History of the Far East, vol. 1. Rome: Jesuit Historical Institute, 1968.

Fontenelle, Bernard le Bovier de. *La Comète: comédie.* 1681.

———. *Conversations on the Plurality of Worlds.* Trans. H. A. Hargreaves. Berkeley: University of California Press, 1990.

———. *Entretiens sur la pluralité des mondes.* 1686. Ed. Robert Shackleton. Oxford: Clarendon Press, 1955.

———. *L'histoire des oracles.* 1686.

———. "Relaçion de l'île de Bornéo." *Nouvelles de la république des lettres* (January 1686). In vol. 1 of Fontenelle, *Oeuvres complètes,* ed. Alain Niderst. 7 vols. Paris: Fayard, 1989–.

Formigari, Lia. *Language and Experience in Seventeenth-Century British Philosophy.* Trans.

William Dodd. Studies in the History of the Language Sciences, ser. 3, vol. 48. Amsterdam: J. Benjamins, 1988.

Forsyth, Donald W. "The Beginnings of Brazilian Anthropology: Jesuits and Tupinamba Cannibalism." *Journal of Anthropological Research* 39 (2) (1983): 147–78.

Foster, Martha Harroun. "Lost Women of the Matriarchy: Iroquois Women in the Historical Literature." *American Indian Culture and Research Journal* 19 (3) (1995): 121–40.

Foucault, Michel. *Discipline and Punish: The Birth of the Prison*. 1975. Trans. Alan Sheridan. New York: Vintage Books, 1995.

———. "Of Other Spaces." Trans. Jay Miskowiec. *Diacritics* 16 (Spring 1986): 22–27.

———. *The Order of Things*. 1966. New York: Pantheon Books, 1971.

Fox, Richard G., ed. *Recapturing Anthropology: Working in the Present*. Santa Fe, N.M.: School of American Research Press, 1991.

Foxon, David. *Libertine Literature in England, 1660–1745*. New Hyde Park, N.Y.: University Books, 1965.

Fracastoro, Girolamo. *Syphilis, sive morbus gallicus*. Veronica, 1530.

Freud, Sigmund. *Civilization and Its Discontents*. Ed. and trans. James Strachey. 1930. New York: W. W. Norton, 1962.

———. "The Resistances to Psychoanalysis." 1925. In *Character and Culture*, ed. Phillip Rieff. New York: Collier Books, 1963.

Friedman, Jerome. *The Battle of the Frogs and Fairford's Flies*. London: Routledge, 1993.

Friedman, John Block. *The Monstrous Races in Medieval Art and Thought*. Baltimore: Johns Hopkins University Press, 1979.

Frisch, Max, ed. *Joannis Kepleri astronomi opera omnia*. 8 vols. Frankfurt and Erlangen, 1858–71.

Frost, Robert. "Desert Places." In *A Further Range*. New York: H. Holt, 1936.

Fry, Gladys-Marie. *Stitched from the Soul: Slave Quilting in the Ante-Bellum South*. New York: E. P. Dutton, 1990.

Frye, Northrop. *Anatomy of Criticism*. 1957. New York: Atheneum, 1966.

Gabbey, Alan. "Innovation and Continuity in the History of Astronomy: The Case of the Rotating Moon." In *Revolution and Continuity: Essays in the History and Philosophy of Early Modern Science*, ed. Peter Barker and Roger Ariew. Washington, D.C.: Catholic University of America Press, 1991.

Galilei, Galileo. Letter to Benedetto Landucci, 29 August 1609. In *Opere*, ed. Favaro (see entry), vol. 10.

———. *Sidereus nuncius*. 1610. In *Opere*, ed. Favaro (see entry), vol. 3, pt 1.

———. *Siderius nuncius: Le messager céleste*. (See entry under Pantin.)

———. *Sidereus Nuncius, or, The Sidereal Messenger*. Trans. Albert Van Helden (see entry). Chicago: University of Chicago Press, 1989.

Gallagher, Catherine. "Embracing the Absolute: The Politics of the Female Subject in Seventeenth-Century England." *Genders* 1 (1) (1988): 24–39.

———. *Nobody's Story: The Vanishing Acts of Women Writers in the Marketplace, 1670–1820*. Berkeley: University of California Press, 1994.

Garber, Marjorie. "Spare Parts: The Surgical Construction of Gender." In *The Lesbian and*

Gay Studies Reader, ed. Henry Abelove, Michele Aina Barale, and David M. Halperin. New York: Routledge, 1993.

Geertz, Clifford. *Works and Lives: The Anthropologist as Author*. Stanford: Stanford University Press, 1988.

Gelbart, Nina Rattner. "Introduction." In Bernard le Bovier de Fontenelle, *Conversations on the Plurality of Worlds*. (See entry under Fontenelle.)

Geoffrey of Monmouth. *Historia regnum Brittaniae*. 1136.

Gerard, l'abbé de. *Entretiens de Philomon et de Théandre sur la philosophie de cour*. 1680.

Gilbert, Sandra M., and Susan Gubar. *The Madwoman in the Attic: The Woman Writer and the Nineteenth-Century Literary Imagination*. New Haven: Yale University Press, 1979.

Gilman, Sander L. "Black Bodies, White Bodies: Toward an Iconography of Female Sexuality in Late Nineteenth-Century Art, Medicine, and Literature." *Critical Inquiry* 12 (1) (1985): 205–43.

———. *Difference and Pathology: Stereotypes of Sexuality, Race, and Madness*. Ithaca: Cornell University Press, 1985.

Glück, Louise. "Vita Nova" (I). In *Vita Nova*. New York: Ecco Press, 1999.

Godard de Donville, Louise. "La femme dans le discours sur la Mode au XVIIe siècle." *Travaux de littérature* 3 (1990): 417–28.

Godwin, Francis [Domingo Gonsales]. *The Man in the Moon*. 1638.

———. *Nuncius Inanimatus*. Utopia, 1629.

Godwin, Joscelyn. *Athanasius Kircher: A Renaissance Man and the Quest for Lost Knowledge*. London: Thames and Hudson, 1979.

Goldberg, Jonathan. *Sodometries: Renaissance Texts, Modern Sexualities*. Stanford: Stanford University Press, 1992.

Goodman, Nelson. *Ways of Worldmaking*. Sussex: Harvester Press, 1978.

Gove, Phillip. *The Imaginary Voyage in Prose Fiction: A History of Its Criticism and a Guide to Its Study*. New York: Columbia University Press, 1941.

Grafton, Anthony, with April Shelford and Nancy Siraisi. *New Worlds, Ancient Texts: The Power of Tradition and the Shock of Discovery*. Cambridge: The Belknap Press of Harvard University Press, 1992.

Grant, Douglas. *Margaret the First: A Biography of Margaret Cavendish, Duchess of Newcastle, 1623–1673*. Toronto: University of Toronto Press, 1957.

Greenblatt, Stephen. "Invisible Bullets: Renaissance Authority and Its Subversion." *Glyph* 8 (1981): 40–61.

———. *Marvelous Possessions: The Wonder of the New World*. Chicago: University of Chicago Press, 1991.

———. *Renaissance Self-Fashioning*. Chicago: University of Chicago Press, 1979.

Greene, Robert. *Friar Bacon and Friar Bungay*. 1594.

Grew, Nehemiah. *The Anatomy of Plants*. 1682.

———. *Musaeum Regalis Societatis, or a Catalogue and Description of the Natural and Artificial Rarities Belonging to the Royal Society And preservedd at Gresham Colledge*[sic]. 1681.

Grossman, Allen. "The Passion of Laocoön, or the Warfare of the Religious against the Poetic Institution." Unpublished paper.

———. *Poetry: A Basic Course*. [Videotapes]. 2 tapes. Prod. Jay Reed. Springfield, Va.: Teaching Company, 1993.

Grotius, Hugo. *De origine gentium americanarum dissertatio*. Paris, 1643. Trans. Edmund Goldschmidt as *On the Origin of the native races of America*. Edinburgh: Priv. print, *Biblioteca curiosa*, 1884.

Grove, Richard H. *Green Imperialism: Colonial Expansion, Tropical Island Edens, and the Origins of Environmentalism, 1600–1860*. Cambridge: Cambridge University Press, 1995.

Gutch, J. M., ed. *Caraboo: A Narrative of a Singular Imposition*. Bristol, 1817.

Guthke, Karl S. *The Last Frontier: Imagining Other Worlds from the Copernican Revolution to Modern Science Fiction*. Ithaca: Cornell University Press, 1990.

Gutiérrez, Ellen Turner. *The Reception of the Picaresque in the French, English, and German Traditions*. New York: Peter Lang, 1995.

Hahn, Roger. *The Anatomy of a Scientific Institution: The Paris Academy of Sciences, 1666–1803*. Berkeley: University of California Press, 1971.

Hakluyt, Richard. *Divers Voyages touching the discovery of America*. 1582.

———. *Principle Navigations of the English Nation*. . . . 1589.

Hall, Joseph. *Mundus alter et idem*. 1607.

Hall, Kim. *Things of Darkness: Economies of Race and Gender in Early Modern England*. Ithaca: Cornell University Press, 1995.

Hallyn, Fernand. *The Poetic Structure of the World: Copernicus and Kepler*. Trans. Donald M. Leslie. 1987. New York: Zone Books, 1990.

Handler, Richard, and Daniel Segal. *Jane Austen and the Fiction of Culture*. Tucson: University of Arizona Press, 1990.

Hanke, Lewis. *Aristotle and the American Indians*. Bloomington: University of Indiana Press, 1959.

Haraway, Donna. "Cyborg Manifesto." In *Simians, Cyborgs, and Women*. New York: Routledge, 1991.

Harbsmeier, Michael. "Spontaneous Ethnographies: Towards a Social History of Travellers' Tales." *Studies in Travel Writing* 1 (1) (1997): 216–38.

Hargreaves, H. A. "New Evidence of the Realism of Mrs. Behn's *Oroonoko*." *Bulletin of the New York Public Library* 74 (1970): 437–44.

Hariot, Thomas. *A Briefe and true report of the new founde land of Virginia*. 1588.

Harrisse, Henry. *Découverte et évolution cartographique de Terre Neuve*. Paris: H. Welter, 1900.

Hart, John. *Orthographie*. 1569.

Harth, Erica. *Cyrano de Bergerac and the Polemics of Modernity*. New York: Columbia University Press, 1970.

Harwood, John T. "Rhetoric and Graphics in *Micrographia*." In *Robert Hooke: New Studies*, eds. Michael Hunter and Simon Schaffer. Woodbridge, Suffolk: Boydell Press, 1989.

Headley, John. "Geography and Empire in the Late Renaissance: Botero's Assignment, Western Universalism, and the Civilizing Process." Paper delivered to the Committee on Renaissance Studies, UNC-Chapel Hill, February 1998.

———. "The Sixteenth-Century Venetian Celebration of the Earth's Total Habitability." *Journal of World History* 8 (1) (1997): 1–27.

———. *Tommaso Campanella and the Transformation of the World*. Princeton: Princeton University Press, 1997.

Hebdige, Dick. *Subculture: The Meaning of Style*. London: Methuen, 1979.

Heilbron, John. "Experimental Natural Philosophy." In *The Ferment of Knowledge: The Historiography of Eighteenth-Century Science*, ed. G. S. Rousseau and Roy Porter. Cambridge: Cambridge University Press, 1980.

Helms, Mary W. "Essay on Objects: Interpretations of Distance Made Tangible." In *Implicit Understandings: Observing, Reporting, and Reflecting on the Encounters between Europeans and Other Peoples in the Early Modern Era*, ed. Stuart B. Schwartz. Cambridge: Cambridge University Press, 1994.

Hendricks, Margo, and Patricia Parker, eds. *Women, "Race," and Writing in the Early Modern Period*. London: Routledge, 1994.

Heninger, S. K. *Touches of Sweet Harmony: Pythagorean Cosmology and Renaissance Poetics*. San Marino: The Huntington Library, 1974.

Herbert, Christopher. *Culture and Anomie: Ethnographic Imagination in the Nineteenth Century*. Chicago: Chicago University Press, 1991.

Herrick, Robert. "Upon Julia's Clothes." In *Poetical Works*, ed. L. C. Martin. Oxford: Clarendon Press, 1956.

Hervier, Julien. "Cyrano de Bergerac et le voyage spatial: de la fantaisie á la science-fiction." *Proceedings of the Tenth Congress of the International Comparative Literature Association*, eds. Anna Balakian and James J. Wilhelm. New York: Garland, 1985.

Hodgen, Margaret. *Early Anthropology in the Sixteenth and Seventeenth Centuries*. Philadelphia: University of Pennsylvania Press, 1964.

Hooke, Robert. "Discourse concerning Telescopes and Microscopes." In *Robert Hooke: Philosophical Experiments and Observations*, ed. W. Derham. 1726. Library of Science Classics, vol. 8. London: Cass, 1967.

———. *Micrographia: or Some Physiological Descriptions of Minute Bodies Made by Magnifying Glasses*. 1665. New York: Dover Publications, 1961.

Huet, Marie Hélène. *Monstrous Imagination*. Cambridge: Harvard University Press, 1993.

Hulme, Peter. *Colonial Encounters: Europe and the Native Caribbean, 1492–1797*. 1986. Rev. ed. London: Routledge, 1992.

Hulton, Paul. *America 1585: The Complete Drawings of John White*. London: British Museum Publications, 1984.

Hunt, Lynn. "Introduction: Obscenity and the Origins of Modernity, 1500–1800." In *Invention of Pornography*, ed. Hunt (see entry).

———, ed. *The Invention of Pornography: Obscenity and the Origins of Modernity*. New York: Zone Books, 1993.

Hunter, J. Paul. *Before Novels: The Cultural Contexts of Eighteenth-Century English Fiction*. New York: W. W. Norton, 1990.

Hutchinson, Keith. "What Happened to Occult Qualities in the Scientific Revolution?" *Isis* 73 (267) (1982): 233–53.

Huygens, Christiaan. *The Celestial Worlds Discovered: or, Conjectures Concerning the Inhabitants, Plants and Productions of the Worlds in the Planets*. [*Cosmotheoros*.] 1698.

Impey, Oliver, and Arthur MacGregor, eds. *The Origins of Museums: The Cabinet of Curiosities in Sixteenth- and Seventeenth-Century Europe*. Oxford: Clarendon Press, 1985.
Itard, Jean Marc Gaspard. "Of the First Developments of the Young Savage of Aveyron." (See entry under Malson, *Wolf Children*.)
Ivins, William Mills. *On the Rationalization of Sight, with an Examination of Three Renaissance Texts on Perspective*. New York: Da Capo Press, 1973.
Jacob, James, and Margaret Jacob. "The Anglican Origins of Modern Science: The Metaphysical Foundations of the Whig Constitution." *Isis* 71 (257) (1980): 251–67.
Jakobson, Roman. "On Realism in Art." In *Language in Literature*, ed. Krystyna Pomorska and Stephen Rudy. Cambridge, Mass.: Belknap Press, 1987.
Jameson, Fredric. *The Political Unconscious: Narrative as a Socially Symbolic Act*. Ithaca: Cornell University Press, 1981.
Jane, Cecil, ed. *Select Documents Illustrating the Life of Columbus*. 2 vols. Hakluyt Society, 1930, 1934. Reprint, New York: Dover Publications, 1988.
Janowitz, Anne. "Caraboo: A Singular Imposition." Unpublished paper, 1998.
Janson, H. W. *Apes and Ape Lore in the Middle Ages and Renaissance*. London: Warburg Institute, University of London, 1952.
Jardine, Lisa. *Francis Bacon: Discovery and the Art of Discourse*. Cambridge: Cambridge University Press, 1974.
Jayne, Sears R. *Library Catalogues of the English Renaissance*. Berkeley: University of California Press, 1956.
[Jeamson, Thomas]. *Artificiall Embellishments or Arts Best Directions How to Preserve Beauty or Procure It*. Oxford, 1665.
Jordanova, Ludmilla. *Sexual Visions: Images of Gender in Science and Medicine between the Eighteenth and Twentieth Centuries*. Madison: University of Wisconsin Press, 1989.
Jowett, Claire. *Real and Imaginary Worlds: English Politics, Utopianism, and Travel Literature, 1589–1667*. Manchester: Manchester University Press, 1999.
Kant, Immanuel. *Observations on the Feeling of the Beautiful and the Sublime*. Trans. John T. Goldthwaite. Berkeley: University of California Press, 1960.
Kargon, Robert. *Atomism in England from Hariot to Newton*. Oxford: Clarendon Press, 1966.
Kegl, Rosemary. "'This World I Have Made': Margaret Cavendish, Feminism, and the Blazing World." In *Feminist Readings of Early Modern Culture: Emerging Subjects*. Ed. Valerie Traub, M. Lindsay Kaplan, and Dympna Callaghan. Cambridge: Cambridge University Press, 1996.
Keller, Evelyn Fox. *Reflections on Gender and Science*. New Haven: Yale University Press, 1985.
——. "Baconian Science." In *Reflections on Gender and Science* (see entry).
Kelly, Joan. "Did Women Have a Renaissance?" In *Women, History, and Theory: The Essays of Joan Kelly*. Chicago: University of Chicago Press, 1984.
Kenseth, Joy. "'A World of Wonders in One Closet Shut.'" In *Age of the Marvelous*, ed. Kenseth (see entry).
Kenseth, Joy, ed. *The Age of the Marvelous*. Hanover, N.H.: Hood Museum of Art, 1991.
Kepler, Johannes. *Dioptrice*. Augsburg, 1611.

———. *Kepler's "Somnium": The Dream, or Posthumous Work on Lunar Astronomy*. Trans. Edward Rosen. Madison: University of Wisconsin Press, 1967.

———. *Mysterium cosmographicum*. Frankfurt, 1587.

———. "Preface" to *Dioptrice*. In *"The Sidereal Messenger" of Galileo Galilei and a part of the preface to Kepler's "Dioptrice"*. Trans. E. S. Carlos. London: Rivingtons, 1880.

———. *Somnium, seu Astronomia Lunae*. . . . 1634. In *Opera*, ed. Frisch (see entry), vol. 8.

Keynes, Geoffrey, ed. *The Works of Sir Thomas Browne*. 4 vols. London: Faber and Faber, 1964.

Khanna, Lee Cullen. "The Subject of Utopia: Margaret Cavendish and Her *Blazing-World*." In *Utopian and Science Fiction by Women: Worlds of Difference*, ed. Jane L. Donawerth and Carol A. Kolmerten. Syracuse: Syracuse University Press, 1994.

Kiefer, Frederick. "The Dance of the Madmen in *The Duchess of Malfi*." *Journal of Medieval and Renaissance Studies* 17 (2) (1987): 211–33.

King, Thomas. *Queer Articulations: Men, Gender, and the Politics of Effeminacy*. Madison: University of Wisconsin Press, 2000 (forthcoming).

Kircher, Athanasius. *Oedipus Aegyptiacus*. 4 vols. 1652–54.

Knight, David. "Science Fiction of the Seventeenth Century," *Seventeenth-Century* 1 (1986): 69–79.

Knowlson, James R. "A Note on Bishop Godwin's *Man in the Moone*: The East Indies Trade Route and a 'Language' of Musical Notes." *Modern Philology* 65 (4) (1968): 357–61.

———. *Universal Language Schemes in England and France, 1600–1800*. Toronto: University of Toronto Press, 1975.

Knox, Dilwyn. "John Bulwer." In *Lexicon Grammaticorum: Who's Who in the History of World Linguistics*, vol. 20, ed. Harro Stammerjohann. Tübingen: Niemeyer, 1996.

Koestler, Arthur. *The Sleepwalkers: A History of Man's Changing Vision of the Universe*. London: Hutchinson, 1959.

Kristeva, Julia. *The Powers of Horror: An Essay on Abjection*. Trans. Leon S. Roudiez. New York: Columbia University Press, 1982.

La Bruyère, Jean de. *The "Characters" of Jean de La Bruyère*. Trans. Henri van Laun. London: John C. Nimmo, 1885.

Lacouperie, Terrien de. "Formosa Notes on MSS, Races, and Languages . . . Including a Note on *Nine Formosan MSS*. by E. Colbourne Baker, H. B. M. Chinese Secretary, Peking." *Journal of the Royal Asiatic Society*, n.s., 19 (1887): 413–75.

Laet, Johann de. *Notae ad dissertationem Hugonis Grotii*. Paris, 1643.

Lafayette, Madame de [Marie-Madeleine Pioche de la Vergne]. *La Princesse de Clèves*. 1678.

Lafitau, Joseph. *Customs of the American Indians Compared with the Customs of Primitive Times*. Ed. William N. Fenton and Elizabeth Moore, with introduction. Publications of the Champlain Society, nos. 48 and 59. 2 vols. Toronto: Champlain Society, 1977.

———. *Moeurs des sauvages amériquaines, comparées aux moeurs des premiers temps*. 2 vols. 1724.

Lafond, Jean. "Le Monde á l'envers dans les 'Etats et Empires de la Lune' de Cryano de Bergerac." In *L'Image du monde renversé et ses représentations littéraires et para-littéraires*, ed. Lafond and Augustin Redondo. Paris: Librairie Philosophique J. Vrin, 1979.

Lahontan, Baron de. *New Voyages to North America*. 1703. Ed. Reuben Gold Thwaites. 2 vols. Chicago: A. C. McClurg, 1905.

La Lande, Jérôme de. *Bibliographie astronomique.* Paris, 1803.
Laqueur, Thomas. *Making Sex: Bodies and Gender from the Greeks to Freud.* Cambridge: Harvard University Press, 1991.
Las Casas, Bartolomé de. *The Spanishe Colonie [Brevissima relaçión].* Trans. M. M. S. London, 1583.
Latour, Bruno. *We Have Never Been Modern.* Trans. Catherine Porter. 1991. Cambridge: Harvard University Press, 1994.
Laugaa, Maurice. "Lune ou l'Autre." *Poètique: Revue de Théorie et d'Analyse Littéraires* 3 (1970): 282–96.
Lear, John. *Kepler's "Dream"; With the full text and notes of "Somnium, sive Astronomia lunaris."* Trans. Patricia Frueh Kirkwood. Berkeley: University of California Press, 1965.
Lebowitz, Fran. "A Humorist at Work." Interview. *Paris Review* 127 (1993): 160–88.
Leeuwenhoek, Antony van. "Letter 25." In Clifford Dobell, *Antony van Leeuwenhoek and His "Little Animals."* 1932. New York: Dover, 1960.
Legati, Lorenzo. *Museo Cospiano Annesso a quello del Famoso Ulisse Aldrovandi.* Bologna, 1677.
Lehmann-Haupt, Hellmut. "The Microscope and the Book." In *Festschrift für Clau Nissen.* Weisbaden: Pressler, 1973.
Le Jeune, Paul. "Relation de ce qui s'est passée en la Nouvelle France, en l'année 1635," and "Relation . . . 1636." In *Jesuit Relations,* ed. Thwaites (see entry), vols. 8–10.
Léry, Jean de. *History of a Voyage to the Land of Brazil.* Trans. Janet Whatley. Berkeley: University of California Press, 1990.
Leslie, Marina. "Gender, Genre, and the Utopian Body in Margaret Cavendish's *Blazing World.*" *Utopian Studies* 7 (1) (1996): 6–24.
———. *Renaissance Utopias and the Problem of History.* Ithaca: Cornell University Press, 1998.
Lestringant, Frank. *André Thevet: Cosmographe des derniers Valois.* Geneva: Druz, 1991.
———. "Travels in Eucharista: Formosa and Ireland from George Psalmanaazaar to Jonathan Swift." *Yale French Studies* 86 (1994): 109–25.
———. *Cannibals: The Discovery and Representation of the Cannibal from Columbus to Jules Verne.* Trans. Rosemary Morris. The New Historicism: Studies in Cultural Poetics, vol. 37. Berkeley: University of California Press, 1997.
Li Yu. *A Tower for the Summer Heat.* Trans. Patrick Hanan. New York: Columbia University Press, 1992.
Lilley, Kate. "Blazing Worlds: Seventeenth-Century Women's Utopian Writing." In *Women, Texts, and Histories, 1575–1760,* ed. Clare Brant and Diane Purkiss. New York: Routledge, 1992.
———, ed. *The Description of a New World Called the Blazing-World and Other Writings,* by Margaret Cavendish. New York: New York University Press, 1992.
Linebaugh, Peter. *The London Hanged: Crime and Civil Society in the Eighteenth Century.* Cambridge: Cambridge University Press, 1992.
Linnaeus, Carolus. *Systema Naturae, sive Regna tria naturae systematice propositae per classes, ordines, genera & species.* Leiden, 1635.
Lipovetsky, Gilles. *The Empire of Fashion: Dressing Modern Democracy.* 1987. Trans. Catherine Porter. Princeton: Princeton University Press, 1994.
Liu, Lydia. "Robinson Crusoe's Earthenware Pot: Science Aesthetics, and the Meta-

physics of True Porcelain in the Eighteenth Century." *Critical Inquiry* 25 (4) (Summer 1999).

Locke, John. *An Essay concerning Human Understanding.* 1689.

Longino, Michèle Farrell. "The Staging of Exoticism in Seventeenth-Century France." Book in progress.

"L. S." *Nature's Dowrie: or, The peoples native liberty asserted.* 1652.

Lucian. *Certain Select Dialogues of Lucian, Together with His True Historie.* Trans. "T. H. Mr. of Arts of Christ-Church in Oxford." Oxford, 1634.

Mackenthun, Gesa. *Metaphors of Dispossession: American Beginnings and the Translation of Empire.* Norman: University of Oklahoma Press, 1996.

Mackie, Erin. *Market à la Mode: Fashion, Commodity, and Gender in the "Tatler" and the "Spectator".* Baltimore: Johns Hopkins University Press, 1997.

Malinowski, Bronislaw. *Argonauts of the Western Pacific.* London: G. Routledge and Sons, 1922.

Malson, Lucien. *Wolf Children and the Problem of Human Nature.* Trans. Edmund Fawcett, Peter Ayrton, and Joan White. 1964. Includes text of 1802 translation of Jean Marc Gaspard Itard's report "Of the First Developments of the Young Savage of Aveyron" (1801). New York: Monthly Review Press, 1972.

Mandel, Siegfried. "From the Mummelsea to the Moon: Refractions of Science in Seventeenth-Century Literature." *Comparative Literature Studies* 9 (4) (1972): 407–415.

Manning, Patrick. *Slavery and African Life: Occidental, Oriental, and African Slave Trades.* Cambridge: Cambridge University Press, 1990.

Manuel, Frank E., and Fritzie P. Manuel. *Utopian Thought in the Western World.* Cambridge, Mass.: Belknap Press, 1979.

Marana, Jean Paul. *Lettre d'un Sicilien à un de ses amis.* Introduction and notes by l'abbé Valentin Dufour. Paris: A. Quantin, 1883.

Marguerite de Navarre. *Haeptameron.* 1558.

Martin, Henri-Jean. "Culture écrite et culture orale, culture savant et culture populaire dans la France de l'Ancien Régime." *Journal des Savants* (1975): 246–47.

Marx, Karl. *Capital.* Trans. Samuel Moore and Edward Aveling. New York: International Publishers, 1967.

Massachusetts, Commonwealth of. Task Force on Human Subject Research. *A Report on the Use of Radioactive Materials in Human Subject Research That Involved Residents of State-Operated Facilities within the Commonwealth of Massachusetts from 1943 through 1973.* Boston, 1994.

Massinger, Philip. et al. *The Old Law.* 1599.

McColley, Grant. "The Date of Godwin's *Domingo Gonsales.*" *Modern Philology* 35 (1) (1937): 47–60.

McDermot, Murtagh. *A trip to the Moon* (Dublin, 1728). In *Memoirs concerning the Life and Adventures of Captain Mackheath* [etc.], ed. Josephine Grieder. Foundations of the Novel. New York: Garland Press, 1973.

McKendrick, Neil. "The Commercialization of Fashion." In *Birth of a Consumer Society*, ed. McKendrick, Brewer, and Plumb (see entry).

McKendrick, Neil, John Brewer, and J. H. Plumb, eds. *The Birth of a Consumer Society: The Commercialization of Eighteenth-Century England*. Bloomington: Indiana University Press, 1982.

McKeon, Michael. *Origins of the English Novel, 1600–1740*. Baltimore: Johns Hopkins University Press, 1987.

McMahon, Elise-Noël. "'Le corps sans frontiers': The Ideology of Ballet and Molière's *Le Bourgeois gentilhomme*." *Papers on French Seventeenth Century Literature* 20 (38) (1993): 53–72.

Mead, Margaret. *Coming of Age in Samoa*. New York: William Morrow, 1923.

Menant, Sylvain. "Les Modernes et le 'style à la Mode.'" *Cahiers de l'Association internationale des études francaises* 38 (May 1986): 145–56.

Mendelson, Sara. *The Mental World of Stuart Women*. Brighton: Harvester Press, 1987.

Menzel, Donald H. "Kepler's Place in Science Fiction." In *Kepler: Four Hundred Years: Proceedings of Conferences Held in Honour of Kepler*, ed. Arthur Beer and Peter Beer. Oxford: Pergamon Press, 1975.

Merchant, Carolyn. *The Death of Nature*. San Francisco: Harper and Row, 1980.

Middleton, Anne. "'New Men' and the Good of Literature in the *Canterbury Tales*." In *Literature and Society*, ed. Edward Said. Selected Papers from the English Institute. Baltimore: Johns Hopkins University Press, 1978.

Mignolo, Walter D. *The Darker Side of the Renaissance: Literacy, Territoriality, and Colonialization*. Ann Arbor: University of Michigan Press, 1995.

[Millot, Michel, and Jean L'Ange]. *L'Ecole des filles*. 1655.

"Miso-Spillus." *Wonder of Wonders, or, A Metamorphosis of Fair Faces Voluntarily Turned into Foul, or, An Invective against Black-Spotted Faces*. 1662.

Molière, Jean Baptiste Poquelin. *Le Bourgeois gentilhomme*. 1670.

Molinet, Claude de. *Le Cabinet de la bibliothèque de Saint-Geneviève*. 1692.

Montaigne, Michel Eyquem de. "Of Cannibals." 1588. In *Essays and Selected Writings*, ed. and trans. Donald M. Frame. Bilingual ed. New York: St. Martin's Press, 1963.

Montrose, Louis. "The Work of Gender in the Discourse of Discovery." *Representations* 33 (Winter 1991): 1–41.

Moore, Marianne. "Poetry." 1921. In *Complete Poems*, Note to "Poetry" (rev. ed.). New York: Macmillan, 1981.

Moorshead, John. *Scientific Dream: Voyage to the Moon, &c*. Sydney, 1845.

Morgan, Lewis Henry. *League of the Ho-de-no-sau-nee, or Iroquois*, Rochester: Sage and Brother; New York: M. H. Newman and Co. [etc.], 1851.

Moss, Jean Dietz. "The Interplay of Science and Rhetoric in Seventeenth-Century Italy." *Rhetorica* 7 (1) (1989): 23–43.

Muldoon, James. *The Americas in the Spanish World Order*. Philadelphia: University of Pennsylvania Press, 1994.

Mulligan, Lotte. "Puritans and English Science: A Critique of Webster." *Isis* 71 (258) (1980): 456–69.

Mulligan, Lotte, and Glenn Mulligan. "Reconstructing Restoration Science." *Social Studies of Science* 11 (3) (1981): 327–64.

Münster, Sebastian. *Cosmographia universalis*. Basel, 1544.
Nashe, Thomas. *Unfortunate Traveler*. 1594.
Nelson, William. *Fact or Fiction: The Dilemma of the Renaissance Storyteller*. Cambridge: Harvard University Press, 1973.
Nerlich, Michael. *Ideology of Adventure: Studies in Modern Consciousness, 1100–1750*. Vol. 1. Trans. Ruth Crowley. Minneapolis: University of Minnesota Press, 1987.
Neville, Henry. *The Isle of Pines, or, A late Discovery*. 1668.
Newton, Isaac, Sir. *Philosophiae naturalis principia mathematica*. London, 1687.
Nicholl, Charles. *The Creature in the Map: A Journey to El Dorado*. New York: William Morrow, 1995.
Nicolson, Marjorie Hope. *The Breaking of the Circle*. Evanston: Northwestern University Press, 1950.
———. *Voyages to the Moon*. New York: Macmillan, 1948.
Noyes, Nicholas. "Reasons against the Wearing of Periwigs." In chap. 24 of *Remarkable Providences, 1600–1760*, ed. John Demos. New York: Braziller, 1972.
Nuñez Cabeza de Vaca, Alvar. *La relación de la jornada*. Zamora, 1542.
O'Banion, John D. *Reorienting Rhetoric: The Dialectic of List and Story*. University Park, Pa.: Pennsylvania State University Press, 1992.
Oldys, William. *British Librarian*. 1738.
Ong, Walter J. *Ramus, Method, and the Decay of Dialogue: From the Art of Discourse to the Art of Reason*. Cambridge: Harvard University Press, 1958.
Ovenell, R. F. *The Ashmolean Museum, 1683–1894*. Oxford: Clarendon Press, 1986.
Pagden, Anthony. *The Fall of Natural Man: The American Indian and the Origins of Comparative Ethnology*. Cambridge Iberian and Latin American Studies. Cambridge: Cambridge University Press, 1982.
Pantin, Isabelle, ed. and trans. *Sidereus nuncius: Le Messager céleste* by Galileo Galilei. Science and Humanism. Paris: Les Belles Lettres, for l'Association Guillaume Budé, 1992.
Paradis, James. "Montaigne, Boyle, and the Essay of Experience." In *One Culture: Essays in Science and Literature*, ed. George Levine. Madison: University of Wisconsin Press, 1987.
Paré, Ambroise. *On Monsters and Marvels (Des monstres)*. 1573. Trans. Janet L. Pallister. Chicago: University of Chicago Press, 1982.
Paris, James. *Prodigies and Monstrous Births of Dwarfs, Sleepers, Giants, Strong Men, Hermaphrodites, Numerous Births, and Extreme Old Age, etc*. British Library, MS Sloane 5246.
Park, Katharine, and Robert A. Nye. "Destiny Is Anatomy." Review essay on *Making Sex*, by Thomas Laqueur. *New Republic*, 18 February 1991, 53–57.
Parker, Patricia. "Rhetorics of Property." Chap. 7 in *Literary Fat Ladies: Rhetoric, Gender, Property*. London: Methuen, 1987.
"Patch, [The]. An Heroi-Comical Poem . . . in Three Cantos . . . By a Gentleman of Oxford." London, 1724.
Pecora, Vincent. "The Limits of Local Knowledge." In *The New Historicism*, ed. H. Aram Veeser. New York: Routledge, 1989.

Penzer, N. M., ed. *An Historical and Geographical Description of Formosa.* London: Robert Holden, 1926.
Pepys, Samuel. *The Diary of Samuel Pepys.* Ed. Robert Latham et al. 11 vols. Berkeley: University of California Press, 1970–83.
Petty, William. "Quaeries Concerning the Nature of the Natives of Pennsylvania." In *The Petty Papers: Some Unpublished Writings of Sir William Petty,* ed. Marquis of Lansdowne. Vol. 2, nos. 113, 115–19. Boston: Houghton Mifflin, 1927.
Phillips, William D. *Slavery from Roman Times to the Early Transatlantic Trade.* Minneapolis: University of Minnesota Press, 1985.
Philosophical Transactions of the Royal Society. London, 1665–.
Pietz, William. "The Problem of the Fetish, I." *Res* 9 (Spring 1985): 5–17.
———. "The Problem of the Fetish, II: The Origin of the Fetish." *Res* 13 (Spring 1987): 23–45.
———. "The Problem of the Fetish, IIIa: Bosman's Guinea and the Enlightenment Theory of Fetishism." *Res* 16 (Autumn 1988): 105–23.
Pizon, Faith K., and T. Allen Comp, eds. *The Man in the Moone and Other Lunar Fantasies.* New York: Praeger Publishers, 1971.
Plot, Robert. *The Natural History of Oxfordshire.* Oxford, 1677.
Plutarch. *De facie lunae.* c. 100. Trans. Harold Cherniss and William C. Humbold. In *Plutarch's Moralia,* vol. 12, pp. 34–223.
Postel, Guillaume. *De orbis terra concordia.* Basel, 1544.
Powers, Henry. *Experimental Philosophy.* 1664.
Pratt, Mary Louise. *Imperial Eyes: Travel Writing and Transculturation.* London: Routledge, 1992.
———. "Scratches on the Face of the Country, or What Mr. Barrow Saw in the Land of the Bushmen." In *"Race," Writing, and Difference,* ed. Henry Louis Gates. Chicago: University of Chicago Press, 1986.
Prynne, William. *The Unloveliness, of Love-Lockes.* 1628.
Psalmanazar, George. *A Historical and Geographical Description of Formosa.* 1704. Ed. N. M. Penzer. London: Robert Holden, 1926.
———. *Memoirs of ****. Commonly Known by the Name of George Psalmanazar.* 1764.
Purchas, Samuel. *Hakluytus Posthumus, or, Purchas His Pilgrimes.* 1625. 20 vols. Glasgow: James MacLehose and Sons, 1905.
Quevedo, Francisco de. *La vida del Buscón.* Saragosa, 1628.
Quinn, David Beers. *The Roanoke Voyages, 1584–1590.* Hakluyt Society, 2d ser., vol. 114. 1955. Reprint, Nendeln: Kraus Reprints, 1967.
Rabelais, François. *Quart livre.* Trans. J. M. Cohen. In *Gargantua and Pantagruel.* 1552. Harmondsworth: Penguin Books, 1955.
Rafter, Nicole Hahn. *Making Born Criminals.* Urbana: University of Illinois Press, 1997.
Ralegh, Walter. *History of the World.* 1614. In *Works,* ed. Thomas Birch and William Oldys. Vols. 2–7. Oxford: The University Press, 1829.
———. *The Large, Rich and Bewtiful Empyre of Guiana.* 1596.
Ramusio, Giovanni Battista. *Navigationi et viaggi.* Venice, 1563–1606.

Rawson, Claude. "Cannibalism and Fiction: Reflections on Narrative Form and 'Extreme' Situations." *Genre* 10 (4) (1977): 667–711.

Ray, John, and Francis Willughby. *Ornithology*. (See entry under Willughby.)

Razovsky, Helaine. "Popular Hermeneutics: Monstrous Children in English Renaissance Broadside Ballads." *Early Modern Literary Studies* 2 (3) (1996): 1.1–34.

Reeves, Eileen. "Old Wives' Tales and the New World System: Gilbert, Galileo and Kepler." *Configurations* (forthcoming).

Reichler, Claude. "La Cité sauvage: La Figure du cercle dans les images des *Moeurs des sauvages américains* de Lafitau." *Mots et images nomades*, special issue of *Etudes de lettres*, 1–2 (1995): 59–75.

Reiss, Timothy J. *The Discourse of Modernism*. Ithaca: Cornell University Press, 1982.

Rief, Patricia, Sister. "The Textbook Tradition in Natural Philosophy, 1600–1650." *Journal of the History of Ideas* 30 (1) (1969): 17–32.

Roche, Daniel. *The Culture of Clothing: Dress and Fashion in the "Ancien Régime."* Trans. Jean Birrell. Cambridge: Cambridge University Press and Editions de la Maison des Sciences de l'Homme, 1994.

Roche, John J. "Harriot, Galileo, and Jupiter's Satellites." *Archives internationales d'histoire des sciences* 32 (1982): 9–51.

Rochefort, Charles César de. *Histoire naturelle & morale des iles Antilles de l'Amérique*. Rotterdam, 1658.

Rogers, John. *The Matter of Revolution: Science, Poetry, and Politics in the Age of Milton*. Ithaca: Cornell University Press, 1996.

Romm, James. "Belief and Other Worlds: Ktesias and the Founding of the 'Indian Wonders.'" In *Mindscapes: The Geographies of Imagined Worlds*, ed. George E. Slusser and Eric S. Rabkin. Carbondale, Ill.: Southern Illinois University Press, 1989.

———. *The Edges of the Earth in Ancient Thought: Geography, Exploration, and Fiction*. Princeton: Princeton University Press, 1992.

———. "Lucian and Plutarch as Sources for Kepler's *Somnium*." *Classical and Modern Literature* 9 (2) (1989): 97–107.

Rosen, Edward, trans. *Kepler's "Somnium": The Dream, or Posthmuous Work on Lunar Astronomy*. (See entry under Kepler.)

Rossi, Paolo. *The Dark Abyss of Time: The History of the Earth and the History of Nations from Hooke to Vico*. Trans. Lydia G. Cochrane. Chicago: University of Chicago Press, 1984.

Rowe, John. "Renaissance Foundations of Anthropology." *American Anthropologist* 67 (1) (1965): 1–21.

Royal Society of London. *Transactions of the Royal Society of London*. . . . 1665–.

Rubiés, Joan-Pau. "New Worlds and Renaissance Ethnology." *History of Anthropology* 6 (2–3) (1993): 157–97.

Russen, David. *Iter lunare*. London, 1707. In *The Female Critick* [etc.], ed. William Graves. Foundations of the Novel. New York: Garland, 1972.

Ryan, Michael. "Assimilating New Worlds in the Sixteenth and Seventeenth Centuries," *Comparative Studies in Society and History* 23 (1981): 519–538.

Sagard, Gabriel. *The Long Journey to the Country of the Hurons*. Ed. George Wrong. Trans.

H. H. Langton. Publications of the Champlain Society, no. 25. Toronto: Champlain Society, 1939.

Sahagún, Bernardino de. *The Florentine Codex: General History of the Things of New Spain*. Ed. and trans. Arthur J. O. Anderson and Charles E. Dibble. 2d ed. 12 vols. Monographs of the School of American Research, no. 14. Santa Fe, N.M.: The School of American Research, 1970.

Sahlins, Marshall. *Historical Metaphors and Mythical Realities: Structure in the Early History of the Sandwich Islands Kingdom*. ASAO Special Publications no. 1. Ann Arbor: University of Michigan Press, 1981.

Said, Edward. *Orientalism*. New York: Pantheon, 1978.

Saint-Hubert, M. de. *La Manière de composer et faire reussir les ballets*. 1641.

Saliba, George. *A History of Arabic Astronomy: Planetary Theories during the Golden Age of Islam*. New York: New York University Press, 1994.

Sarasohn, Lisa. "A Science Turned Upside Down: Feminism and the Natural Philosophy of Margaret Cavendish." *Huntington Library Quarterly* 47 (1984): 289–307.

Scarry, Elaine. *The Body in Pain*. Oxford: Oxford University Press, 1985.

Schiebinger, Londa. "Feminine Icons: The Face of Early Modern Science." *Critical Inquiry* 14 (Summer 1988): 661–91.

Schlesinger, Roger, and Arthur P. Stabler, eds. *André Thevet's North America: A Sixteenth-Century View*. Kingston: McGill-Queen's University Press, 1986.

Schopp, Gaspar. Letter to Conrad Rittershausen. In Burkhard Gotthelf Struve, ed., *Acta Litteraria* [serial, 1705–13] 3. Jena, 1705.

Schott, Gaspar. *Mechanica hydraulico-pneumatica*. Würzburg, 1657.

Scott, Peter Dale. *Coming to Jakarta: A Poem about Terror*. New York: New Directions, 1989.

Sedgwick, Eve Kosofsky. "The Privilege of Unknowing." In *Tendencies*. Durham, N.C.: Duke University Press, 1995.

———, ed. *Novel Gazing: Queer Readings in Fiction*. Durham, N.C.: Duke University Press, 1997.

Shackleton, Robert, ed. *Entretiens sur la Pluralité des Mondes* by Bernard le Bovier de Fontenelle (see entry).

Shadwell, Thomas. *The Virtuoso. A Comedy*. 1676.

Shakespeare, William. *The Tempest*. 1623.

———. *Titus Andronicus*. 1594.

Shapin, Steven. "'A Scholar and a Gentleman': The Problematic Identity of the Scientific Practitioner in Early Modern England." *History of Science* 29 (3) (1991): 279–327.

———. *The Scientific Revolution*. Chicago: University of Chicago Press, 1996.

———. *A Social History of Truth: Civility and Science in Seventeenth-Century England*. Chicago: University of Chicago Press, 1994.

Shapin, Steven, and Simon Schaffer. *Leviathan and the Air Pump: Hobbes, Boyle, and the Experimental Life*. Princeton: Princeton University Press, 1985.

Sherman, Sandra. "Trembling Texts: Margaret Cavendish and the Dialectic of Authorship." *English Literary Renaissance* 24 (1) (Winter 1994): 184–210.

Shirley, John. *Thomas Harriot: A Biography*. Oxford: Clarendon Press, 1983.

Shumaker, Wayne. "Accounts of Marvelous Machines in the Renaissance." *Thought* 51 (September 1976): 255–70.

———. "George Dalgarno's Universal Language." In *Renaissance Curiosa*. Medieval and Renaissance Texts and Studies, no. 8. Binghamton, N.Y.: Center for Medieval and Early Renaissance Studies, 1982.

———. *Natural Magic and Modern Science: Four Treatises, 1590–1657*. Medieval and Renaissance Texts and Studies, no. 63. Binghamton, N.Y.: Center for Medieval and Early Renaissance Studies, 1989.

Silverman, Kaja. *The Threshold of the Visible World*. New York: Routledge, 1996.

Simons, Sarah, ed. *No One May Ever Have the Same Knowledge Again: Letters to Mount Wilson Observatory, 1915–1935*. Los Angeles: Society for the Diffusion of Useful Information Press, for the Trustees of the Museum of Jurassic Technology, 1993.

Singer, Dorothy Waley. *Giordano Bruno: His Life and Thought, with Annotated Translation of His Work on the Infinite Universe and Worlds*. New York: Henry Schuman, 1950.

Slaughter, Mary M. *Universal Languages and Scientific Taxonomy in the Seventeenth Century*. Cambridge: Cambridge University Press, 1982.

Snow, C. P. *The Two Cultures and the Scientific Revolution*. Cambridge: Cambridge University Press, 1959.

Sober, Elliott. *Conceptual Issues in Evolutionary Biology*. 2d ed. Cambridge, Mass.: MIT Press, 1994.

Solomon, Julie R. *Objectivity in the Making: Francis Bacon and the Politics of Inquiry*. Baltimore: Johns Hopkins University Press, 1998.

———. "To Know, to Fly, to Conjure." *Renaissance Quarterly* 44 (3) (Autumn 1991): 513–58.

Southerne, Thomas. *Oroonoko: A Tragedy*. 1699.

Spedding, James, Robert Leslie Ellis, and Douglas Heath, eds. *The Works of Francis Bacon*. 14 vols. London: Longmans, 1861–1879.

Spenser, Edmund. *Edmund Spenser's Poetry: Authoritative Texts and Criticism*. Ed. Hugh Maclean. Norton Critical Editions. New York: Norton, 1968.

Sprat, Thomas. *History of the Royal Society*. 1667.

Spufford, Margaret. *Small Books and Pleasant Histories: Popular Fiction and Its Readership in Seventeenth-Century England*. Athens, Ga.: University of Georgia Press, 1981.

Stabler, Arthur P. "Rabelais, Thevet, L'Ile des Démons, et les Paroles Gelées." *Etudes Rabelaisiennes* 11 (1974): 57–62.

Stabler, Arthur P., and Roger Schlesinger, eds. *Thevet's North America* (see entry under Schlesinger).

Staden, Hans von. *The Captivity of Hans Stade of Hesse*. Ed. and trans. Richard Burton. London: Hakluyt Society, o.s., 51, 1874.

Stafford, Barbara Maria. *Body Criticism: Imaging the Unseen in Enlightenment Art and Medicine*. Cambridge, Mass.: MIT Press, 1991.

Stagl, Justin. "The Man Who Called Himself George Psalmanazar, or The Problems of the Authenticity of Ethnographic Description." In *A History of Curiosity: The Theory of Travel, 1550–1800*. Chur, Switzerland: Harwood Academic Publishers, 1995.

Stewart, Susan. "Antipodal Expectations: Notes on the Formosan 'Ethnography' of George Psalmanazar." *History of Anthropology* 6 (1989): 44–73.

——. *On Longing: Narratives of the Miniature, the Gigantic, the Souvenir, the Collection.* 1984. Durham, N.C.: Duke University Press, 1993.

Stocking, George. *Victorian Anthropology.* New York: Free Press, 1987.

——, ed. *Observers Observed: Essays on Ethnographic Fieldwork.* Vol. 1 of *History of Anthropology.* Madison: University of Wisconsin Press, 1983.

Strachan, Geoffrey, trans. *Other Worlds,* by Savinien Cyrano de Bergerac (see entry).

Sturtevant, William C. "First Visual Images of America." In *First Images of America,* ed. Chiappelli (see entry), vol. 1.

Suvin, Darko. *Metamorphoses of Science Fiction: The Poetics and History of a Literary Genre.* New Haven: Yale University Press, 1979.

Swanson, Roy Arthur. "The True, the False, and the Truly False: Lucian's Philosophical Science Fiction." *Science Fiction Studies* 3 (3) (1976): 228–39.

Swanton, John R. *Indians of the Southeastern United States.* Smithsonian Institution Bureau of American Ethnology, bulletin 137. 1946. Reprint, Washington, D.C.: Smithsonian Institution Press, 1979.

Swiderski, Richard M. *The False Formosan: George Psalmanazar and the Eighteenth-Century Experiment of Identity.* San Francisco: Mellen Research University Press, 1991.

Swift, Jonathan. *Travels into several Remote Nations of the World. In Four Parts. By Lemuel Gulliver.* Ed. Robert A. Greenberg. 1726. 2d ed. New York: W. W. Norton, 1970.

Tambiah, Stanley Jeyaraja. *Magic, Science, Religion, and the Scope of Rationality.* Cambridge: Cambridge University Press, 1990.

Tanner, John. "'And Every Star Perhaps a World of Destined Habitation': Milton and Moonmen," *Extrapolation* 30 (3) (1989): 267–279.

Taussig, Michael T. *Mimesis and Alterity: A Particular History of the Senses.* New York: Routledge, 1993.

——. *Shamanism, Colonialism, and the Wild Man: A Study in Terror and Healing.* Chicago: University of Chicago Press, 1987.

Tax, Sol. "From Lafitau to Radcliffe-Browne: A Short History of the Study of Social Organization." In *Social Anthropology of North American Tribes,* ed. Fred Eggan. Chicago: University of Chicago Press, 1955.

Taylor, F. Sherwood. "Alchemical Papers of Dr. Robert Plot." *Ambix* 4 (1 and 2) (December 1949): 67–76.

Terzago, Paulo Mario. *Museo o Galerie Adunata . . . Manfredo Settala.* Tertona, 1664.

Thevet, André. *La Cosmographie universelle d'André Thevet.* 2 vols. 1575.

——. "Description de plusiers isles par M. André Thevet." 1588, Bibliothèque nationale, MS fr. 17174.

——. "Grand insulaire et pilotage d'André Thevet." 2 vols. 1586, Bibliothèque nationale, MS fr. 15452 and 15453.

——. *Les Singularitez de la France Antarctique.* 1558.

——. *Les Singularitez.* Facsimile of 1558. Ed. J. Baudry. Paris: Le Temps, 1981.

——. *Les vrais pourtraicts et vies des hommes illustres grecz, latins, et payens. . . .* 2 vols. 1584.

Thomas, Keith. *Man and the Natural World: A History of the Modern Sensibility.* New York: Pantheon Books, 1983.

——. *Religion and the Decline of Magic.* New York: Charles Scribner and Sons, 1971.

Thomas, Nicholas. *Entangled Objects: Exchange, Material Culture, and Colonialism in the Pacific*. Cambridge: Harvard University Press, 1991.

Thwaites, Reuben Gold, ed. *The Jesuit Relations and Allied Documents: Travels and Explorations of the Jesuit Missionaries in New France, 1610–1791*. Vols. 8–10. Bilingual edition. Cleveland: Burrows Brothers, 1897.

Todd, Dennis. *Imagining Monsters: Miscreations of the Self in Eighteenth-Century England*. Chicago: University of Chicago Press, 1995.

Todd, Janet. *The Secret Life of Aphra Behn*. New Brunswick, N.J.: Rutgers University Press, 1997.

———, ed. *The Works of Aphra Behn*. Columbus: Ohio State University Press, 1995.

Torgovnick, Marianna. *Gone Primitive: Savage Intellects, Modern Lives*. Chicago: University of Chicago Press, 1990.

Tradescant, John, the Younger. *Musaeum Tradescantianum: Or, A Collection of Rarities Preserved at South-Lambeth neer London*. 1656.

Traherne, Thomas. "Leaping over the Moon." In *Poems: "Centuries" and Three "Thanksgivings,"* ed. Anne Riddler. London: Oxford University Press, 1966.

Trigault, Nicolas. *China in the Sixteenth Century: The Journals of Matteo Ricci*. Trans. L. J. Gallagher. New York: Random House, 1953.

Trouillot, Michel-Rolph. "Anthropology and the Savage Slot: The Poetics and Politics of Otherness." In *Recapturing Anthropology*, ed. Fox (see entry).

Tufayl, ibn. *Hayy ibn Yaqzan*. Trans. Len Evan Goodman. In *Ibn Tufayl's "Hayy ibn Yaqzan," A Philosophical Tale*. New York: Twayne, 1972.

Tuffal, Jacqueline. "Les Recueils de costumes gravés au XVIe siècle." In *Actes du 1er Congrès international d'histoire du costume*, by Congrès international d'histoire du costume. Venice: Centro internazionale delle arti e del costume, 1952.

Tylor, E. B. *Religion in Primitive Culture*. Chaps. 11–19 of *Primitive Culture*. 1910. New York: Harper and Brothers, 1958.

U.S. Congress. House. Committee on Government Operations. Legislation and National Security Subcommittee. *Cold War Era Human Subject Experimentation Hearing before the Legislation and National Security Subcommittee*. . . . 103d Cong., 2d sess., 28 September 1994.

Van Helden, Albert. *The Invention of the Telescope*. Transactions of the American Philosophical Society 76, pt. 4. Philadelphia, 1977.

———, trans. *Sidereus Nuncius, or, The Sidereal Messenger*, by Galileo Galilei. (See entry under Galilei.)

Varenius, Bernhardus. *Descriptio regni Japoniae et Siam*. Amsterdam, 1673.

Vickers, Ilse. *Defoe and the New Sciences*. Cambridge: Cambridge University Press, 1966.

Vickers, Nancy. "The Body Re-Membered: Petrarchan Lyric and the Strategies of Description." In *Mimesis: From Mirror to Method, Augustine to Descartes*, ed. John D. Lyons and Stephen G. Nichols Jr. Hanover, N.H.: University Press of New England for Dartmouth College, 1982.

Vida de Lazarillo de Tormes. Burgos, Alcalá, Antwerp, 1554.

Wagner, Roy. *The Invention of Culture*. 2d ed. Chicago: University of Chicago Press, 1981.

Waquet, Françoise. "La Mode: De la folie à l'usage." *Cahiers de l'Association internationale des études francaises* 38 (May 1986): 91–104.
Warren, George. *An Impartial Description of Surinam*. 1667.
Washington, Edward. "'At the Door of Truth': The Hollowness of Signs in Othello." In *Othello: New Essays by Black Writers*, ed. Mythili Kaul. Washington, D.C.: Howard University Press, 1997.
Watt, Ian. *The Rise of the Novel: Studies of Defoe, Richardson, and Fielding*. Berkeley: University of California Press, 1957.
Weatherford, Jack. *Indian Givers: How the Indians of the Americas Transformed the World*. New York: Fawcett Columbine, 1988.
Webster, Charles. *The Great Instauration: Science, Medicine, and Reform, 1626–1660*. London: Duckworth, 1975.
Wells, Christopher. "Dwindling into a Subject: George Psalmanazar, the Exotic Prodigal." Unpublished paper.
Wethey, Harold E. *The Paintings of Titian: Complete Edition*. 3 vols. London: Phaidon Press, 1975.
Whaples, Miriam K. "Exoticism in Dramatic Music, 1600–1800." Ph.D. diss., Indiana University, 1958.
White, Hayden. "The Value of Narrativity in the Representation of Reality." In *On Narrative*, ed. W. J. T. Mitchell. Chicago: University of Chicago Press, 1981.
Wilkins, John. *A Discourse Concerning a New World and Another Planet: The First Book, Discovery of a New World, or A Discourse tending to prove, that 'tis probable there may be another habitable World in the Moon*. 1638.
William of Rubruck. "The Journey of William of Rubruck." In *The Mission to Asia*, ed. Christopher Dawson. Toronto: University of Toronto Press in association with Medieval Academy of America, 1980.
Williams, Arnold. *The Common Expositor: An Account of the Commentaries on Genesis, 1527–1633*. Chapel Hill: University of North Carolina Press, 1948.
Williams, Brackette. "Dutchman Ghosts and the History Mystery: Ritual, Colonizer, and Colonized Interpretations of the 1763 Berbice Slave Rebellion." *Journal of Historical Sociology* 3 (2) (June 1990): 133–65.
Williams, Eric. *Capitalism and Slavery*. Chapel Hill: University of North Carolina Press, 1944.
Williams, Raymond. "Utopia and Science Fiction." *Science Fiction Studies* 5 (16) (1978): 203–14.
Williams, Roger. *A Key into the Language of America*. 1643.
Williamson, James A. *English Colonies in Guiana and on the Amazon, 1604–1668*. Oxford: Clarendon Press, 1923.
Willughby, Francis. *The Ornithology of Francis Willughby, . . . translated into English, and enlarged with many additions . . . to which are added Three considerable discourses . . . by John Ray*. 1678.
Wilson, Catherine. *The Invisible World: Early Modern Philosophy and the Invention of the Microscope*. Princeton: Princeton University Press, 1995.

———. "Visual Surface and Visual Symbol: The Microscope and the Occult in Early Modern Science." *Journal of the History of Ideas* 49 (1988): 85–108.

Wilson, Dudley. *Signs and Portents: Monstrous Births from the Middle Ages to the Enlightenment*. London: Routledge, 1993.

Wollock, Jeffrey. "John Bulwer's (1606–1656) Place in the History of the Deaf." *Historiographia Linguistica* 23 (1–2) (1996): 417–33.

Woolf, Virginia. *A Room of One's Own*. London: Harcourt, Brace, Jovanovich, 1929.

Worm, Olaus. *Museum Wormianum, seu Historia rerum rariorum, adornata ab Olao Wormio*. Leyden, 1655.

Wotton, Sir Henry. *The Life and Letters of Sir Henry Wotton*. Ed. Logan Pearsall Smith. 2 vols. Oxford: Clarendon Press, 1907.

Yates, Frances. *Giordano Bruno and the Hermetic Tradition*. London: Routledge and Kegan Paul, 1964.

———. *The Rosicrucian Enlightenment*. London: Routledge and Kegan Paul, 1972.

INDEX

Names and titles not quoted or individually discussed are not indexed.
Page numbers in italic refer to figures.

Abbot, George, 145
Académie des Sciences, 6, 98n, 144, 145
Acosta, Joseph, 135, 287
Adam and Eve, *54*, 63, 288
Adams, Percy, 288
Africans, 90, 91, 94, 95, 188, 269; represented as Europeans, *93*, 188
Agricola, Georg, 44
air, 196
Alexander the Great, 146
Algonquins, 51–53, 57n, *58*, *59*, 309; engravings of, 52, 57–59; reaction to Bible, 52–53, 60
Amadas, Philip, 1
America, 49, 131, 324; appropriation of, 48–50; imaginative opportunity, 155, 180; Moon and, 136, 152–53, 154–56, 159, 166–67, 169, 172 (*see also* New World; *Somnium*); narrative of, 1, 63, 204; as *topos*, 49, 308
America, Part 1, 51–67; early printing history, 53n; exoticism, 57–59
Americans, 29, 42, 46, *57–59*, 268, 308; absence of, 36–38; "naked," 260, 263, *263*, 267; nonhuman, 253; origins of, 288, 292. *See also* Africans; Algonquins; Iroquois; Tupinamba
Americas, 41–42. *See also* Brazil; Canada; New World; Surinam; Virginia
Anderson, Benedict, 113n
animal, as category, 173, 175, 179
anthropology, 9–13, 176, 208, 251, 291, 298; and Browne, 88, 89, 93–94, 96; goals of, 66, 173, 287; and Lafitau, 285–310; and microanthropology, 317, 323; predisciplinary, 176, 248–49, 289, 289n; seventeenth-century term, 264n; universalizing tendency of, 18, 211n, 223. *See also* anthropology, physical; colonial expansion; ethnography; ethnology; human
anthropology, cultural. *See* ethnology
anthropology, physical: human origins, 43–44, 288; and Moon, 143, 167; and sentient species, 205. *See also* Laet, de; primates; species
apes. *See* primates
astronomy, 112, 143; and fiction, 158, 212;

astronomy (*continued*)
and narrative, 128–30; and pleasure, 149
Arcadia, 161; space as, 129
Aretino, Pietro, 183, 196
Ariosto, Ludovico, 153
Aristotle, 95n, 119, 121–22, 122–23; on slavery, 10n; on women, 243
Asquacoqoc, razing of, 61
Aubrey, John, *Brief Lives*, 60n
Austen, Jane, 75, 166
automata, 254, 282. See also robot
autre monde, L' (Cyrano de Bergerac), 12, 171–80; and Cavendish, 202–3; publication history, 172, 172n
Auzout, Adrian, 76–77n
Axtell, James, 30n, 48, 226n

Bachelard, Gaston, 6
Bacon, Sir Francis, 72, 74, 75, 80n, 82, 82–83n, 96; *Advancement of Learning*, 5, 74, 78, 87, 139n; and figurative language, 73–78; *New Atlantis*, 62, 170n; *Novum Organum*, 16, 69, 71–80, 72n, 86, 109n. See also Spedding
Bacon, Roger, 103, 123
ballet, 18, 229–30
Barlowe, Arthur, 1
Bateson, Gregory, 266
Bath, 323; *Bath Herald*, 319
Battigelli, Anna, 206n
bees, as extraterrestrials, 10n
Behn, Aphra, 230, 258, 265, 267, 276; *Emperor of the Moon*, 258, 282–83; translator of Fontenelle, 115, 257, 287; *The Widow Ranter*, 261. See also *Oroonoko*
Belleforest, François de, 31, 37n
Bender, John, 6n
Benjamin, Walter, 139
Benzoni, Girolamo, 28
Bible, 52–53, 287
Bibliothèque bleue, 133n
Bierce, Ambrose, 188

Bishop, Elizabeth, 36, 166
Blackwell, Thomas, 294n
Blaeu, Willem, 90
Blake, William, 112
blazon, 211–12, 212n, 215, 217; of violence, 246, 301
Blondel, James, 246–47n
Blount, Thomas, *Glossographia*, 260
Boccaccio, Giovanni, 49, 264
body: and adornment, 17, 228; as costume, 253; as costume model, 230; fashions, 18, 228, 236, 258 (*see also* tattoo); figure for nation, 242; mutability of, 232–33, 258. See also pain; sensation
Boissard, Jean-Jacques, *Habitus variarum orbis gentium*, 228, *229*
Borges, Jorge Luis, 33
Bora, Renu, 194n
Bose, Georg Matthias, and *Venus electrificata*, 67n
Bossy, John, 122
Bowen, Emanuel, 312n, 314
Boyle, Robert, 77
Brazil: and Crusoe, 320; and Léry, 69–70; and *Receuil de la diversité des habits*, 229, 253; and Thevet, 30n, 47, 69
Brébeuf, Jean de, 285, 294, 304–5, 306
Breiner, Laurence A., 41n
Brief and true reporte of . . . Virginia (Hariot), 51, 53, 55, 123, 140n. See also *America, Part I*
Browne, Sir Thomas: "Blackness of Negroes," 89–96, 240; on class, 97–101; *Musaeum Clausum*, 87n; on plagiarism, 31n; and racism, 93–96; "Urne Buriall," 88. See also *Pseudodoxia epidemica*
Bruno, Giordano, 10, 113–14; cosmology, and Galileo's, 131; as Henry Fagot, 122n; and Godwin, 156n; *De l'infinito universo et mondi*, 113, 114n, 118–23, 182; *De innuberalis immenso et infigurabili*, 114n; and J. Kepler, 114, 134–35, 143
Bucher, Bernadette, 48

Bulwer, John, 233, 241n, 263; *Anthropometamorphosis*, 221, 228, 233–50; *View of the People of the Whole World*, 234n, 235
Burton, Robert, 81, 228
Bynum, Carolyn Walker, 3

Cabeza de Vaca, Alvar Nuñez, 23, 261
cabinets: Ashmolean, 23, 70, 80n, 81; Browne's, 87; collections, 227; of Royal Society ("Antiquaries"), 277; Tradescant's Ark, 80, 101. *See also* museums
Cabot, John, 31
Calvino, Italo, 33
Campanella, Tommaso, 250n
Campbell, Mary Baine, 46, 282
Canada ("New France"): and Cyrano de Bergerac, 152, 176, 180; and Lafitau, 49, 285, 285–86n, 289; Matter of, 306; and Thevet, 30n, 31–32, 35, 47, 49, 50, 156
Candidius, George, 312n
cannibalism, 210, 227; in *L'autre monde*, 173; Christian, 309–10; in Formosa, 312; in New World, 173n, 309n; in Old World, 309n
"cannibals" (label), 264, 320
capitalism: and atheism, 294; compared to evangelism, 309; expansion of, 295. *See also* colonial expansion; empire
Caraboo, Princess, 19, 319, 323–24, *324*
Carroll, Margaret, 125–26, 129n
Cartier, Jacques, 31, 33–34
catalogue: and detail, 183, 186; and narrative, 80–82, 84; and pornography, 183; of unwritten books, 79
catalogues: of collections, 79–85, 227; list of, 83–84n; parody of, 87n; of private libraries, 15, 15n; and property, 80, 82
categories: anthropological, 254; "Blackness" as, 93; concept of, 80; social, 251; and wonder, 250–51
Cavendish, Margaret, Duchess of Newcastle, 209, 222, 324; and atomism, 182n, 205n; *Blazing-World*, 12, 39n, 112, 179, 202–18, 270, 308; and Hooke, 182, 203, 213–15, 217–18; *Nature's Pictures*, 208n; *Observations upon Experimental Philosophy*, 203, 206, 207, 216–18; *Poems and Fancies*, 181, 214; *Sociable Letters*, 204
Cavendish, William, Duke of Newcastle, 203, 208, 213; as character, 207–9
Centlivre, Susan, 225
Certeau, Michel, 26, 226, 290–95
Chappuys, Gabriel, 253
character: 13, 121–22, 128, 129, 178, 211, 281; and agency, 280–82; avatar(s) of narrator, 272; banished, 178; duplicated, 208; enslaved, 279–82; "lunatic," 178–80; narrator as, 50, 101, 128, 172. *See also* La Bruyère
Charles II (of England), 192, 195, 276
Chibka, Robert, 266–68
China, 162, 163, 294; and Defoe, 317, 321; languages of, 162; and *Man in the Moon*, 152, 158, 159; and utopia, 159
Christianity, 299–300, 304, 309–10, 313; and Church, 123; and colonialism, 60, 309, 320; and ethnology, 295, 310
Civil War (English), 213
Clarke, Arthur C., 120
Clifford, James, 62, 271
Codex Mendoza, 31, 31n
colonial conquest: activities of, 75; as allegory, 74, 190, 282n, 295; pain of, 303; *Sidereus nuncius* and, 129–31
colonial expansion, 117, 145, 176, 187, 222, 256, 309; and *Blazing-World* (M. Cavendish), 204; and moon voyages, 152, 176; and *Somnium*, 135, 140, 142
colonial report, 12, 51, 185, 272. See also *America, Part 1*; *Brief and true reporte*; Candidius; Warren
Columbus, Christopher: *Epistola de insulis nuper repertis* (*Letter*), 133, 156, 182; on islands, 40n; as narrator, 23, 49; as

New World writer, 4n, 9, 9n, 251; as pseudonym, 1
Copernicus, Nicolas, 7–8, 116, 117n, 143, 153
Coppo, Pietro, world chart, 42, *42*
Cornelius, Paul, 162
Cortés, Hernan, 74, 168, 169, 217
Cosmographie universelle. *See* Thevet
cosmography, 16, 47–50, 287; and acquisition, 30, 47–50; and aesthetic pleasure, 29; commodity, 47–48; and fiction, 12, 38–39, 147–48; and market, 26–29
cosmology, 116, 117, 119, 135; social models of, 118, 125, 129, 132, 135, 205
Cotton, Charles, *Erototopolis*, 203–4n
Couliano, Ioan, 71–72
Courtier, The (Castiglione), 4
Cox, Captain, book collector, 28n
Cross, the, in America, 299–300
Crusades, 49n
Crusoe, Robinson (character), 19, 36, 38, 39, 158, 166, 198, 206, 319–23
cryptography. *See* stenographia
culture: construction of, 174, 176, 320 (*see also* scientific communities; utopia); history of concept, 234–45, 236, 245, 294, 322; as performance, 18 (*see also* performance); one person as, 321, 323; as term, 94, 245n. *See also* anthropology; ethnography; ethnology
Cunningham, J. V., 3
Cyclops, 229, 251–54, 263
Cyrano de Bergerac, Savinien de, 12n, 49. *See also autre monde, L'*

Daemon from Levania (character), 139, 141, 143
Dampier, William, 222
Danae, 125, 126
Darwin, Charles, 7
Daston, Lorraine, 75n; and Katharine Park, 3, 5
de Bry, Theodor: and *America, Part 1*, 51, 52, 53, 55; as engraver, 51, 55, figs. 3, 5,

7, 8, 9; *Les grands voyages*, 16, 27, 51, 53; New World engravings of, 262, 266; publisher and editor, 16, 26, 51, 53, 55. *See also* Hariot
Dee, John, and Euclid, 124, 127n, 128n
Defoe, Daniel, 317, 319, 322. See also *Robinson Crusoe*
Dejean, Joan, 180
Delft, Louis van, 231
Delgarno, George, 88n
demon of Socrates (character), 172, 175
depopulation, 146, 147–48; and repopulation, 166–67
Descartes, René, 77, 148, 309; Lafitau and, 298; and machines, 103, 104; *Meditations*, 69, 105, 299; and pain, 306; and robots, 296
Desprez, François. See *Receuil de la diversité des habits*
detail, 26, 188, 215, 317; and attention, 201; and commerce, 187–88; and sadism, 186; sensational, 278; and visual representation, 183, 184, 192, 196–99, 201–2, 214; and writing, 183–202, 300–304
dialogue, 12, 191; philosophical, 118–23, 175, 177–78
Dictionnaire de botanique chrétienne, 191, 192n
Diego (Domingo's slave), 164–65
Digges, Leonard, 110, 116n
Digges, Thomas, *110*, 116, 116n
dismemberment: European practice of, 264; as getting "inside," 272; of Iroquois, 302; of objectified persons, 270; of Oroonoko, 258, 263, 265
distinction, loss of, 232–33, 243, 268, 278
Don Quixote (Cervantes), 201, 280
Donne, John, 40, 323
Douglas, Mary, 235n
drone-fly, gray, 196–99, *197*, 213–15
Dryden, John: and Robert Howard, *Indian Queen* (play), 18n, 261; and Henry Purcell, *Indian Queen* (opera), 230
Dumée, Jeanne, 148–49

Durocotus (character), 137–38, 143
Dyrcona (character), 175; as female, 154, 173, 179–80

Earle, John, *Microcosmographie*, 231
Earth: as ethnos, 145; as home, 164; as moon, 140–41, 175; as ornament, 141; as world, 141
East India Company, 186, 256
ecole des filles, L', 184–88
Eden: paradigm, 62n; and present tense, 65–66; trope, 1, 61, 273. *See also* paradise
Egmond, Florike, and Peter Mason, 65n
El Dorado, 26n
empire: colonial, 1, 2, 51, 62, 120; mercantile, 22, 238
enargeia, 174, 186n
England's Vanity, 228, 241
Entretiens sur la pluralité des mondes (Fontenelle), 6, 10n, 143–49; as dialogue, 144; as fiction, 115, 144, 147–49; and *Somnium*, 145
error, 73, 85, 85n, 88–89, 92–94, 97
escape: of feral child, 322; in *Oroonoko*, 259, 277, 281; and pleasure, 169; as theme, 169, 171, 176
escapist fiction: *Blazing-World*, 204; medieval romance, 169; moon voyages, 170, 204
ethnography, 9, 25, 66, 189; defined, 51; commercial, 52–62; and fiction, 29–67, 147–48; history of, 26–29, 28n, 51 (see also *America, Part 1*; ethnology); and illustrations, 51–52, 57–59; and language, 11, 51; lunar, 158, 159, 205; scientized, 16, 59; and time, 47, 62–63
ethnology, 17, 25–27, 65, 221; and ancient world, 45–46, 289, 290; and commercial publishing, 26–29, 30–31, 41n, 48, 295; and dance, 229–30, 246–48; defined, 51; didactic, 231, 290, 313 (*see also* Tylor); and fashion, 17, 225–33, 236, 242–50, 254–56; and fashion plates, 228–29; and fiction, 26–67, 34, 143, 287; history of, 289; lunar, 143, 145, 179; and natural history, 58–103
eucharist, 173–74, 312–13
Evelyn, John, 86
experience, 72, 156; and experiment, 73, 128n; and facts, 62, 70; female, 230, 274; hidden, 272; non-subjective, 106; prosthetic, 126, 130; and scientific authority, 28; shared by reader, 128. *See also* identification; reading, as experience; witnessing
experiment: and artifice, 106; and experience, 73, 128n; and play, 199–200; and pleasure, 189; and imaginative literature, 215; reports on, 104, 109, 127–28, 303; thought experiment, 319

Fabian, Johannes, 62–63n
Fanon, Frantz, 301n
fashion, 18, 225, 226, 233, 236; and disease, 231, 232–33; fashion plates, 228–30, *229*, 254, 260, 263; ladies of, 244, 245; and Tradescant, 81n
Fénelon, François de Salignac de la Mothe, *Télémaque*, 287
Fenton, William, 289
feral children ("wolf children"), 320, 321, 322–23. *See also* Victor
Ferguson, Margaret, 270
Fernandez de Oviedo, Gonsálo de, 27, 48n
fetish, and detail, 187–88
fieldwork, 18, 23, 51; as intervention, 65–66. *See also* ethnography
Fiolxholde (character), 134n, 137, 138–39. *See also* Kepler, K.; Libussa
Fitelieu, M. de, *La contre-mode*, 231–32, 246n
Flecknoe, Richard, 13, 185
Fletcher, Angus, 242n
Fontenelle, Bernard le Bovier: and Bruno, 144, 145, 149; *La Comète*, 143–44; on extra-terrestrials, 119; *L'histoire des oracles*, 145, 287, 298; "Relaçion curieuse

Fontenelle, Bernard le Bovier (*continued*) de l'île de Bornéo," 145, 147; on universe as theatre, 144–45. See also *Entretiens de la pluralité des mondes*
Formosa (Taiwan), 310, 312, 314; alphabet of, 314, *315*; map of, 313
Foucault, Michel, 14n, 93, 117, 215; *heterotopias*, 304n
Freud, Sigmund, 27, 29, 76
Friday (character), 198, 320
Frost, Robert, "Desert Places," 32

Galileo Galilei, 124, 127, 130; and *L'autre monde*, 171, 175; *Dialogue Concerning the Two Chief World Systems*, 115, 134n. See also *Sidereus nuncius*
Gallagher, Catherine, 209, 270, 271n
gallants, 236, 239, 242, 243, 244, 245
Gardiner, Chauncy, 171
Garland, Judy, 164
Gelbart, Nina Rattner, 144
gender, 179, 271; and identification, 305; and monstrosity, 242–43, 244n; of narrator, 179–80; and sodomy, 242. See also experience: female; hermaphrodites; hybrid; hybridity; parthenogenesis; women
Genesis 1:2, 253
genres, 12, 14, 202; antipseudodoxical, 85, 85–86n, 100; autoethnography, 270; colonial report, 185; Greek novel, 282; heroic drama, 269, 280, 281; ethnography, 51, 185; memoir, 317; mirror for princes, 290, 309; moon voyage, 202–3; *receuil de costumes*, 228–29; science fiction, 65; utopia, 14, 204. *See also individual genres*
Geoffrey of Monmouth, 100n
Glück, Louise, "Vita Nova," 2
Godwin, Francis: *Nuncius inanimata*, 158; pseudonym of, 154, 155. *See also* Gonsales, Domingo; *Man in the Moon*
Gonsales, Domingo (character): lineage of, 167–68; and Cortés, 169; character in *L'autre monde*, 172, 173, 175, 179; as midget, 158, 160, 165, 167–68, 170; protagonist of *Man in the Moon*, 152, 158, 178, 179, 180; and Ralegh, 156n
Goodman, Nelson, 9, 10, 15, 324
Greenblatt, Stephen, 60
Greene, Robert, *Friar Bacon and Friar Bungay*, 126n
Grew, Nehemiah, 190
Grossman, Allen, 118n, 265n
Grotto at Enston, 106, *108*
Grove, Richard, 42n, 165n
Guiana, 269. *See also* Ralegh
Guinea, gold trade, 187
Gulliver, Lemuel (character), 23, 188, 198; and *Man in the Moon*, 158
Gutch, J. M., *Caraboo*, 323, 324
gypsies, 96, 324

Hackluyt, Richard, 1, 37, 163
hairy girl of Pisa, 244–45
Hall, Joseph, *Mundus alter et idem*, 203
Hallyn, Fernand, 117
Hariot, Thomas, 52–53, 60, 61n, 62, 124, 257; captions for de Bry, 61; and ethnographic discourse, 51–66, 226, 279; magician, 60n, 61; on Virginia, 16, 51–66, 279. See also *America, Part 1*; *Brief reporte*; White
Harrisse, Henry, 45
Harth, Erica, 12n
Harwood, John, 186n
headdress from Surinam, 261
Heart of Darkness (Conrad), 267, 269
heliocentrism, 115, 116, 118. *See also* Copernicus
hermaphrodites, 208n, 246–48
Herodotus, 45, 289
Herrick, Robert, 194
heterocosmos, 118n
Hodgen, Margaret, 86–87
Homer, 287, 294n; *Odyssey*, 33, 232, 289

Hooke, Robert, 186, 187, 188, 189. See also *Micrographia*
Huet, Marie-Hélène, 244, 246–47n, 250
Hulme, Peter, 50
Hulton, Paul, 53n, 57
human: as category, 19, 119, 142, 173, 174, 175, 176, 251–54, 262, 264; as collectivity, 205, 305; as commodity, 27; experimental objects, 91, 215; *Homo americanus*, 293; *Homo sapiens*, 321; *Homo sapiens ferus*, 321, 322 (*see also* Victor; Wild Man; Wild Woman); as spectrum, 253; subcategories of, 293
Hunt, Lynn, 186
Hunter, Paul, 7n, 12n
Huygens, Christiaan, 130n, 205, 222
hybrid, 13, 235; fops, 227, 239; and J. Kepler, 138; masculine-feminines, 227; transexuality, 242; unmoralized, 246
hybridity, 20, 208, 227, 254
hyperbole, 120, 169, 203, 198, 303

identification, 13, 65, 127, 178, 108–9, 210–13, 213n, 217, 261; and idealization, 210–11, 217; with the feminine, 210–11; with lunars, 140; with *pícaro*, 158; with slaves, 280
idols: and F. Bacon, 74; and Cortés, 74
illustration, scientific, 52n, 56, 58, 84; defined, 55–56; ethnographic, 245–46; ethnological, 57–59. See also *Micrographia*
image: dissemination of, 245–46; as event, 213; "Spectacle of a Mangled King," 263–64, 273, 275, 276–77, 281, 308
Indians. *See* Americans
information, 133, 171, 229, 215; and fiction, 112, 171, 230; rhetoric of, 292
instrument, 190, 209, 215; as prosthetic eye, 111, 269–70; as prosthetic hand, 189; scientific, 75, 84, 189–90
interior: concept of, 182, domestic, 198; and exterior, 206, 321; knowability of, 13, 92, 192; of machines, 105–9; and monstrosity, 240, 240n; of person, 182, 192; unknowability of, 215
interiority, 304, 309; of characters, 206, 207–10, 213, 308; of Iroquois captives, 304, 306
Iroquois, 285, 292, 298–99, 310; dreams of, 299; kinship system, 289; war captives, 301–2, 304–6, *307*
islands, 16, 32–34, 42n, 46–47, 101–2, 135–36; artificial, 105–9, *107*; and biogeography, 42–44; celestial bodies as, 116, 121, 135–36, 156–57; desert, 36–40, 42–44, 319–22; Hong Kong, 152; Iceland, 136; imaginary, 135–36, 159, 286; Isle of Demons, 35–40, 42–43; Isle of Thevet, 37, 40, 165; and self, 40, 164–65, 170; St. Helena, 152, 159, 164–66, 168–70, 321 (*see also* paradise); as *topos*, 37, 41–42, 155; Trinidad and Tobago, 317, 321. See also Formosa; Javasu
Itard, Jean Marc Gaspard, 19, 322–23; constructivist, 322; and Hooke, 322

Jakobsen, Roman, 184
James, Henry, 194n
James I (of England), 85
Jameson, Fredric, 141
Janowitz, Anne, 324n
Javasu, 19, 323; alphabet of, 324, *324*
Jeamson, Thomas, 228
Jesuits, 152, 158, 159, 169, 310, 312. *See also* Brébeuf; journals: *Mémoires de Trévoux*; Lafitau; Pantoja; Ricci
Jesus Christ, 304, 314
Johnson, Samuel, 187
journals, scientific, 278; *Journal des sçavans*, 145; *Journal of the Royal Asiatic Society*, 314; *Mémoires de Trévoux*, 285n, 286, 294n; *Philosophical Transactions*, 59–60, 277
Jupiter, moons of, 114, 125, 127, 128, 145

Kansas, 135, 168
Kant, Immanuel, 188
Kepler, Johannes, 72, 140n; *Astronomica Copernicana*, 139; and Bruno, 114, 134–35, 143; "*Conversation with the* Starry Messenger," 134–35; *Dioptrice*, 123, 126, 200–201; on extraterrestrial beings, 119, 180, 203; and Galileo, 134n, 134–35; *Harmonica Mundi*, 115; *Mysterium cosmographicum*, 115, 117n. See also *Somnium*
Kepler, Katerina, 134, 136, 138
Kircher, Athanasius, 142
knowledge, 77–78, 96, 139–40, 154; alternative, 138–39; and gender, 138–39, 148–49; hermetic, 99–100; management of, 23, 72–85, 98–99, 100, 112, 221; of persons, 310; Tree of, 54, 55, 176, 189
Koestler, Arthur, 113
Kristeva, Julia, 235

La Bruyére, *Caractères*, 231, 232
Laet, Johann de, 288
Lafayette, Madame de. See *Princesse de Clèves, La*
Lafitau, Joseph, 17, 18, 49, 223, 285, 288, 312; as Christian, 288n, 298–99, 293; and fieldwork, 18, 289–90, 292; ginseng, 285–86n; lost history, 292–93n; and study of religion, 290, 293–95, 298. See *Moeurs des sauvages amériquains*
lang, k. d., 202
Lahontan, Baron de, 287, 306, 310
Las Casas, Bartolomé, 10n, 175, 299
Latour, Bruno, 7, 7n, 8n, 13, 20
Laudonniére, René de, 31
Laugaa, Maurice, 179
Lazarillo de Tormes, 156
Lebowitz, Fran, 191
Leeuwenhoek, Antony van, 174, 213n
Léry, Jean de, 70; *Histoire d'un voyage*, 27–28, 69–70, 304; and Lafitau, 296–98
Lestringant, Frank, 29, 32n, 310, 313

Levania (Moon), 135, 136, 140, 142
Lévi-Strauss, Claude, 65, 168
Li Yu, 126n
Libussa (of Bohemia), 137, 140
Linnaeus, Carolus, 44n, 222, 293, 320–21
Lipovetsky, Gilles, 226, 232n
list, 80–81, 80n, 106, 142, 228; and anatomy, 228, 277; and narrative, 80–81n; rhetorical device, 30, 79, 202
Liu, Lydia, 320n
Locke, John, 181
Longino, Michèle, 230n
Lopez de Gómara, Francisco, 31
Lord's Prayer, 314, *316*
louse, 195, 213, 213n
Lucian, 152, 202–3
Lucretius, 116
lunar color, 160–61, 170, 175

magic, 154, 165, 174; and Hariot, 60, 61n; and K. Kepler, 134, 138; and language, 138–39, 140n; as "magia," 96, 99; natural, 69; pneumatic, 139; and representation, 139–40, 140n
magus: Bruno as, 117–18, 139; Dee as, 124
Malinowski, Bronislaw, 26–27, 66n
Malson, Lucien, 319, 321
Man in the Moon (Godwin), 151–71, 202–3; and *L'autre monde*, 171–80; Spanish protagonist, 156n
Mandeville, Sir John, 31, 174, 313
manners and customs, 227, 281, 283; and Lafitau, 289, 291
Marguerite de Navarre, 37n, 264
Martyr, Peter, 4n, 30
Marx, Karl, 32n
Mead, Margaret, 66n, 266
Medici, Cosimo de, 125–26, 132
Melville, Herman, *Moby Dick*, 87, 281
Micrographia, 111, 181, 183–202, 213–15, 217n, 225; and *cosmographia*, 190; and novel, 191; as romance, 189, and utopia, 189

microscope, 181, 182, 188, 189, 208; critique of, by M. Cavendish, 213–18; deconstructive, 215, 230; and reification, 216, 217
Milton, John, 73, 168
mimesis: and magic, 139–40, 140n; and revolution, 180, 208n
miscegenation: and fashion, 233, 234n; on Moon, 167; on stage, 269
"Misospillus," *Wonder of Wonders*, 234n
"la Mode," 225, 227, 236, 249
Moeurs des sauvages amériquains (Lafitau), 222, 286–310, 313; and historicity, 291, 309, 310; reviews of, 285, 286n, 294; as scientific ethnology, 287
Molière, Jean Baptiste Poquelin, 230n
monkeys, 174n
monsters, 92, 175, 247, 248, 248n
monstrosity, 227, 238, 240n; and culture, 236–56; and the foreign, 239–40, 242, 245, 246–49; and gender, 242–43
monstrous races, 92, 248n, 254
Montaigne, Michel de, 174, 297
Moon, 115, 135, 159, 164, 177, 205, 321; without affinities, 161–63, 166; conquest of, 136, 146; fictional object, 115, 131; fictional setting, 135–43; language of, 161–62, 172; and satire, 161, 177; uselessness of, 130–31, 169
moon voyages, 134, 151–55, 169, 170. See also *autre monde, L'*; Lucian; *Man in the Moon*; *Somnium*
Moore, Elizabeth, 289, 290n
Moore, Marianne, "Poetry," 151
Moors, 96, 147, 164, 229
More, Saint Thomas, 232, 313
Morgan, Lewis Henry, 289
Morris, John, library of, 28
Mt. Wilson Observatory, 5n
Münster, Sebastian, 44, 190
museums, 85, 270; catalogues of, 80, 80n, 81n, 83, 101. *See also* cabinets
music: and language, 161–62; and transport, 296–98

mutilation, 262, 268; punitive, 258; as representation, 262–63; ritual, 258; of self, 262–63, 269, 308

nakedness, as trope, 151–53, 260, 260–61n, 262–63, 263, 267, 272n
narrator, 204, 205, 306; female, 179–80, 262, 267, 270–72, 274–75; as protagonist, 13, 23, 53, 153, 172, 292; as observer, 70, 101; as witness, 302 (see also *Oroonoko*; *Sidereus nuncius*)
Nashe, Thomas, 50n
National Geographic, 59
natural history, 30, 55, 56, 76, 85, 102, 308; and F. Bacon's reformation, 72–85; and novel, 12, 281
natural philosophy, 12, 62, 76, 79, 189, 290–91; exclusion of subjectivity, 308
nature, 85, 234, 239; artificial, 106, 109; v. culture, 236–56; and gender, 73, 75, 84–85; as Daphne, 85, 86; as female love interest, 191; as opera house, 144–45; and supernatural, 70
Neoplatonism, 116
New World, 4, 25, 43, 46, 135, 155; and human origins, 43–44, 288 (*see also* Laet, de); metaphorical, 190, 190n, 203; parallels with old, 285, 289, 290, 292, 293, 297, 309; and *Somnium*, 135, 136, 142; as trope, 1, 29. *See also* Behn; Bucher; de Bry; Lafitau; Hariot; *Moeurs des sauvages amériquains*; *Oroonoko*; Thevet
New World writing, 135, 136, 153, 232, 261, 273
Newton, Isaac, 72, 112, 222
Nicholl, Charles, 26n
Nicolas of Cusa, 116, 130
Nicolson, Marjorie Hope, 154n, 159
nightingale, artificial, 106
No One May Ever Have the Same Knowledge Again, 5n
novel, 11–12, 132, 211–12n, 259, 264n, 267; and catalogue, 80–81; and detail,

novel (*continued*)
81; and dialogue, 191; and disenfranchised protagonists, 280; and fashion, 226; and gender, 182–83, 230, 274–75; history of, 186, 201, 206, 230; and New World, 279; picaresque, 133, 151, 158, 171 (see also *Lazarillo de Tormes*, *vida del Buscón, La*); realistic, 75, 206, 232; and *Sidereus nuncius*, 132–33; space in, 190–91

Noyes, Reverend Nicholas, 232

O'Banion, John D., 80–81n
objectivity: in fiction, 282; in Lafitau, 293, 310; rhetoric of, 293, 306
observation, 194, 195, 322; and pleasure, 195, 201; sadistic, 276; and style, 278. See also detail; microscope; surveillance; telescope
objectification, 196, 226; of Algonquins, 63; language of, 75, 77–78, 244; of women, 85, 244
Oldenburg, Henry, 277
Olmos, Andrés de, 28n, 31, 96
Ong, Walter, 82–84, 97n
"Oroonoko" (Carib word), 268–69
Oroonoko (Behn), 12, 18, 201, 221, 258–83, 287; and anthropology, 260, 264; and historical ethnography, 271; historicity of, 266–67; and Lafitau's torture passages, 301; and moon voyages, 259; as pornographic novel, 273; publication history, 259n; as realistic fiction, 260; as romance, 258
other worlds, 1, 10, 75, 78; actuality of, 114–15; America as, 49, 309; Europe as, 308; experience of, 130; imaginary, 203; list of examples, 78; moon as, 153
Oxford, 101, 117, 189
Oxfordshire, 70

Pagden, Anthony, 51
pain, 221, 309; and the aesthetic, 27, 29;
ritualized, 300–310; and the sublime, 283; vicarious, 308
Pantoja, Father, 163
Paracelsus, 242
paradise: as Browne's house, 86; as island, 40, 40n; on Moon, 159, 176; as St. Helena, 159; as trope, 1
Paré, Ambroise, *Des monstres*, 246n
Paris, James, 248n
Park, Katharine, 3, 5
Parker, Patricia, 214
parthenogenesis, 174, 242
"Patch, The: An Heroi-Comical Poem," 234n
Pepys, Samuel, 184
performance: of imaginary foreignness, 18, 285–86, 310–17, 323–24; of lunar language, 162; queer, 18n
perspective glass. See telescope
petit prince, Le (Saint-Exupéry), 147
Petty, William, 77
Phillip II (of Spain), 126
picaresque, the, 133, 156, 157, 158
Picts (and Britons), 63, 65; in ethnographic illustration, 52, 55, 63, *64*
planets. See Earth; Jupiter; Moon; other worlds; Venus
pleasure, 52, 100, 146, 148, 254; and ownership, 27–28, 148, 186; and deception, 105; and early modern ethnography, 26–29, 51–67; and illustration, 27, 52, 59, 66–67, 235; and marketplace, 26–29, 51, 55, 59, 61; and scientific knowledge, 76, 127, 133, 146, 189
Pliny the Elder, 97–98, 289
plot, 279, 280; in *Micrographia*, 190, 191, 192, 201; erotic, 191, 192; of object loss, 201; of romance, 267, 278, 282; structured like dreams, 149, 170
Plot, Robert, 17, 70, 101, 103, 105; *Natural History of Oxfordshire*, 70, 101–9
plurality of worlds, 111, 113–49; aestheticized, 115; and difference, 147

Plutarch, *De Facie lunae*, 152, 153, 180, 291
poetry, 3, 84, 183, 216; and anthropology, 21, 287; epistemological role, 264–65; exotic, 297–98; song, 304n, 309. *See also* music
Polo, Marco, 33
popularization, 6, 12n, 134n, 154n; and Fontenelle, 115, 143, 144, 149
pornography, 183, 191; and comparative anatomy, 280n; and ethnographic illustration, 26–27, 66–67; "philosophical literature," 15, 187; and verisimilitude, 183–86, 191. *See also* rhetoric: of pornography
Postel, Guillaume, 34
Pratt, Mary Louise, 10n, 46, 136, 146, 147, 270
present tense, 23, 62–63, 65–66, 201
primates, 205, 240. *See also* monkeys
Princesse de Clèves, La (Madame de Lafayette), 115, 147, 201, 206, 259n, 288
printing, 71, 83, 97; and monstrosity, 245–46
protagonist, 13, 207, 217; African, 270; cybernetic, 189, 190; female, 267; instrument as, 127; as spectacle, 206, 212; and unstable subjectivity, 177–80, 217
Prynne, William, 256
Psalmanazar, George: and Formosa, 285–86, 310–17; *The Geographical and Historical Description of Formosa*, 18, 286, 310–18; later writings of, 312n, 314, 317; performance, 18, 317
pseudodoxia, and fiction, 100–101
Pseudodoxia epidemica, 16–17, 85–101; and anthropology, 89–96; as catalogue, 86–87; quoted in *Anthropometamorphosis* (Bulwer), 240; and *Man in the Moon*, 161; as wonder book, 86–87
Purchas, Samuel, 32, 163
Putumayo, 265–66

Rabelais, François, 49; and islands, 33–35; and plagiarism, 34–35; *Quart livre*, 35–36, 39
racism, 93–96, 166
Ralegh, Sir Walter, 25; *Empire of Guiana*, 30, 46, 156, 161, 169. *See also* School of Night
Ramism, 81–84
Ramus, Peter, 81–82
Ramusio, Giovanni, 31
rape imagery, 125–26, 129
Rawson, Claude, 301n
Ray, John, 101–2n. *See also* Willughby
readers, 15–16, 26–29, 94, 98, 132–33; of Behn, 274; Caraboo as, 323–24; early modern, 15n; ideal, 33, 128; implied, 13; of Lafitau, 302n (see also *Moeurs des sauvages amériquains*: reviews of); leisured, 133; of *Sidereus nuncius*, 127, 132–33. *See also* Medici; Pepys
readership, of travels and cosmography, 26–29, 53–55, 304n, 317
reading, and sensation, 178, 198, 302, 303
reading, as experience: 50, 75–76, 100, 128, 132–33, 191, 196, 198, 253, 273; history of, 28–29, 132–33; paranoid, 2; and pleasure, 59, 76, 133, 154, 254; sensational, 188, 196–99, 300–310
real character, 17, 77, 77n, 88, 161; and denotation, 159, 161–62; Mandarin (written) as, 159, 162
Receuil de la diversité des habits, 228–29, 230n, 250–56, 263
Reformation, 71
Reichler, Claude, 306–8
Reiss, Timothy, 19–20, 135n
rhetoric: of fiction, 16; of the imaginary, 88; and pornography, 183, 185–86, 201; of science, 28, 59–60, 79, 103–7, 188, 190–91, 196–202; of the sublime, 192, 196–99; of wonder, 4, 76n, 106
Restoration, 192, 195, 202, 213, 274, 278
Ricci, Matteo, 159
Richardson, Samuel, 201, 206

Roberval, Marguerite de la Rocque de, 35, 38–40, 43; as castaway, 22; as female hero, 39n
Roberval, Sieur de, 35
Robinson Crusoe (Defoe), 36, 38–39, 201, 259n, 322. *See also* Crusoe, Robinson
robot, 279; and Descartes, 109, 283; mechanical scientist, 104; and Christopher Wren, 14n
Rochefort, Charles César de, 300
Rogers, John, 182
romance, 185, 191, 281, 282; and F. Bacon, 74; and closure, 201; heroic, 278, 279; medieval, 169–70, 275
Romano, Giulio, *I modi*, 183
Rosen, Edward, 134, 134n, 135n
Rossi, Paolo, 44
Royal Society, 74, 182, 214; membership, 98, 98n54. *See also* journals, scientific: *Philosophical Transactions*
Rudolf II (of Holy Roman Empire), 121, 137

Sade, Marquis de, 301
Sagan, Carl, 6
Sagard, Gabriel, 295n, 304, 305, 312
Sahagún, Bernardino de, 28n, 96
Sahlins, Marshall, 100n
Saintonge, Alfonse de, 31, 40n
Salamon House, 74
Sarasohn, Lisa, 203–4
Scarry, Elaine, *Body in Pain*, 304n
Schaffer, Simon, and Steven Shapin, 14n, 277n
Schlesinger, Robert, 31–32; and Arthur P. Stabler, 31n
School of Night, 114n, 145
Schopp, Gaspar, 114
science fiction, 65, 67, 115n, 116, 134; as reading practice, 144; and realistic novel, 112, 320, 320n
scientific communities, 145, 290. *See also* Académie des Sciences; journals, scientific; Royal Society; School of Night

Scientific Revolution, 24; and Cortés, 74
scintillation, 199, 217, 250
Scott, Peter Dale, 266
Sedgwick, Eve Kosofsky, 2, 2n, 7n, 250, 266
self-made man, 155, 233, 234, 234nn, 248
Selkirk, Alexander, 320, 322
sensation, 160, 177, 277; and reading, 178, 198; evacuation of, 277, 297
sensational, the: and ethnography, 66–67, 272, 300–310; and fictionality, 111, 272 (*see also* mimesis: and magic); and *Micrographia*, 17n, 182, 187–88, 189, 195–97; and natural philosophy, 17, 189–91; and novel, 13; and the sublime, 182, 303
sensationalism, 259, 304; and dismemberment, 272–73; in history of fiction, 100; and Lafitau, 303–8; and Psalmanazar, 286
Shackleton, Robert, 144
Shadwell, Thomas, *The Virtuoso*, 225
Shirley, John, *Thomas Hariot*, 60n
Sidereus nuncius (Galileo), 111, 115, 124–33; as colonial report, 129–30; and Columbus, 133; erotic imagery in, 128–31; illustrations, 126, 130; and novel, 132–33; and picaresque, 133; printing of, 133, 133n
Silverman, Kaja, 209n, 210–11, 210n, 211n, 217
Singer, Dorothy, 114, 156
skin color, 57, 161; Browne on, 89–96; and environment, 89–92; in *Man in the Moon*, 161, 161n. *See also* lunar color
slave trade, 90n, 94n, 95, 279
slavery: American, 10n, 90n, 95, 258, 264, 265, 267–68, 271, 275; Aristotle on, 10n; and fictional agency, 278–83; plantation, 275, 279; and reification, 91; and representation, 267–68, 280; on St. Helena, 164–65; of women, 90n
Snow, C. P., 7. *See also* "Two Cultures"
sociology, 199–201, 227, 249; and *L'autre monde*, 173, 175, 179; and *Man in the*

Moon, 160, 166, 170. *See also* ethnology; manners and customs
Solomon, Julie R., 52n, 61n
Somnium (J. Kepler), 67, 133–43; audience of, 134; and cosmographical context, 135–36, 140, 142; as heterotopia, 14n; as parody, 142; as voyage, 136, 142
sorcery: and Dyrcona, 152, 180; and Fiolxhilde, 137–39; and Libussa, 137; and Godwin, 180
Southerne, Thomas, *Oroonoko*, 259n, 269
species, 80, 82–83, 83n; etymology of, 83n; lunar, 142
Spedding, James, 72n
Sprat, Thomas, 36, 71n, 74, 161, 215, 183, 184, 188, 277, 303
Stabler, Arthur P., 34. *See also* Schlesinger
Staden, Hans von, 27, 173n, 264n, 305n
Stein, Gertrude, 14n
stenographia, 84, 158, 162, 165
Stewart, Susan, 187, 214, 314n
Stubbe, Henry, 17
sublime, the, 131, 141, 196, 214, 265n; arithmetic, 149; and category collapse, 254; and ecstatic religion, 293–94; and helter-skelter, 265; and cosmology, 116, 119; feminine, 217–19; and fictionality, 111; and the sensational, 182; relative to wonder, 5, 103. *See also* pain
Sun Ra, 128
Surinam: and Behn, 221; as fictional setting, 158, 268, 271, 273, 278, 308; and Maria Sybilla Merian, 16n, 274; plantation culture of, 261, 265, 282
surveillance, 190, 192, 276
Swanton, John R., 57n
Swift, Jonathan, 168; *Gulliver's Travels*, 194, 201, 217, 222, 259n, 261, 287, 317. *See also* Gulliver, Lemuel

Tacitus, 289
tattoo, *64*, 258, 262
Taussig, Michael: *Mimesis and Alterity*, 139–40, 143, 249–50; *Shamanism, Colonialism, and the Wild Man*, 264, 265–66, 303
taxonomy, 62, 65, 81–84
technical terms, 79
telescope, 120, 123–24, 127; and R. Bacon, 103, 123; disapproval of, 125, 212–13; erotic dimension of, 126, 148; and Galileo, 124, 126–31; and Hariot, 120, 123; and J. Kepler, 123, 126
Tempest, The (Shakespeare), 36
terror, 258, 265–70; absent in *Moeurs des sauvages amériquains*, 308
theatre, 123, 130; anatomy, 85, 280n; as Behn's venue, 262; fashion as, 226; folkways as, 306; as trope, 306; universe as, 144–45
Theophrastus, 232
Thevet, André, 26, 30, 31n, 52, 149, 190, 257, 287, 290, 320; *Cosmographie universelle*, 22, 27, 35–40, 41n, 44, 46, 52; cosmographies of, 155, 156, 270, 287; "Descriptions de plusiers isles," 32n; and ethnography, 45–50, 293, 304; and Galileo, 127; "Grand insulaire et pilotage," 31n, 34, 35n, 37n, 39–40, 42, 44, 46–47; and Léry, 69–70; on New World, 25–50; *Singularitez de la France Antarctique*, 27, 30, 32, 41n, 45–48, *48*; *Les vrais pourtraicts*, 41, 41n, 231
Thirty Years' War, 136, 137, 140
Thomas, Nicholas, 233
Titian, 126
Titus Andronicus (Shakespeare), 194
Torgovnik, Marianna, 26–27
torture, 258, 275, *307*; F. Bacon's tropes of, 72–73, 85; and dissociation, 300, 302, 306; and empathy, 308–10
Tradescant, John (the younger), *Musaeum Tradescantium*, 80–82, 85
Traherne, Thomas, 322
travel writing, 12–13, 51, 156, 324; and catalogues, 82; and colonial expansion, 21, 148; and difference, 89; and ethnology, 26, 287, 291; and pleasure, 25–26

Trouillot, Michel-Rolphe, 248
Truffaut, François, *The Wild Child*, 19
Tufayl, ibn, *Hayy ibn Yaqzan*, 38n
Tupinamba: cannibals, 173n, 264n, 305n; and Greek lyric, 297; music of, 296–98
"Two Cultures" (Snow), 7, 206, 216
Tylor, E. B., 60n, 74n, 245n
Tyresias, 180

universal language schemes. *See* real character
Utopia, 253
utopia, 12, 14, 15, 204, 287, 320; China and, 159; Moon and, 153, 203; parallel world, 120, 160n, 204n, 313; West as, 248

Vallodolid, 10n, 175
Van Helden, Albert, 123–24
Venus, 125, 126
Venus Hottentot (Sarah Bartmann), 85, 198, 280n
"Venuses," 272n
verisimilitude, 163, 180, 185, 209; and prose fiction, 13, 185, 186; and scientific writing, 9–10, 184, 191
Vespucci, Amerigo, 4n
Victor (of Aveyron), 19, 38n, 174n, 322–24; and speech, 322n
vida del Buscón, La, 133
violence, 61–62, 63–65, 279; and Baconian science, 73n, 74n, 84–85, 91; colonial, 65, 74; as liberation, 274. *See also* colonial conquest; pain; torture
Virginia, 65; in *America, Part 1*, 27, 51–65, 261; Roanoke, 52, 54; White's watercolors of, 55–59, 261
virtuosi, 127, 202, 211, 225, 231, 323. *See also* Académie des Sciences; Cavendish, W.; Evelyn; Hooke; Huygens; Fontenelle; Moliére; Royal Society; School of Night; Shadwell; Swift

Voltaire, 223
voyeurism, 187, 272

Warren, George, *Surinam*, 278, 281
waterworks, at Enston, 102, 103–4, 105–9, *107, 108*
Watt, Ian, 258
White, John, 27, 51; and Hariot, 26, 51, 55, 62; watercolors engraved, 57–59, *58, 59, 64*; watercolors of Picts, 55, 64–65; watercolors of Virginia, 55, *56, 58*
Wild Boy of Aveyron. *See* Victor
Wild Man, 38n, 174, 319
"Wild People," 323
Wild Woman, 251
Wilkins, John, 6, 146, 154, 185
William of Rubruck, 11n, 52n
Williams, Brackette, 7n, 49n, 208n
Williams, Roger, 30n
Williamson, James, 271
Willughby, Francis, 84, 101, 101n
witnessing: and authority, 29, 69–70; and detachment, 276–77; internal, 213; and Lafitau, 292; of others' narratives, 275–78, 310; virtual, 130, 277n
Wizard of Oz, The, 164, 168–69
women: as aliens, 149, 206; as authors, 261, 271, 271n, 278; and imagination, 244–47, 250; as mediating category, 244; and monstrosity, 239, 242–43; and transmission of culture, 244; utility of, 247
Woolf, Virginia, 211n
world: defined by commerce, 186; only possible, 222; owned, 199; and self, 119n, 171, 178, 321; as theatre, 119; unified, 222–23
world-upside-down, 195, 201, 204, 313
Wren, Christopher, 104, 185

xenophobia, 238, 313

Yates, Frances, 114, 121